Understanding Physics

A Study Guide for

Principles of Physics

Understanding Physics

A Study Guide for Blatt
Principles of Physics

Third Edition

Michael E. Browne
University of Idaho

Allyn and Bacon
Boston • London • Sydney • Toronto

A Division of Simon & Schuster
160 Gould Street
Needham Heights, Massachusetts 02194

ISBN 0-205-11785-6

Printed in the United States of America

10 9 8 7 6 5 4 3 2 1   94 93 92 91 90 89

Contents

Chapter

Chapter

Preface

The study of physics can be very rewarding. It is intellectually stimulating and challenging and it is fun. You will gain a better understanding of the universe you live in, and through it you will learn a way of thinking logically and systematically which will help you all through your life, no matter where your path takes you. Gaining a mastery of physics can do wonders for one's self-confidence. You may be undertaking the study of physics because you are required to do so, and you may have some doubts about your aptitude for science and mathematics. Don't worry. You can do it if you are willing to do some work. Mostly you are going to be learning a way of thinking, a way of piecing together lots of observations of what happens around us so that they form the basis for a beautifully simple pattern of ideas called the "laws of nature". Learning to think will help you no matter what your profession.

We can express the laws of nature most clearly by using mathematics, which is sometimes called the "language of science". Studying physics will do much to help you improve your "fluency" in mathematics.

Much of the beauty and power of science stems from the fact that it is based on observation and measurement in quantitative terms, i.e. using numbers. Thus to truly understand physics you must be able to apply the basic concepts to solve problems. Solving problems will do much to increase your understanding of the basic principles, and you should devote a significant part of your study time to this effort. All problem solving has certain features in common, and an outline of the steps in solving physics problems is given on the next page. If you have difficulty with the mathematics you should first study the math review in this guide and in the appendix of the text.

Learning physics can be easy if you apply some self-discipline. Here is a system which works:

1. Make a study schedule and stick by it. Try to allot your physics study time in one to three hour blocks. All-night grinders are not worth much. You can't learn physics the night before an exam. Physics will be easy if you don't fall behind, but goof off for a week or two and the coyotes will soon be chewing on your bones. When studying, keep a "talking paper" on which you list points which are not clear to you. Periodically go to see your instructor or tutor and take along your list. Often you will find that you can answer your own questions after a few days.

2. Make the best use of your time. Don't spend hours hitting your head against the wall on a tough problem or concept. Leave it and return to it later or get help from your teacher or from other students.

3. Read quickly through the appropriate section of the text before your teacher lectures about it. Don't get stuck on details or complications, but try to get an overview of what the chapter is all about.

4. Always attend the lecture. Try to record important points discussed, but don't make your notes so comprehensive that you don't have time to listen. You can always refer to the text for details you may omit.

5. Read the text a second time, more carefully. Note on a piece of paper important definitions, units and equations. Write out in your own words any important principles or laws.

6. Try to work assigned problems and questions at the end of the chapter. Refer back to the text for help. Look at relevant examples in the text. If you encounter any difficulty, turn to the study guide. If your math background is weak first review the mathematical appendix in the text and in the first section of the study guide.

7. Look at the study guide chapter corresponding to your assignment. Read it and first answer the qualitative questions to make sure you understand the terminology and underlying concepts. Next try to work some of the example problems, looking at the solution only after you have made a fair effort on your own (perhaps 5-10 minutes).

8. In solving problems, first read the question several times to be certain that you understand what is happening, what information you are given, and what it is you are trying to find. It is helpful to draw a little pictorial sketch first. Next, draw a careful diagram which shows force or velocity vectors, distances, masses, etc. Label each quantity with a symbol, such as F, v, m, x, etc. Below the diagram list any numerical values for these quantities. Next decide what basic principles of physics to apply to the problem in order to deduce the desired result. Finally, set up the corresponding mathematical equations and solve them. Be sure to express your answers in the correct units, and check you work so that you don't make careless arithmetic errors.

9. Once again try to work the assigned text problems. You will find that similar problems are worked out in the study guide, and you can refer to them again as needed.

10. After you have studied three or four chapters of the text and are preparing for an examination, go through the study guide chapters and summarize on a sheet of paper important equations, terms and concepts. It may be helpful to go over this sheet with your teacher to ascertain that you have not overlooked any important points. If you have difficulty, review the guide and work additional problems from the text.

The material in each study guide chapter is organized as follows:

Summary of important ideas, principles and equations.
Qualitative questions, with answers.
Sample problems, with solutions.

You may find it helpful to study some of the above material, particularly the qualitative questions, with other students. In studying physics you should not try to memorize a lot of equations. You will soon forget them, and they will be of little value in your life or in your course. However, just as it is convenient to know the multiplication table by heart, it is also desirable to know the fundamental physical laws. There are only a few of these, and if you work lots of problems you will automatically remember them without any special effort.

Review of Mathematics

An inadequate math background is the main source of trouble for students studying physics. Study this review carefully if you are weak in this area. You are expected to know the following material from previous studies.

1. Arithmetic

We can multiply out an expression like 2(3 + 2) to obtain 10.

$$2(3 + 2) = 6 + 4 = 10$$

Similarly, $2(3x + 2x) = 6x + 4x = 10x$

Also, $3(5x + 1) = 15x + 3$.

The reverse process is called factoring.

$$15x + 3 = 3(5x + 1)$$

This idea is useful when adding fractions.

2. Fractions

Consider the fraction 2/3. The number upstairs, 2, is called the numerator. The number downstairs, 3, is the denominator. To add two fractions first change them so that they have the same denominators, like this:

$$\frac{2}{3} + \frac{1}{4} = (\frac{2}{3})(\frac{4}{4}) + (\frac{1}{4})(\frac{3}{3}) = \frac{8}{12} + \frac{3}{12} = \frac{1}{12}(8 + 3) = \frac{11}{12}$$

Be careful when you encounter fractions in the denominator. Thus,

$$\frac{1}{\frac{2}{5}} = \frac{1}{\frac{2}{5}} \cdot \frac{5}{5} = \frac{5}{2}$$

$$\frac{1}{\frac{1}{2} + \frac{1}{6}} = \frac{1}{\frac{3}{6} + \frac{1}{6}} = \frac{1}{\frac{4}{6}} = \frac{6}{4} = \frac{3}{2}$$

3. Exponent notation is useful when multiplying a number by itself.

Thus $2 \times 2 = 2^2$

$2 \times 2 \times 2 \times 2 \times 2 = 2^5$

To multiply two such numbers we add exponents:

$$2^2 \times 2^5 = (2 \times 2)(2 \times 2 \times 2 \times 2 \times 2) = 2^7$$

Negative exponents are used for reciprocals.

$$2^{-1} = \frac{1}{2}$$

$$2^{-4} = \frac{1}{2^4} = \left(\frac{1}{2}\right)^4$$

4. Powers of Ten

Very large and very small numbers are conveniently written using the following scheme:

$$\begin{aligned} 10{,}000 &= 10^4 \\ 1{,}000 &= 10^3 \\ 100 &= 10^2 \\ 10 &= 10^1 \\ 1 &= 10^0 \\ 0.1 &= 10^{-1} \\ 0.01 &= 10^{-2} \\ 0.001 &= 10^{-3} \\ 0.0001 &= 10^{-4} \end{aligned}$$

To multiply two such numbers add exponents:

$$10^4 \times 10^2 = 10^6 = 1{,}000{,}000$$

$$10^3 \times 10^{-2} = 10^1 = 10$$

$$10^2 \times 10^{-5} = 10^{-3} = 0.001$$

Other numbers may be written this way as well.

$$2,153 = 2.153 \times 1000 = 2.153 \times 10^3$$

$$0.0067 = 6.7 \times 0.001 = 6.7 \times 10^{-3}$$

$$(5 \times 10^4)(3 \times 10^3) = (5 \times 3)(10^4 \times 10^3) = 15 \times 10^7 = 1.5 \times 10^8$$

$$(4 \times 10^{-5})(2 \times 10^3) = 8 \times 10^{-2}$$

To divide two such numbers subtract the exponents.

$$\frac{6 \times 10^9}{3 \times 10^6} = 2 \times 10^3$$

$$\frac{8 \times 10^3}{4 \times 10^6} = 2 \times 10^{3-6} = 2 \times 10^{-3}$$

$$\frac{5 \times 10^4}{2 \times 10^7} = 2.5 \times 10^{-3}$$

5. Logarithms

Exponents need not be integers. Consider, for example, the number $4^{\frac{1}{2}}$. To see what this means, multiply this number by itself, remembering that when multiplying we add exponents.

$$4^{\frac{1}{2}} \times 4^{\frac{1}{2}} = 4^1 = 4$$

Thus $4^{\frac{1}{2}}$ is what is called $\sqrt{4}$, the "square root" of 4, which has the value 2.

In studying radioactivity or bacterial growth or energy consumption we often encounter expressions of the form

$$e^a \text{ or } e^{-a}$$

Here e is a special number which, like π, pops up in many interesting places in mathematics. It has the value 2.718... .

Suppose you want to evaluate $e^{3.5}$. Merely punch 3.5 on your calculator, then punch "INV" and then "Ln" and presto, you have the answer. (Press 3.5 and then e^x on some machines.) The exponent to which e is raised is called the "natural logarithm" of the number $e^{3.5}$, written in this case as

$$\ln e^{3.5} = 3.5$$

Sometimes you meet an expression like $14 = e^a$ and you wish to determine a. To do this punch 14 on your calculator, then punch Ln. The answer for a will appear.

$\ln 14 = \text{Ln } e^a = a$

The logarithm of a number less than 1 is negative.

Another logarithm system is based on raising 10 to different powers. These logarithms are called common logs and written log x.

Thus if $\log x = a$, $x = 10^a$

We won't make use of these. Be careful to punch the "Ln" button, and not the "log" button on your calculator if you want to evaluate natural logs.

6. Geometry and Trigonometry

A. Angles are measured in degrees or in radians. A full circle is 360° or 2π radians.

Thus $360^\circ = 2\pi$ radians, or 1 radian $= 57.3^\circ$

We often label an angle with the Greek letter θ, theta.

B. The angles inside a triangle add up to 180° (or π radians).

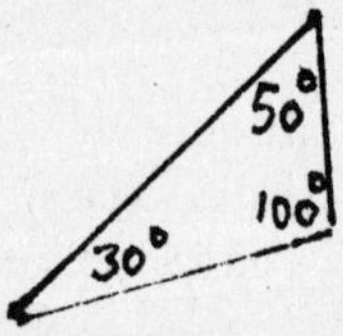

If one angle is 90° the triangle is called a right triangle. This means the other two angles must add up to 90°, since all three add to 180°. The two small angles are called complementary angles.

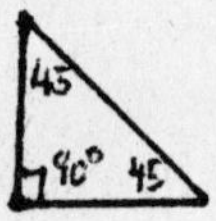

C. Two triangles which have the same shape (i.e. the same angles) are called similar triangles. Their sides are all proportional.

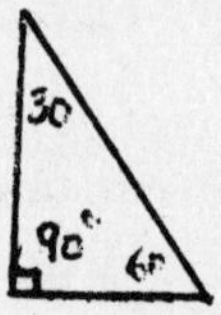

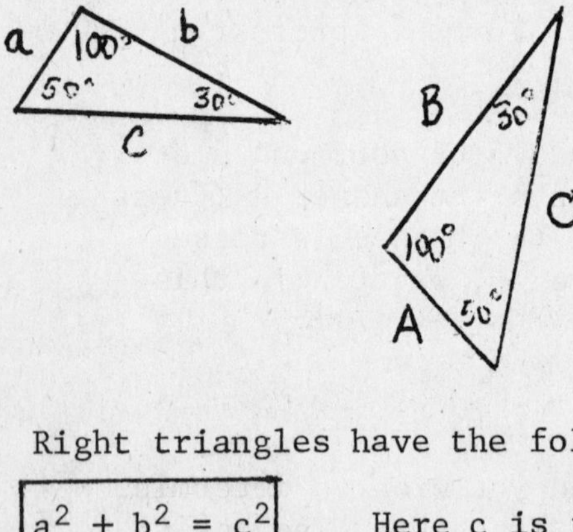

$$\boxed{\frac{a}{A} = \frac{b}{B} = \frac{c}{C}}$$

D. Right triangles have the following properties:

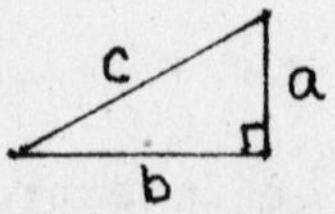

$$\boxed{a^2 + b^2 = c^2}$$ Here c is the hypotenuse (the longest side).

Long ago some diligent soul tabulated the ratios of the sides of a right triangle for all possible shapes of right triangles, i.e. for all different acute angles θ.

These ratios are called the sine of θ, the cosine of θ and the tangent of θ.

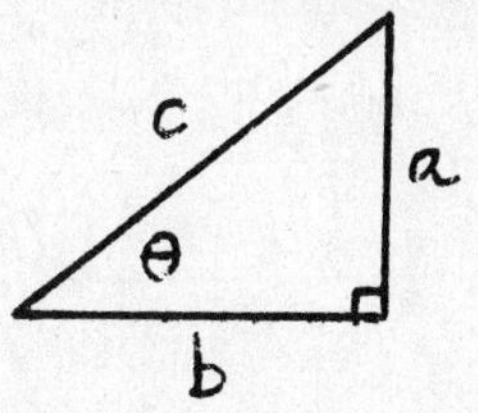

$$\sin\theta = \frac{a}{c} = \frac{\text{opposite side}}{\text{hypotenuse}}$$

$$\cos\theta = \frac{b}{c} = \frac{\text{adjacent side}}{\text{hypotenuse}}$$

$$\tan\theta = \frac{a}{b} = \frac{\text{opposite side}}{\text{adjacent side}}$$

If you have trouble keeping these straight, remember that b, the side next to the angle θ, involves the <u>co</u>sine, just as the person who works next to you is your <u>co</u>-worker.

One can also write these relations as

$$a = c\sin\theta$$

$$b = c\cos\theta$$

The values of the functions sine, cosine and tangent of any angle are readily found with a calculator. Be sure your calculator is set on "degrees".

When solving problems make a careful diagram using a straightedge. Draw a little square in the corner of 90° angles.

The following are useful trigonometric identities:

$$\sin^2\theta + \cos^2\theta = 1$$

$$\sin(\alpha + \beta) = \sin\alpha\cos\beta + \sin\beta\cos\alpha$$

$$\cos(\alpha + \beta) = \cos\alpha\cos\beta - \sin\alpha\sin\beta$$

$$\sin 2\theta = 2\sin\theta\cos\theta$$

$$\cos 2\theta = \cos^2\theta - \sin^2\theta$$

When θ is a small angle the following approximations may be used:

$$\sin\theta \simeq \theta$$

$$\cos\theta \simeq 1 - \tfrac{1}{2}\theta^2$$

$$\tan\theta \simeq \sin\theta$$

θ here must be expressed in radians.

$(\theta \ll 1)$

Formulas which are useful in finding the sides of triangles are

$$c^2 = a^2 + b^2 - 2ab \cos \theta$$ ← Law of Cosines

$$\frac{\sin \alpha}{A} = \frac{\sin \beta}{B} = \frac{\sin \gamma}{C}$$

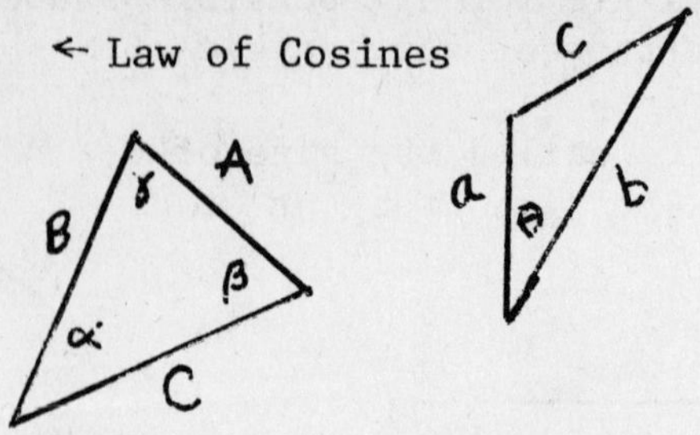

E. The following formulas are useful:

circumference of a circle = $\pi D = 2\pi R$

R = radius
D = diameter

area of a circle = πR^2

surface area of a sphere = $4\pi R^2$

volume of a sphere = $\frac{4}{3}\pi R^3$

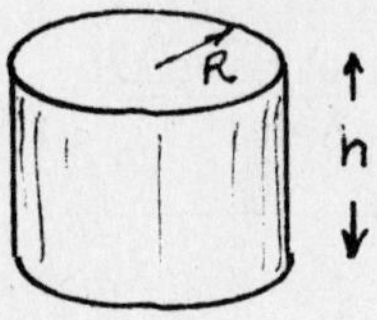

surface area of curved surface of a cylinder (sides) = $2\pi Rh$

volume of a cylinder = $\pi R^2 h$

7. Algebra

To solve a single equation involving one unknown x, first collect together all terms in x:

Example: $2x + 16 = 10x - 8$

Subtract 2x from each side and add 8 to each side. $24 = 8x$

$$x = \frac{24}{8} = 3$$

To solve for two unknowns one must have two equations. One easy way to solve two such simultaneous equations is to solve for one unknown in the first equation and then substitute this in the second equation.

Example: $2x - y = 8$ (1)

$x - 5y = 13$ (2)

From (1), $y = 2x - 8$. Substitute this in (2):

$$x - 5(2x - 8) = 13$$

$$x - 10x + 40 = 13$$

$$-9x = -27 \qquad x = \frac{-27}{-9} = +3 \quad \text{(Answer)}$$

To find y substitute $x = 3$ in (1): $2(3) - y = 8$

$$6 - 8 = y, \quad y = -2 \quad \text{(Answer)}$$

Sometimes one encounters two equations involving an unknown F and the two unknown (but related) functions $\sin\theta$ and $\cos\theta$.

Example: $F \sin\theta = 18$

$F \cos\theta = 24$

To solve divide one equation by the other: $\dfrac{F \sin\theta}{F \cos\theta} = \dfrac{18}{24} = 0.75$

or $\dfrac{\sin\theta}{\cos\theta} = \tan\theta = 0.75 \qquad \theta = \tan^{-1} 0.75 = 36.9^{\circ}$

To find θ when you know $\tan\theta$, punch 0.75 on your calculator, then punch INV (for "inverse"), then punch TAN and the value of θ will appear in the display. Be sure your calculator is set on "degrees". On some calculators arctan is written instead of $\tan^{-1}$.

We also encounter equations in which the unknown appears as x in some terms and as x^2 in others. These are called quadratic equations.

Such an equation is of the form $ax^2 + bx + c = 0$

Here a, b and c are numbers which we are given.

Two values of x, which I call x_1 and x_2, satisfy this equation (i.e. make it hold true). They are given by the following formulas:

$$x_1 = \frac{-b + \sqrt{b^2 - 4ac}}{2a} \quad \text{and} \quad x_2 = \frac{-b - \sqrt{b^2 - 4ac}}{2a}$$

Example: $2t^2 + 4t - 6 = 0$

$$t_1 = \frac{-4 + \sqrt{4^2 - 4(2)(-6)}}{(2)(2)} = \frac{-4 + \sqrt{16 + 48}}{4} = 1$$

$$t_2 = \frac{-4 - \sqrt{4^2 - 4(2)(-6)}}{(2)(2)} = \frac{-4 - 8}{4} = -3$$

1 Units, Dimensions, and Other Preliminaries

• Summary of Important Ideas, Principles, and Equations

1. Units must be used to give meaning to any measurement. We will use the SI (metric) system in which the fundamental units are the meter (for length), the kilogram (mass) and the second (time). There is another old fashioned system, called the "British" system, which is best avoided like the plague. Unfortunately it is sometimes necessary to convert from one system to another. The method for doing this is best explained with examples as given later in this chapter. Note that conversion factors are given in an appendix of your text.

   Sometimes combinations of the fundamental units occur so often that they are given special names. For example, the unit of force is 1 kg-m $/s^2$. This combination is called a newton. Other such derived units are those for energy (the "joule"), power (the "watt") and pressure (the "pascal").

   A particular combination of fundamental units gives the dimension of the quantity. If we let L = length, M = mass and T = time, we see that force has dimensions of ML $/T^2$.

   All of the terms in an equation must have the same dimension. You can sometimes use this fact to check your work for errors.

2. Order of magnitude estimates are often useful in order to gain a feeling for the size of an effect or phenomenon. One can make such an estimate by rounding off each factor in a numerical expression to the nearest multiple of ten. The result is then readily calculated.

3. The result of a calculation should not contain more significant figures (i.e. digits) than the least accurate factor in the calculation. In our work we will not write more than three significant figures for a given value.

4. A <u>vector</u> is a quantity characterized by a <u>magnitude</u> and a <u>direction</u>.

A vector is not changed if it is slid parallel to itself.

A vector may be represented by an arrow, where the length of the vector is proportional to the magnitude of the vector and the direction of the arrow is the direction of the vector.

Two vectors may be added by sliding one parallel to itself until the vectors are aligned tail-head-tail-head. The resultant vector is then drawn from the first tail to the final head, as shown here.

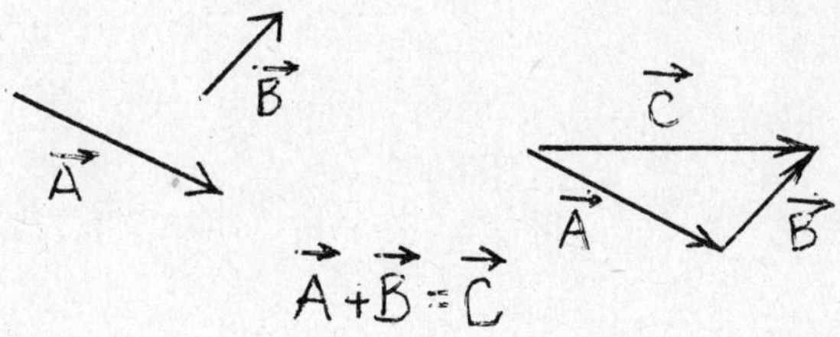

$-\vec{A}$ is a vector of the same magnitude as $\vec{A}$. It points in the direction opposite to $\vec{A}$.

The vector $\vec{C}$ is called the <u>resultant</u> of $\vec{A}$ and $\vec{B}$.

The process of combining two vectors to form a resultant can be reversed. A given vector can be replaced by two smaller vectors, called <u>components</u>, such that when the components are added they yield the original vector. It is useful to choose the components along the x and y axes in most problems. In this case the components of the vector $\vec{C}$ are called $\vec{C}_x$ and $\vec{C}_y$.

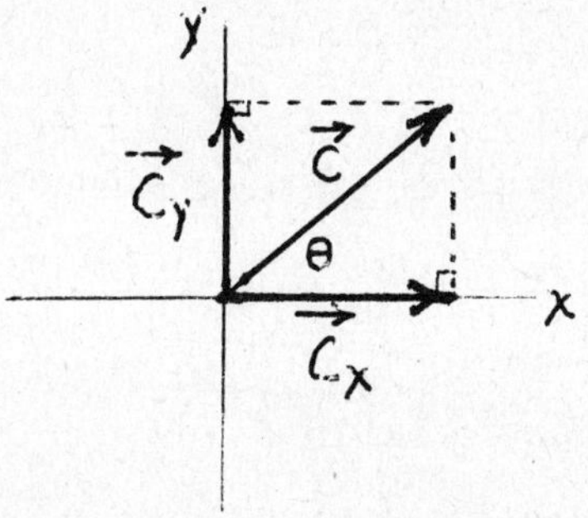

$$C_x = C \cos \theta \qquad (1.1)$$

$$C_y = C \sin \theta \qquad (1.2)$$

$$C^2 = Cx^2 + Cy^2 \qquad (1.3)$$

Be sure to draw diagrams like this very carefully, making certain that the dashed construction lines are perpendicular to the x and y axes.

In the drawing above where $\vec{A}$ and $\vec{B}$ are added to yield $\vec{C}$ one could measure the length of $\vec{C}$ with a ruler and thereby obtain an approximate value for the magnitude C. A better method is first to break $\vec{A}$ and $\vec{B}$ into components, then combine the components of $\vec{A}$ and $\vec{B}$ to yield the components of $\vec{C}$. This is illustrated in the drawing on the next page.

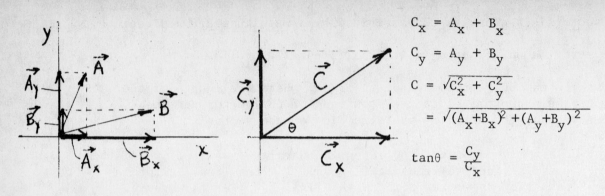

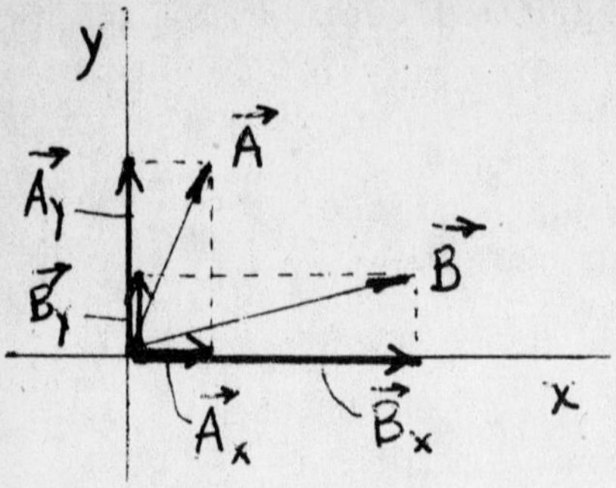

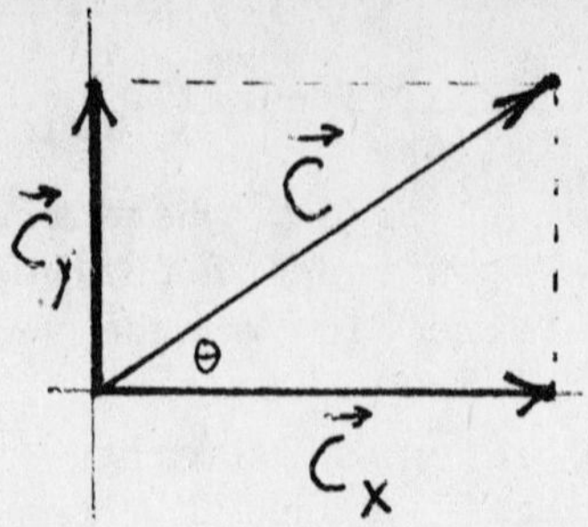

$$C_x = A_x + B_x$$

$$C_y = A_y + B_y$$

$$C = \sqrt{C_x^2 + C_y^2}$$

$$= \sqrt{(A_x+B_x)^2 + (A_y+B_y)^2}$$

$$\tan\theta = \frac{C_y}{C_x}$$

• Qualitative Questions

1M.1 If you multiply 27.0 by 9411 on your calculator you will obtain 254,097. We recognize that not all of the digits in this result are significant (i.e. meaningful), and hence this result should be written

A. 25.4090 ± x, where x is an unknown error.
B. 2.5409×10^5
C. 2.54×10^5
D. 2.55×10^5
E. 2.50×10^5

1M.2 Which of the following operations will not change a given vector?

A. Rotating it.
B. Making it longer or shorter.
C. Sliding it parallel to itself.
D. Multiplying it by a constant scalar factor.
E. Adding a constant vector to it.

1M.3 When two vectors of magnitude 3 Glorks and 4 Glorks are added, the resultant will be a vector of magnitude

A. 7 Glorks.
B. 1 Glork.
C. 5 Glorks.
D. between 1 and 7 Glorks, but we can't deduce the exact value.
E. which can't be determined, since Glorks are not defined in the SI system of units.

IM.4 If three vectors add up to zero, then

A. at least one of them must be itself zero.
B. they must all lie in one plane.
C. all three of the vectors must be of equal magnitude.
D. all three vectors must lie along the same line.
E. None of the above is true, since it is impossible for the resultant of three vectors to be zero.

1M.5 Two displacement vectors of magnitudes 3 meters and 4 meters are drawn here to scale. What is the magnitude of their resultant when added?

A. 7 m
B. 6.4 m
C. 5 m
D. These vectors cannot be added because they are situated at different points in space.

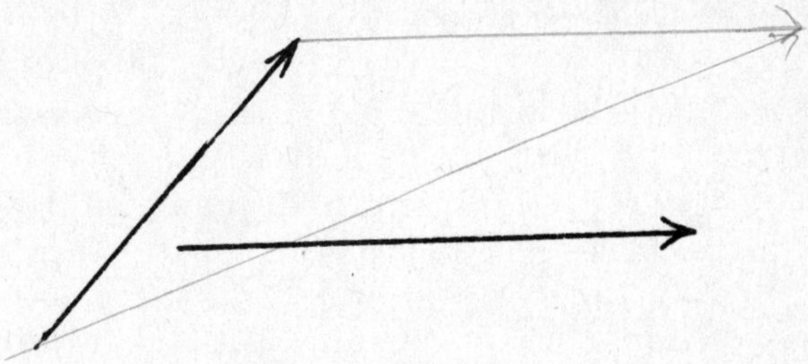

• Multiple Choice: Answers and Comments

1M.1 C is correct. There are only three significant figures in the factor 27.0, and 4 in the factor 9411, thus there should be only 3 significant figures in the answer. We round off 254,097 to 254,000, which is most easily written 2.54×10^5.

1M.2 C is correct. A vector is characterized by magnitude (length) and direction, and these are unchanged when a vector is slid parallel to itself. All of the other choices change the vector.

1M.3 D is correct. If the vectors are parallel their resultant will be 3 + 4 = 7 Glorks. If they are antiparallel the resultant will have magnitude 4 - 3 = 1. If they are at an angle to each other the resultant will have some intermediate magnitude.

1M.4 B is correct. Imagine you first add two of the vectors. Their resultant must lie in their plane. Call this resultant R. The third vector must just cancel R, i.e. it must be antiparallel to R (i.e. along the same line but pointing in the opposite direction). In this case the third vector also lies in the plane of the first two.

1M.5 Slide one vector parallel to itself until they are aligned as shown here. Measure the length of the resultant with a ruler, using the two given vectors to determine the scale. The shortest one must be 3 meters long.

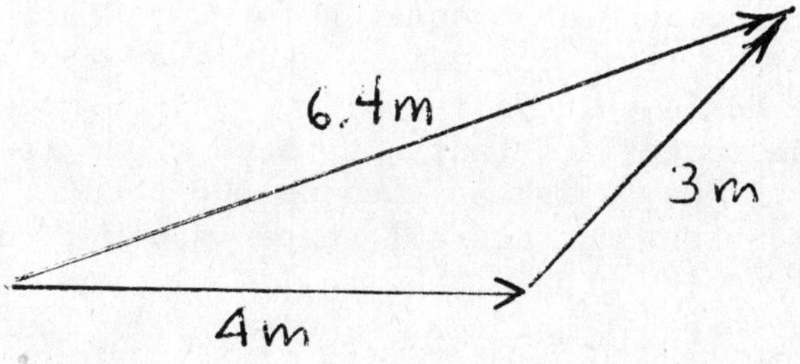

• Problems

1.1 Express 7 hours in seconds, given 1 hr = 3600 s.

1.2 Express 50 meters in feet, given 1 ft = 0.305 m.

1.3 A world class sprinter can run 100 m in 10.2s. Express this speed in mi/hr.

1.4 Suppose that you buy 24 yd^2 of carpet. How many square meters is this?

1.5 Some gas stations are now selling gasoline by the liter (this makes it sound cheaper). What is the price per gallon of gas which sells for \$0.40 per liter? 1 gallon = 231 in^3 and 1 inch = 2.54 cm.

1.6 Recently I helped a friend paint a big cylindrical water tank on his ranch. It was 4 meters in diameter and 3 meters tall. We painted the curved sides and the flat top (but not the bottom). The paint we bought was supposed to cover 200 ft^2 per gallon. How many gallons should we have bought to do the job?

1.7 Evaluate the following numerical expressions:

(a) $(7.1 \times 10^4)(3.0 \times 10^{-9})$

(b) $\dfrac{(0.061)(2.0 \times 10^{-7})(4.1 \times 10^{12})}{7.2 \times 10^5}$

(c) $\dfrac{(6.4 \times 10^3)^{\frac{1}{2}}(2 \times 10^{-4})^{-2}}{(9 \times 10^8)^2}$

1.8 Gravitational potential energy may be expressed as E = mgh. Here h is the height through which a mass m has been lifted. Referring to the table above problem 1.16 in the text, deduce the dimension of the quantity g. What kind of quantity do you suppose g represents?

1.9 Out where I live people tramping through the forest are always on the lookout for Sasquatch (or Big Foot, as he is also called). This critter reputedly looks like your average physics professor, but he is quite a lot bigger and a whole lot hairier and uglier. From footprints that have been found it is deduced that a sasquatch may weigh as much as 600 pounds. If we assume that a sasquatch has the same shape as a man, and that a man 6 feet tall weighs 180 pounds, how tall would you estimate a sasquatch is? (Sasquatches hang around with Artesians, and I know a guy whose brother-in-law's cousin seen 'em just a hobnobbin' together.)

1.10 I've always had fantasies about finding a suitcase stuffed full of $20 bills. This suitcase has been stashed away by dope smugglers who are long gone, and anyway, the money isn't really theirs. I'm not dead certain what I'd do with all that money. Cross that bridge when I come to it. Meanwhile, I've worried a lot about the following question. Just how much money in 20's could you stuff in a big suitcase? See if you can make an order of magnitude estimate. While you're at it, here's a related question that I've pondered. How much would a million dollars in 20's weigh? Could we even lug it away?

1.11 Physiologists sometimes try to obtain a measure of lung capacity by having a person breathe out into a plastic sack. Suppose you did and filled a spherical bag of radius 20 cm . What volume of gas did you exhale? Express your result in cm^3 and in m^3.

1.12 In a race a yacht sails 100 km due north and then 50 km toward the southeast. How far is the vessel then from its starting point? Estimate graphically and calculate analytically.

1.13 Force is a vector which is measured in a unit called a newton (1 newton is a little less than ¼ pound). Now that you know how to add vectors, see if you can answer the following question. Once while exploring a cave a kid in my Boy Scout troop got stuck. We tied two ropes to him and tried to pull him up. Suppose that one rope was pulled with a force of 200 n and the other was pulled with 100 n, as shown here. What was the magnitude and direction (with respect to vertical) of the resultant force exerted on the boy?

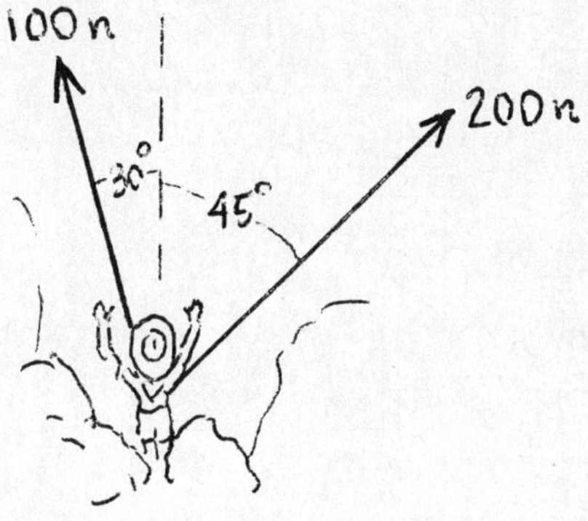

• Problem Solutions

1·1 TO EXPRESS $t = 7\,Hr.$ IN SECONDS GIVEN $1\,Hr. = 3600\,s$
WHEREVER YOU SEE "Hr" WRITE "3600 s."
THUS $t = 7\,Hr = 7(3600\,s) = 25{,}200\,s = 2.52 \times 10^4\,s$.

1·2 TO EXPRESS $s = 50.0\,m$ IN FEET, GIVEN $1\,FT = 0.305\,m$ WE WANT TO REPLACE "m" BY ITS EQUIVALENT IN FEET.

UNFORTUNATELY WE ARE GIVEN A CONVERSION FACTOR WHICH TELLS HOW MANY METERS ARE IN 1 FT, NOT HOW MANY FEET ARE IN 1 METER. THUS WE FIRST OBTAIN THIS VALUE BY DIVIDING BY 0.305.

THUS $0.305\,m = 1\ ft$

$$1\,m = \frac{1}{0.305}\ ft$$

NOW SUBSTITUTE THIS IN THE EXPRESSION FOR DISTANCE.

$$s = 50.0\,m = 50.0\left(\frac{1}{0.305}\ ft\right) = \underline{\underline{164\ ft}}$$

1.3 VELOCITY $V = \dfrac{100\,m}{10.2\,s}$

FROM APPENDIX A IN TEXT, 1609 m = 1 mi., 3600 s = 1 Hr

THUS $1\,m = \frac{1}{1609}\,mi$, $1\,s = \frac{1}{3600}\,Hr.$

$$V = \frac{(100)\left(\frac{1}{1609}\,mi\right)}{(10.2)\left(\frac{1}{3600}\,Hr.\right)} = \frac{(100)(3600)}{(10.2)(1609)}\ \frac{mi}{Hr.} = \underline{\underline{21.9\ mi./Hr.}}$$

1.4 GIVEN $A = 24\,yd^2$. FROM TABLES $1\,yd = 3\,ft$, $1\,ft = 0.305\,m$.

THUS $A = 24(3\,ft)^2 = (24)(9)(ft^2)$

NOTE CAREFULLY THAT "3" IS SQUARED AS WELL AS "ft".

FAILURE TO DO THIS IS A COMMON ERROR.

$A = (24)(9)(0.305\,m)^2 = (24)(9)(0.305)^2\,m^2$

AGAIN, BE SURE TO SQUARE 0.305 AS WELL AS "m".

FINALLY, $\underline{\underline{A = 20\ m^2}}$

1.5 $P = 0.40\ \dfrac{\text{DOLLARS}}{\text{LITER}}$

1 LITER = 1000 cm^3, 1 GAL = 231 in^3, 1 in = 2.54 cm, SO $1\,cm = \frac{1}{2.54}\,in$

$$P = 0.40\ \frac{\text{DOLLAR}}{1000\,cm^3} = \frac{0.40}{1000}\ \frac{\text{DOLLAR}}{\left(\frac{1}{2.54}\,in\right)^3} = \frac{(0.40)(2.54)^3}{1000}\ \frac{\text{DOLLAR}}{in^3}$$

$$= \frac{(0.40)(2.54)^3}{1000}\ \frac{\text{DOLLAR}}{\left(\frac{1}{231}\right)\text{GAL}}$$

$$P = \frac{(0.40)(2.54)^3(231)}{1000}\ \frac{\text{DOLLAR}}{\text{GAL}} = \underline{\underline{\$1.51\ \text{PER GAL}}}$$

1.6 AREA OF TOP IS $A_T = \pi R^2$

AREA OF SIDES IS $A_S = 2\pi Rh$

TOTAL AREA IS $A = A_T + A_S = \pi R^2 + 2\pi Rh$

FROM TABLES $1 \text{ m} = 3{\cdot}28 \text{ ft}$

$$A = 50{\cdot}3(3{\cdot}28\,\text{ft})^2 = (50{\cdot}3)(3{\cdot}28)^2\,\text{ft}^2 = 541\,\text{ft}^2$$

VOLUME OF PAINT NEEDED IS V, WHERE

$$\frac{V}{541\,\text{ft}^2} = \frac{1\,\text{GAL}}{200\,\text{ft}^2}, \quad V = \frac{541\,\text{ft}^2}{200\,\text{ft}^2}\,\text{GAL} = \underline{\underline{2{\cdot}7\,\text{GAL}}}$$

THUS WE HAD TO BUY THREE 1-GALLON CANS OF PAINT.

1·7 (a) $2{\cdot}1 \times 10^{-4}$
(b) $6{\cdot}9 \times 10^{-2} = 0{\cdot}069$
(c) $2{\cdot}5 \times 10^{-9}$

1·8 FROM THE TABLE WE FIND E HAS DIMENSION $\frac{ML^2}{T^2}$
m HAS DIMENSION M.
h HAS DIMENSION L.
THUS $g = \frac{E}{mh}$ HAS DIMENSION $\dfrac{\frac{ML^2}{T^2}}{ML} = \frac{L}{T^2}$

FROM THE TABLE WE SEE THAT g HAS THE SAME DIMENSION AS ACCELERATION, AND IN FACT g IS CALLED THE "ACCELERATION DUE TO GRAVITY."

1·9 A PERSON'S WEIGHT IS PROPORTIONAL TO HIS VOLUME V, WHICH IN TURN IS PROPORTIONAL TO THE CUBE OF HIS LINEAR DIMENSION h.

THUS $\frac{W_{SAS\,Q}}{W_{MAN}} = \frac{V_{SAS\,Q}}{V_{MAN}} = \frac{h^3_{SAS\,Q}}{h^3_{MAN}}$

$$h^3_{SAS\,Q} = \left(\frac{W_{SAS\,Q}}{W_{MAN}}\,h^3_{MAN}\right) = \frac{600\,\text{lbs}}{180\,\text{lbs}}(6\,\text{ft})^3$$

$$h_{SAS\,Q} = \left(\frac{600}{180}\right)^{1/3}(6\,\text{ft}) = \underline{\underline{9\,\text{ft TALL}}}$$

THE SEATTLE SEAHAWKS SHOULD SIGN HIM!

1·10 A REAM OF PAPER (500 SHEETS) IS ABOUT 6 CMS. THICK. A \$20 BILL IS ABOUT 6 CMS × 16 CMS. THUS 500 BILLS OCCUPY A VOLUME

$$V_1 = 6\,\text{cms} \times 6\,\text{cms} \times 16\,\text{cms} \simeq 576\,\text{cm}^3 \simeq 600\,\text{cm}^3$$

A BIG SUITCASE IS ABOUT 75 CMS × 60 CMS × 20 CMS, SO ITS VOLUME IS ABOUT $V_S \simeq 9 \times 10^4\,\text{cm}^3$

THUS THE NUMBER OF "PACKETS" OF 500 20'S WHICH WILL FIT IN THE SUITCASE IS ABOUT

$$N = \frac{9 \times 10^4\,\text{cm}^3}{600\,\text{cm}^3} = 150$$

500 - \$20 BILLS IS \$10,000, SO THE SUITCASE HOLDS A SUM $S = (150)(\$10{,}000) = \$1{,}500{,}000$

\$1·5 MILLION AIN'T HAY, EVEN IF MY ESTIMATE IS OFF A LITTLE BIT. I WOULD ESTIMATE THAT I AM OFF BY AT LEAST A FACTOR OF 2, SINCE I DOUBT IF BILLS WOULD BE PACKED AS TIGHTLY AS A REAM OF TYPING PAPER.

BY LIFTING MY PAPER TABLET I ESTIMATE THAT A SHEET OF PAPER IS ABOUT 2 GM AND THE SIZE OF ABOUT 10 BILLS, SO ONE BILL HAS MASS 0·2g.

\$1 MILLION IS $\frac{10^6}{20} = 5 \times 10^4$ \$20 BILLS

THE TOTAL MASS WOULD THUS BE ABOUT

$$5 \times 10^4 \times 0.2\,g = 10{,}000\,g = 10\,kg$$

THIS WOULD WEIGH ABOUT 22 lbs. NOT BAD. I'D BE OFF AND AWAY LIKE A BIG BIRD.

1·11 VOLUME OF A SPHERE IS $V = \frac{4}{3}\pi R^3$ $\quad 1\,m = 10^2\,cm$

$V = \frac{4}{3}\pi (0.2m)^3 = \underline{\underline{3.4 \times 10^{-2}\,m^3}}$ $\quad$ so $V = 3.4 \times 10^{-2} (10^2\,cm)^3 = \underline{\underline{3.4 \times 10^4\,cm^3}}$

1·12

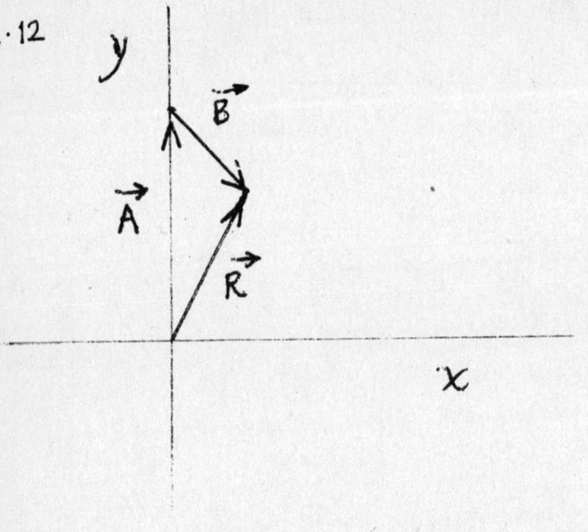

GRAPHICALLY, I MEASURE LENGTH OF R TO BE $\underline{\underline{75\,km}}$.

ANALYTICALLY,

$A_x = 0 \qquad A_y = 100\,km$

$B_x = B\cos 45° = (50\,km)(\cos 45°) = 35.4\,km.$

$B_y = -B\sin 45° = -(50\,km)(\sin 45°) = -35.4\,km.$

$R_x = A_x + B_x = 0 + 35.4\,km$

$R_y = A_y + B_y = 100\,km - 35.4\,km = 64.6\,km.$

$R = \sqrt{R_x^2 + R_y^2} = \sqrt{(35.4)^2 + (64.6)^2}$

$= \underline{\underline{73.7\,km}}$

YACHT IS THUS 73·7 km FROM STARTING POINT.

1·13

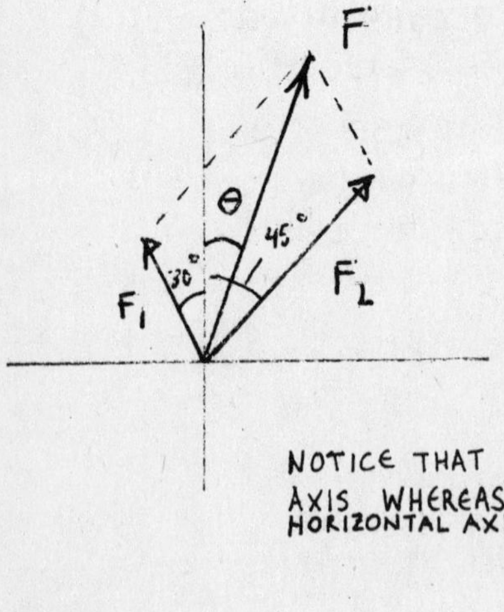

$F_x = F_{1x} + F_{2x} = -F_1 \sin 30° + F_2 \sin 45°$

$= -100 \sin 30° + 200 \sin 45° = 91.4\,n$

$F_y = F_{1y} + F_{2y} = F_1 \cos 30° + F_2 \cos 45°$

$= 100 \cos 30° + 200 \cos 45° = 228\,n$

$F = \sqrt{F_x^2 + F_y^2} = \sqrt{(91.4)^2 + (228)^2}$

$\underline{F = 246\,n}$

$\tan\theta = \frac{F_x}{F_y} = \frac{91.4\,n}{228\,n} = 0.4, \quad \underline{\underline{\theta = 21.8°}}$

NOTICE THAT ANGLES ARE GIVEN WITH RESPECT TO VERTICAL AXIS WHEREAS VISUALLY WE STATE THEM WITH RESPECT TO HORIZONTAL AXIS.

Kinematics in One Dimension

2

• Summary of Important Ideas, Principles, and Equations

Kinematics is the study of motion of objects. Suppose an object moving along the x-axis is at position x_1 at time t_1 and a short time later, at time t_2, it is at position x_2. The average velocity during this time interval is defined to be

$$\text{Average velocity} = \bar{v} = \frac{\text{distance moved}}{\text{time in motion}} = \frac{x_2 - x_1}{t_2 - t_1} = \frac{\Delta x}{\Delta t} \qquad (2.1)$$

Here Δx is pronounced "delta x" and stands for the "difference in x".

Δt is the "difference in time, t".

By Multiplying both sides of eq. 2.1 by Δt we obtain the distance travelled in time Δt,

When an object is moving to the right along the x-axis we take its velocity to be positive, and when it is moving to the left its velocity is negative.

Note that the displacement Δx is measured in meters, the time interval Δt is measured in seconds, and the velocity (and speed) are measured in m/s.

In general the average velocity of an object is different from its velocity at any instant because the object may speed up or slow down at various times. It is useful to define the velocity at a given instant.

$$\text{Instantaneous velocity} \qquad v = \lim_{\Delta t \to 0} \frac{\Delta x}{\Delta t} \qquad (2.2)$$

The reason for considering the limit where the limit interval vanishes is that we want a sufficiently small time interval so that the velocity cannot change during this interval. It will then have a definite value, e.g. the value your speedometer reads at any instant.

The ratio of Δx to Δt in the limit of vanishing Δt is called, in the terminology of calculus, the "time derivative of x with respect to t" and written $v = dx/dt$.

Frequently the velocity of an object is changing, and the quantity that measures how fast velocity is changing is acceleration. If velocity changes by an amount Δ in a time Δt, we define the average acceleration as

$$\text{Average acceleration} = \quad \bar{a} = \frac{\Delta v}{\Delta t} \qquad (2.3)$$

The instantaneous acceleration is

$$\text{Instantaneous acceleration} = \quad a = \lim_{\Delta t \to 0} \frac{\Delta v}{\Delta t} \qquad (2.4)$$

Acceleration is measured in meters per second per second, written as m/s^2.

Observe that acceleration (the "rate of change of velocity") is related to velocity in the same way that velocity (the "rate of change of displacement") is related to displacement.

For motion in one dimension, the slope of the graph of position as a function of time (x vs. t) is the velocity. Similarly, the slope of the graph of velocity as a function of time (v vs. t) is the acceleration.

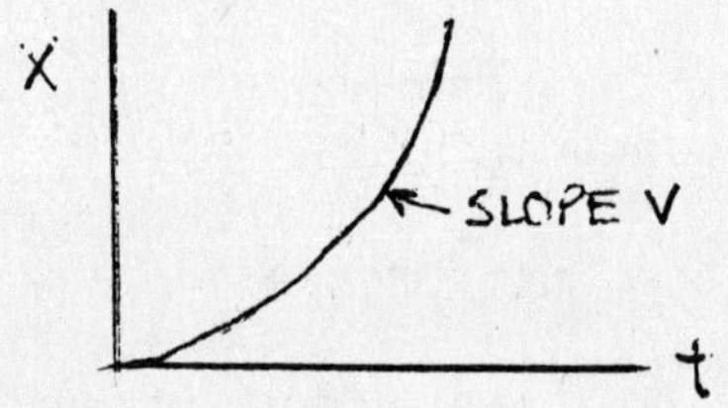

A positive slope (positive velocity) means object is moving to the right. A negative slope (negative velocity) means object is moving to the left.

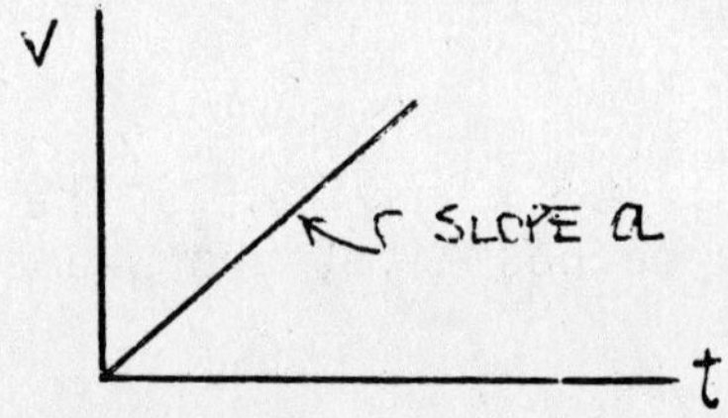

For motion in one dimension two important cases arise.

Case 1. Zero acceleration.

$$a = 0 \qquad (2.5)$$

$$v = v_o = \text{constant} \qquad (2.6)$$

$$x = x_o + vt \qquad (2.7)$$

Here x_o is the position at $t = 0$. Usually we choose the origin such that $x_o = 0$.

Case 2. Uniform (i.e. constant) acceleration.

$$a = \text{constant} \qquad (2.8)$$

$$v = v_o + at \qquad v_o = \text{initial velocity at } t = 0. \qquad (2.9)$$

$$x = x_o + v_o t + \tfrac{1}{2}at^2 \qquad x_o = \text{initial position} \qquad (2.10)$$

$$2as = v^2 - v_o^2, \text{ where } s = x - x_o = \text{distance travelled.} \qquad (2.11)$$

If an object starts from rest ($v_o = 0$) and accelerates to speed v with acceleration a, it will travel a distance s given by

$$2as = v^2 \qquad (2.12)$$

If an object travelling with speed v_o decelerates with acceleration $-a$ ($a > 0$) it will come to rest ($v = 0$) in a distance s given by

$$2as = v_o^2 \qquad (2.13)$$

At the surface of the earth gravity causes all objects, large and small, to experience a downward acceleration of 9.8 m/s when falling in the absence of friction or other forces. We call this acceleration "g".

For vertical motion,

$$a_y = -g \text{ (choosing y positive upward)} \qquad (2.14)$$

$$v_y = v_{yo} - gt \qquad (2.15)$$

$$y = v_{yo}t - \tfrac{1}{2}gt^2 \text{ (choosing } y_o = 0) \qquad (2.16)$$

• Qualitative Questions

Neglect air friction in order to simplify matters.

2M.1 If a car is moving at constant speed, we can say that

A. its velocity is constant.
B. its acceleration is zero.
C. it is going in a straight line.
D. it is travelling in a circle.
E. We can make none of the above statements with certainty.

2M.2 The graph of velocity versus time for an elevator is shown here. For the time interval plotted

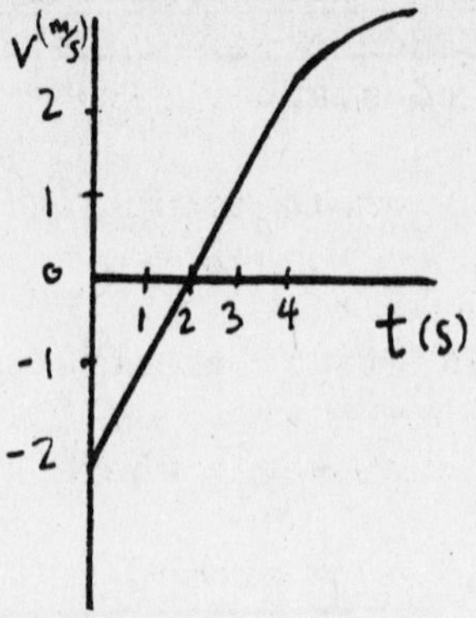

A. the elevator did not stop.
B. the elevator was at rest at t = 0.
C. the elevator did not change direction.
D. the elevator was always going up.
E. the elevator returned to its starting point.

2M.3 The position of a particle moving in one dimension is shown here. From this we can see that the particle

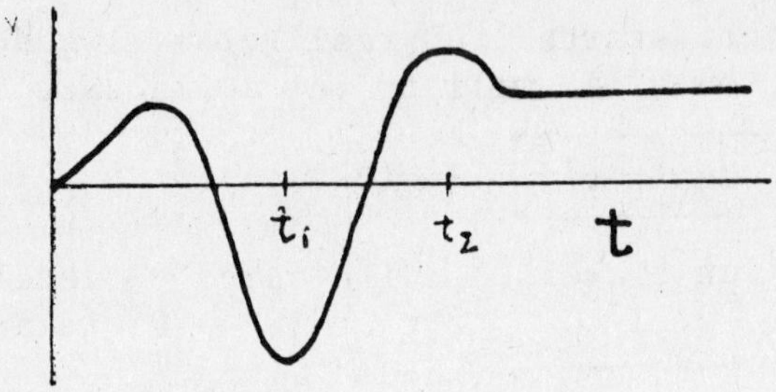

A. had maximum speed at time t_2.
B. had maximum speed at time t_1.
C. stopped twice.
D. turned around twice.
E. sometimes had negative acceleration.

2M.4 The displacement of an object moving in one dimension is sketched here for a 10 second time interval. With respect to this period of time

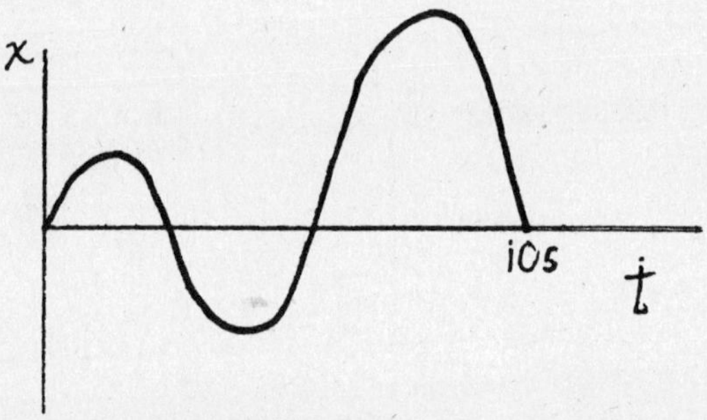

A. the average velocity was positive.
B. the average velocity was negative.
C. the average velocity was zero.
D. the average velocity cannot be determined without knowing the numerical values plotted on the vertical axis.
E. None of the above is correct.

2M.5 The distance moved by an object moving in one dimension is plotted here as a function of time. From this curve we can determine that the average speed of the object between 4 and 8 seconds is

A. 2.5 cm/sec
B. 3.75 cm/sec
C. 5 cm/sec
D. 10 cm/sec
E. 15 cm/sec

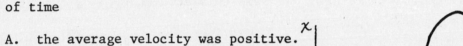

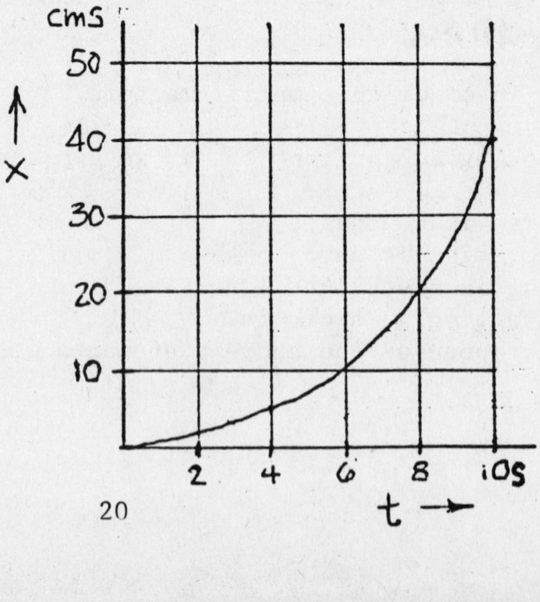

2M.6 Two objects are dropped from rest in a vacuum chamber. One is released one second after the first. Thus the distance separating them as they fall

A. will remain constant.
B. will decrease.
C. will increase.
D. may increase or decrease or remain the same depending on the relative masses of the objects.
E. None of the above is true.

2M.7 I like to mess around in caves and old mines. To figure out how deep a mine shaft is I drop a rock down it. When I release it I start counting "one alligator, two alligators, three alligators, . . .". ("Mississippi's work O.K., too.) If it never hits, I don't go down. How deep would you estimate a 3 alligator shaft is? (Neglect air effects.)

A. 15 m
B. 22 m
C. 44 m
D. 60 m
E. 88 m

2M.8 A motorist travels 60 miles at 30 mph and 60 miles at 60 mph. His average speed for the trip is

A. 32.5 mph
B. 40 mph
C. 45 mph
D. 50 mph
E. 5715 mph

2M.9 Suppose that you take an automobile trip of 300 km. For the first 150 km you travel at a speed v_1 and for the second 150 km you travel at a greater speed v_2.

A. Your average speed will be $\frac{1}{2}(v_1 + v_2)$.
B. Your average speed will be closer to v_1 than to v_2.
C. Your average speed will be closer to v_2 than to v_1.
D. The concept of average speed does not apply here since acceleration is involved in changing from v_1 to v_2.

2M.10 A car travelling 36 km/hr is able to stop with uniform deceleration in a distance of 40 m. The time required for it to stop is

A. 2.22×10^{-3} sec.
B. 1.11 sec.
C. 2.22 sec.
D. 4 sec.
E. 8 sec.

2M.11 How long would it take a drag racer to increase her speed from 20 m/s to 30 m/s if her car can accelerate 15 m/s^2?

A. 0.25 sec.
B. 0.5 sec.
C. 0.67 sec.
D. 1.00 sec.
E. 1.2 sec.

2M.12 You may have heard the "rule of thumb" that when following another car, you should stay back one car length for every 10 mph of your speed. Thus, when driving 50 mph you should stay back about five car lengths, or about 30 meters. To check this out, you would need to measure your reaction time and know the braking deceleration of your car (you might assume other cars are the same for a first approximation). From the laws of physics you could then compute just how far back you should be at any speed. The National Safety Council says that a car should be able to decelerate at about one-fifth g. Assuming this is the case for your car and the one ahead of you, and that your reaction time is 0.8 sec. between the time you see the driver ahead hit his brakes (his brake light goes on) and when you can hit your brakes, what is the minimum following distance (from the back of his bumper to your front bumper allowing no safety factor) if you are travelling 20 m/sec (48 mph)? (Hint: This problem calls for thinking, not calculating.)

A. 12 m
B. 15.36 m
C. 16 m
D. 24 m
E. 31.36 m

• Multiple Choice: Answers and Comments

2M.1 E A is false, since direction of velocity vector could be changing.
B is false, since changing velocity direction means acceleration.
C is false; the object could be moving at constant speed on a curved path.

2M.2 E A is false, since when v = 0 (at t = 2s), the elevator stopped.
B is false; the elevator had velocity of 2 m/s downward at t = 0.
C is false since initially velocity is negative (going down) and later velocity is positive (going up).
D is false. Initially the elevator has negative velocity, i.e. going down.
E is true. The distance travelled is proportional to the area between the curve and the horizontal axis. The elevator descended, stopped at t = 2s, and then travelled upward, passing its starting point at t = 4s.

2M.3 E A and B are false. Maximum speed means maximum slope.
C is false. Stopping means zero slope, and this occurs four times.
D is false. The object turned around three times.
E is true, since negative acceleration means decreasing speed, i.e. decreasing slope.

2M.4 C Average velocity = total displacement/time elapsed. Since the object returned to its starting point the total displacement is zero, so the average velocity is also zero. Do not confuse average speed (not zero here) with average velocity.

2M.5 B At t = 4s, x = 5 cms and at t = 8s, x = 20 cms. Thus Δx = 20 cms - 5 cms = 15 cms and t = 8s - 4s = 4s.

Average speed $v = \frac{\Delta x}{\Delta t} = \frac{15 \text{ cms}}{4 \text{ sec}} = 3.75$ cm/s.

2M.6 C As an object falls its speed increases, so it falls farther during the second second than during the first second, and farther still during the third second. The position of the two objects at intervals of one second is sketched here. The two circles joined by a line indicate the position at any instant. From the drawing we see that the separation steadily increases.

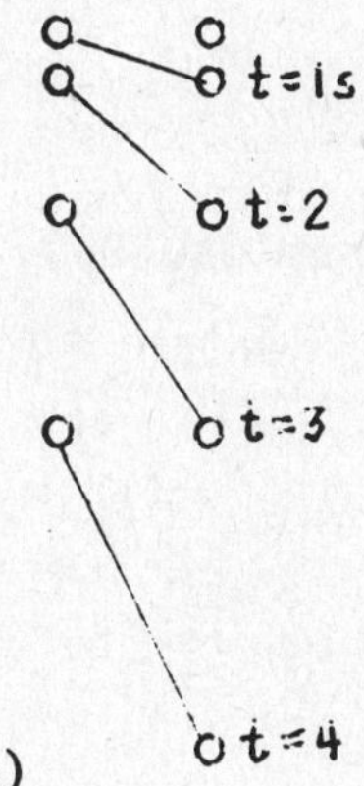

2M.7 C $y = v_o t - \frac{1}{2}gt^2$ $v_o = 0$ (dropped from rest)

t = 3 seconds $g = 9.8$ m/s^2

$y = 0 - \frac{1}{2}(9.8 \text{ m/s}^2)(3\text{s})^2 = -44.1$ m

2M.8 B Average speed = $\frac{\text{total distance}}{\text{total time}}$. Total distance = 120 miles.

Time at 30 mph is $t_1 = \frac{s}{v_1} = \frac{60 \text{ mi}}{30 \text{ mph}} = 2$ hrs

Time at 60 mph is $t_2 = \frac{s_2}{v_2} = \frac{60 \text{ mi}}{60 \text{ mph}} = 1$ hr

Total time = 2 hrs + 1 hr = 3 hrs

Average speed = $\frac{120 \text{ mi}}{3 \text{ hrs}}$ = 40 mph

Note that average speed is <u>not</u> the average of 30 mph and 60 mph (which would be 45 mph) because <u>more time</u> was spent at 30 mph than at 60 mph. Remember that when we say average we mean <u>time average</u>. If a person drove for equal <u>times</u> at 30 mph and 60 mph, then his average speed would be 45 mph.

2M.9 B Using the reasoning of 2M.8, we see that more time is spent travelling at v_1 than at v_2, so the average speed will be closer to this value.

2M.10 E From eq. 2.13, $2as = v^2$. $v_o = 36 \text{ km/hr} = 36 \frac{1000\text{m}}{3600\text{s}} = 10 \text{ m/s}$.

$s = 40 \text{ m}$ $\quad a = \frac{v_o^2}{2s} = \frac{(10 \text{ m/s})^2}{(2)(40 \text{ m})} = 1.25 \text{ m/s}^2$

From eq. 2.9, $v = v_o + at$. When stopped, $v = 0$, so

$0 = v_o + at_s$ $\quad t_s = \frac{-v_o}{a} = \frac{-10 \text{ m/s}}{-1.25 \text{ m/s}^2} = 8\text{s}$

Note that acceleration is negative when decelerating.

2M.11 C From eq. 2.9, $v = v_o + at$.

$v_o = 20 \text{ m/s}$ $\quad at = v - v_o$

$v = 30 \text{ m/s}$ $\quad t = \frac{v - v_o}{a}$

$a = 15 \text{ m/s}^2$ $\quad = \frac{30 \text{ m/s} - 20 \text{ m/s}}{15 \text{ m/s}^2}$

$= 0.67\text{s}$

2M.12 C If your reaction time is 0.8 s, then you will travel a distance of $s = vt = (20 \text{ m/s})(0.8\text{s}) = 16 \text{ m}$ between the time you see the brake light on the car ahead and the time your foot hits your brake pedal. If you were following at this distance behind the car ahead, you would start to brake at a point one car length behind where he did, and you would come to rest just behind him also. Thus this is the minimum distance you should follow under these circumstances (no safety factor).

• Problems

2.1 On a trip you drive 2 hours at 40 km/hr and 3 hours at 60 km/hr. How far do you travel?

2.2 A woman drives from her home in Los Angeles to Laguna Beach, a distance of 70 km. She uses 45 minutes in going to Laguna Beach and one hour on the return trip, and spends 20 minutes on business in Laguna Beach before returning home. What was her average velocity and her average speed for the entire trip, including the stop in Laguna Beach?

2.3 In a 3000 meter race a fast runner laps a slower runner after the fast runner has completed 4 laps and the slow runner has completed 3 laps. Assuming each runner travels at a constant speed, and the fast runner finishes in a time of 9 minutes, what time would you expect the slow runner to register for the race?

2.4 Consider two runners who run the anchor legs for their team's 800 meter relay race. Schneller can run 200 meters in 20.0 seconds (a world class sprinter) and Langsamer can run this distance in 21.4 seconds. Assuming each runs at the corresponding speed in the relay, what is the biggest lead Langsamer can have when Schneller gets the baton if Schneller is just able to catch Langsamer at the finish line?

2.5 A road sign indicates a speed limit of 100 km/hr. Express this speed in miles/hr and in m/s.

2.6 A Chevrolet Cavalier car can accelerate to 30 mph in 6.7 seconds. What is this acceleration in SI units? How far would the car travel from rest in achieving this speed? Assume uniform acceleration.

2.7 A car travelling 100 km/hr tries to overtake and pass a truck travelling 50 km/hr. If the car is 150 meters behind the truck at t = 0, how long will it take to catch up with the truck? How far will the car travel before it catches the truck?

• Problem Solutions

2.1 DISTANCE TRAVELLED IS $x = v_1 t_1 + v_2 t_2 = (40 \frac{km}{hr})(2hrs) + (60 \frac{km}{hr})(3hrs)$

$$= (80 + 180) km = \underline{\underline{260 km}}$$

2.2 AVERAGE VELOCITY = $\dfrac{\text{TOTAL DISPLACEMENT}}{\text{TOTAL TIME}}$

SINCE WOMAN ENDS UP WHERE SHE STARTED, HER TOTAL DISPLACEMENT IS ZERO, AND THUS AVERAGE VELOCITY IS ALSO ZERO.

AVERAGE SPEED = $\dfrac{\text{TOTAL DISTANCE}}{\text{TOTAL TIME}}$

$$v_{AV} = \frac{140\text{ km}}{(45+60+20)\text{ min}} = \frac{140\text{ km}}{125\text{ min}} \cdot \frac{60\text{ min}}{\text{hr}} = \underline{\underline{67.2\text{ km/hr}}}$$

2.3 LET d = LENGTH OF ONE LAP. THEN $4d = v_F t$ AND $3d = v_S t$

THUS $\frac{d}{t} = \frac{v_F}{4} = \frac{v_S}{3}$, $v_S = \frac{3}{4} v_F$

ALSO $v_F = \frac{3000\text{ m}}{9\text{ MIN}} = \frac{3000\text{ m}}{(9)(60\text{s})} = 5.55\text{ m/s}$; $v_S = \frac{3}{4} v_F = \frac{3}{4}(5.55) = 4.17\frac{\text{m}}{\text{s}}$

THUS SLOW RUNNER RUNS 3000 m IN t_S SECONDS, WHERE

$$t_S = \frac{3000\text{ m}}{4.17\text{ m/s}} \qquad t_S = 720\text{s} = \frac{720}{60}\text{ min} \qquad \underline{\underline{t_S = 12\text{ min}}}$$

2.4 $v_S t = 200$, $v_L t = 200 - d$, $v_S = \frac{200\text{ m}}{20\text{ s}} = 10\text{ m/s}$

SO $10t = 200$, $t = 20\text{ s}$ $\qquad v_L = \frac{200\text{ m}}{21.4\text{s}} = 9.35\text{ m/s}$

$d = 200 - v_L t = 200 - (9.35)(20) = 13.1\text{ m}$ $\qquad$ $\underline{\underline{\text{HEAD START} = 13.1\text{ m}}}$

2.5 $1\text{ km} = 1000\text{ m}$, $1\text{ mi} = 1.609\text{ km}$, $1\text{ hr} = 3600\text{ s}$

$$v = 100\frac{\text{km}}{\text{hr}} = 100\left(\frac{1}{1.609}\text{ mi}\right)\left(\frac{1}{\text{hr}}\right) = \underline{\underline{62.2\text{ mi/hr}}}$$

$$v = 100\frac{\text{km}}{\text{hr}} = \frac{100(1000\text{ m})}{3600\text{ s}} = \underline{\underline{27.8\text{ m/s}}}$$

2.6 $v = v_0 + at = 0 + at$, $a = \frac{v}{t} = \frac{13.4\text{ m/s}}{6.7\text{ s}} = \underline{\underline{2\text{ m/s}^2}}$

$2as = v^2$, $s = v^2/2a = (13.4)^2/2(2) = \underline{\underline{44.9\text{ m}}}$

2.7 DISTANCE TRAVELLED BY CAR IS d , $v_c t = d$

DISTANCE TRAVELLED BY TRUCK IS $d - .150$ (IN km)

$v_T t = d - .150$ $\qquad\qquad$ $150\text{ m} = 0.150\text{ km}$

THUS $100t = d$

$50t = d - .150$

$50t = 100t - .150$

$.150 = 50t$

$t = 0.003\text{ hr}$

$d = (100)(t) = (100\frac{\text{km}}{\text{h}})(0.003\text{h}) = \underline{\underline{0.30\text{ km}}}$

$t = 0.003\text{hr} = (0.003)(3600\text{s}) = \underline{\underline{10.8\text{ s}}}$

Kinematics in Two Dimensions; Vectors

3

• Summary of Important Ideas, Principles, and Equations

A vector is a quantity that has magnitude and direction. The location of an object is determined by its position vector, $\vec{r}$. It at time t_1 the object has position vector $\vec{r}_1$ and at a later time t_2 it has position vector $\vec{r}_2$, the displacement of the object during this time interval is $\Delta\vec{r} = \vec{r}_2 - \vec{r}_1$.

Velocity is a vector quantity which measures the rate of change of displacement. The average velocity is the ratio of the displacement to the time interval.

$$\vec{v}_{ave} = \frac{\vec{r}_2 - \vec{r}_1}{t_2 - t_1} = \frac{\Delta\vec{r}}{\Delta t} \tag{3.1}$$

The instantaneous velocity is the limit of the average velocity as the time interval Δt approaches zero:

$$\vec{v} = \lim_{\Delta t \to 0} \frac{\Delta\vec{r}}{\Delta t} \tag{3.2}$$

The speed of an object is the magnitude of the velocity (in meters/sec.)

Speed is a scalar quantity (i.e. it is just a number).

Acceleration is a vector quantity which measures the rate of change of velocity. Observe that acceleration bears the same relationship to velocity that velocity bears to displacement.

If velocity of an object is $\vec{v}_1$ at time t_1 and $\vec{v}_2$ at a later time t_2, then the average acceleration during this time interval is

$$\vec{a}_{ave} = \frac{\vec{v}_2 - \vec{v}_1}{t_2 - t_1} = \frac{\Delta\vec{v}}{\Delta t} \tag{3.3}$$

The <u>instantaneous acceleration</u> is the limit of the average acceleration as the time interval Δt approaches zero.

$$\vec{a} = \lim_{\Delta t \to 0} \frac{\Delta\vec{v}}{\Delta t} \tag{3.4}$$

Acceleration is measured in $\frac{m/s}{s} = m/s^2$. It is best to pronounce this unit as "meters per second per second", rather than "meters per second squared", since the idea of a "squared second" may be confusing.

Observe that for an object moving in two (or three) dimensions two kinds of acceleration can occur. First, the object may speed up or slow down. This kind of acceleration (which we studied in the previous chapter) is called <u>tangential acceleration</u>, because the change in velocity is tangent to the path the object is following.

Second, velocity can change direction without changing magnitude (i.e. speed). In this case the change in the velocity vector $\Delta\vec{v}$ is directed sideways to the velocity vector. Since acceleration is directed in the direction of $\Delta\vec{v}$, it also is directed sideways to the velocity vector. This sideways acceleration is called <u>centripetal acceleration</u>. Thus an object which is turning to the right is always accelerating to the right, even though its speed may be constant. Both kinds of acceleration are measured in m/s^2.

<u>Projectile motion</u> is an important example of motion in two dimensions. A key observation is that vertical and horizontal motion are independent. We can see that this is so by noting that horizontal movement is measured with respect to some reference frame below the object, e.g. the sidewalk. Suppose now that the sidewalk starts moving (like the "people mover" sidewalks in airports). Now the horizontal velocity of the object is changed, but its vertical velocity is not affected by motion of the sidewalk below.

For a projectile, the horizontal motion has zero acceleration. The vertical motion has acceleration $-g = -9.8\ m/s^2$ (i.e. downward acceleration).

If the projectile is launched at angle θ above horizontal with initial velocity v_o,

$$\boxed{v_{ox} = v_o\cos\theta \qquad v_{oy} = v_o\sin\theta} \tag{3.5}$$

The subscript "o" refers to the value of velocity at time $t = 0$.

The distances moved horizontally and vertically are x and y.

$$\boxed{x = v_{ox}t \qquad y = v_{oy}t - \tfrac{1}{2}gt^2} \tag{3.6}$$

The projectile reaches the top of its flight in a time t_1 (at which time $v_y = 0$)

$$v_x = v_{ox} = v_o \cos\theta = \text{constant} \qquad (3.7)$$

$$v_y = v_{oy} - gt = v_o \sin\theta - gt \qquad (3.8)$$

At the top, $v_y = 0 = v_o \sin\theta - gt_1$ so $t_1 = \frac{v_o \sin\theta}{g}$ (3.9)

The object spends as much time going up as coming down, so the total time in the air is $2t_1$. In this time the object travels a horizontal distance R (the "range").

$R = 2t_1 v_x = 2\frac{v_o}{g} \sin\theta \ v_o \cos\theta$

$$R = 2\frac{v_o^2}{g} \sin\theta \cos\theta \qquad (3.10)$$

By eliminating t in the expressions for x and y (equation 3.6) we obtain the equation of the trajectory of a projectile:

$$y = v_{oy} \frac{x}{v_{ox}} - \tfrac{1}{2} g \left(\frac{x}{v_{ox}}\right)^2 \qquad (3.11)$$

Using eq. 3.5 this may be written as

$$y = \frac{\sin\theta}{\cos\theta} x - \frac{g}{v_o^2 \cos\theta} x^2 \qquad (3.12)$$

This is the equation of a parabola.

• Qualitative Questions

3M.1 A boy throws a ball into the air with a speed of 30 m/s. Neglect air effects.

A. The height to which the ball rises depends on the mass of the ball.
B. The time the ball is in the air is independent of the angle at which the ball is thrown.
C. The sum of the horizontal distance travelled plus the vertical height reached will be constant for different angles of throwing.
D. The ball is accelerating all the time it is in the air.
E. The ball is always moving.

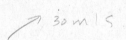

3M.2 Consider two balls which are thrown into the air simultaneously from the same position on a level field. Neglect air friction. If the two balls are in the air the same length of time,

A. they must have been thrown with the same initial speed.
B. they must have been thrown at the same angle.
C. they will both land the same horizontal distance away.
D. they will both reach the same maximum height.
E. they must have had the same masses.

3M.3 Shown here are several trajectories followed by a ball thrown by a baseball outfielder. Assuming that air effects are negligible, which ball stayed in the air the longest?

A. A
B. B
C. C
D. D
E. All were in the air for the same amount of time.
F. Cannot be determined without knowing initial velocities.

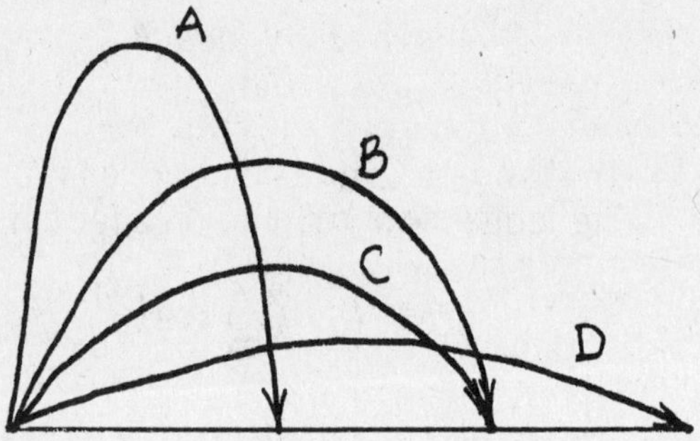

3M.4 A ball rolls off a table with different speeds and follows the trajectories shown here. Neglecting air friction, the time between leaving the table and hitting the floor will be greatest for

A. path A.
B. path B.
C. path C.
D. either path A, B, or C, depending on the mass of the ball.
E. None of the above is correct.

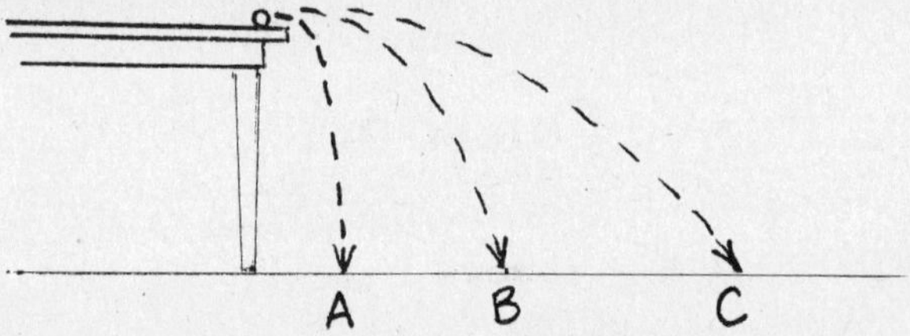

3M.5 Two balls are thrown into the air simultaneously from the same level. They reach the same height. If we neglect air friction, we deduce that

A. they were thrown with the same speed.
B. they were both in the air for the same length of time.
C. they both landed the same horizontal distance from the point from which they were thrown.
D. they were both thrown at the same angle of elevation.

3M.6 A fish jumps up a waterfall with the minimum speed needed to reach the top. He can begin his leap at points A, B, or C. If air friction is negligible, his time in the air will be

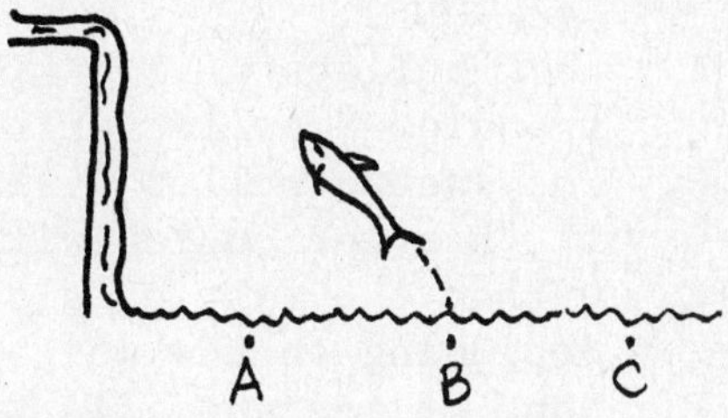

A. greatest if he leaves at point A.
B. greatest if he leaves at point B.
C. greatest if he leaves at point C.
D. independent of the point from which he jumps.

3M.7 Napoleon and the Duke of Wellington, having positioned their cannons on the opposite sides of a gorge as shown here, fire them at each other. If the cannons were initially aimed directly at each other we can reason (neglecting air effects) that

A. the two cannonballs will never strike each other.
B. whether or not the two cannonballs strike each other depends only on the velocities of the balls.
C. whether or not the cannonballs strike each other depends only on whether or not the two cannons were fired simultaneously.
D. the two balls will strike each other only if they have the same initial velocity.
E. None of the above.

3M.8 A rifle is sighted in on level ground at 200 yards so that the bullet will strike the target when the crosshairs are placed right on the target. Suppose now you shoot this rifle downhill at a target 200 yards away. You should aim

A. high.
B. low.
C. right at the target.
D. high for some angles of slope, low for others.

3M.9 Consider two rifle bullets which are fired simultaneously horizontally over a level field. One bullet has very high velocity, the other low velocity. Neglecting friction, we know that if both were fired from the same height,

A. both bullets will strike the ground at the same time.
B. the slower bullet will strike the ground first.
C. the faster bullet will strike the ground first.
D. the heavier bullet will strike the ground first.
E. either the fast or slow bullet may strike the ground first depending on the weights of the bullets.
F. None of the above is true.

3M.10 Two bullets of different weights are fired simultaneously uphill parallel to an inclined plane. The bullets have different muzzle velocities. If we neglect the effect of air friction, we can reason that

A. the heavier bullet will strike the plane first.
B. the lighter bullet will strike the plane first.
C. the faster bullet will strike the plane first.
D. the slower bullet will strike the plane first.
E. More than one of the above are correct.
F. None of the above is correct.

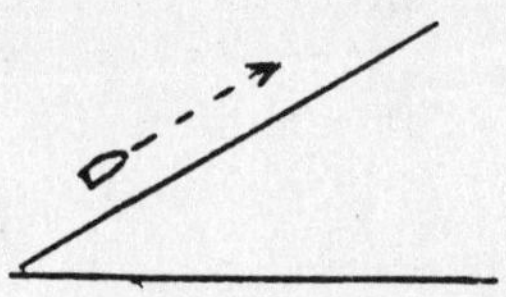

• Multiple Choice: Answers and Comments

3M.1 D Gravity causes the ball to accelerate downward all during its flight.
A is false, since gravity causes an acceleration independent of the mass of the object.
B is false; a ball thrown straight up will be in the air longer than one thrown at an angle, because it has a greater initial upward velocity, and this is what counts.
C is a nonsense answer.
E is false; a ball thrown straight up stops at the top of its trajectory.

3M.2 D Consider the vertical motion independently from the horizontal motion. Both the vertical height and the time in the air depend only on the initial vertical velocity (and on the value of g, which is constant). Thus two balls in the air are the same length of time must have had the same initial vertical velocity and would thus rise to the same height.
A is false; they could have been thrown at different angles.
B is false; they could have been thrown at different angles and different speeds, as long as the vertical component of velocity was the same for each.
C is false; horizontal distance depends on horizontal speed, which need not be the same for both.
E is false; acceleration of gravity doesn't depend on the mass of the ball.

3M.3 A Ignore horizontal motion and we readily see that the ball which reaches the greatest height is in the air longest.

3M.4 E The time of fall will be the same for each path, since it depends only on the vertical distance fallen, not on horizontal motion. Each object takes the same time to fall a given distance down <u>independent</u> of horizontal motion.

3M.5 B Same reasoning as for 2M.14.

3M.6 D Time in air depends only on height jumped, not on horizontal motion.

3M.7 C Suppose there were no gravity. The cannonballs would then strike each other. If gravity is present, the balls will fall as they move sideways. Suppose that their horizontal velocities were such that they would meet after 1 second in the absence of gravity. Then if gravity were present they would still move sideways at the same speed, but they would fall at the same time. However, each would fall the same distance (in the same times), so they would still hit. Thus we require only that the cannons be aimed at each other and that they be fired simultaneously. This is the principle behind the "hunter and the monkey" demonstration which you may have seen in lecture.

3M.8 B Suppose that you are shooting on level ground. You aim the gun barrel at point A so that the bullet will strike the target at point T, having fallen a distance d due to gravity. If you now shoot downhill, your sights will cause you to aim the barrel at point B, as if gravity would move the bullet back to the target T. However, now gravity pulls the bullet downward to point T' (i.e. straight down), instead of to T, and the bullet goes above the target. Thus you must aim low when shooting downhill! See if you can analyze on your own the case of shooting uphill.

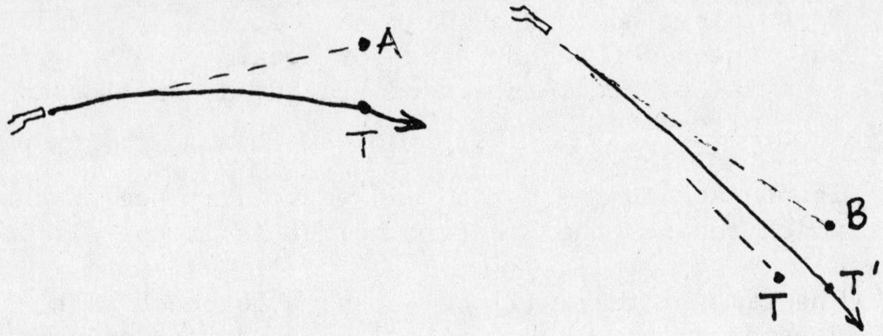

3M.9 A Horizontal motion and vertical motion are independent. If two bullets drop the same vertical distance, they require the same time to do so, independent of thier horizontal velocities (cf. 2M.16).

3M.10 F Both bullets will strike the plane simultaneously, since to do so they must each fall the same distance, and this requires the same amount of time independent of mass or initial speed.

• Problems

3.1 A swimmer can move through the water at a speed of 0.4 m/s. How long would it take her to swim 100 meters in still water? How long would it take to swim 100 m upstream against a current of 0.2 m/s? How long would it take her to swim downstream with a helping current of 0.2 m/s?

3.2 A pilot starts a trip at Memphis, 400 km due south of St. Louis, his destination. He heads his plane due north with an airspeed of 200 km/hr expecting to reach St. Louis in 2 hours. He fails to take account of a 60 km/hr wind from the east. Where in fact will he be located with respect to St. Louis after 2 hours? What is his groundspeed and in what direction is his "ground velocity" vector pointing?

3.3 A sailboat which moves through the water at 20 km/hr wishes to sail from the Ferry Building in San Francisco to Tiburon, a town about 10 km due north of the Ferry Building. An ocean current of 6 km/hr is flowing in through the Golden Gate from west to east, across the port bow of the vessel. In what direction should the sailor head his boat with respect to north in order to reach Tiburon in the shortest time? How long will the voyage take?

3.4 A Coast Guard cutter stationed at Key West picks up an unidentified vessel one morning at 2:00 A.M. They track the boat for an hour and find that it is travelling on a bearing of 330° (i.e. 30° west of north) at a steady speed of 25 km/hr (about 13.5 knots). They suspect the ship is a smuggler, and when it is 300 km due east of them, off the Bahamas, they take pursuit at full speed of 75 km/hr. In what direction should the cutter head in order to intercept the smugglers in the shortest time? How long will it take the cutter to reach them?

3.5 With what speed, expressed in m/s and in mph, would Nolan Ryan have to throw a baseball in order to hit a point 80 meters above his head on the roof of the New Orleans Superdome? Do you think this is physically possible?

3.6 Two sky divers jump out of a plane at high altitude, with the second diver jumping 0.5 seconds after the first. They plan to open their parachutes simultaneously 5 seconds after the first person jumped. Assuming air friction is negligible (a fairly good assumption for the first few seconds) how far apart will the two divers be when they open their chutes? How far will each have then fallen? This is an important consideration to ensure that the chute of the first diver will not be fouled by the person above.

3.7 A bicyclist bets a friend that his Gitane racing bike can beat the friend's 75 Ford Mustang in a 100 meter race if the bike has a running start and the Mustang starts from a stop. The biker can reach and hold a speed of 30 mph, and the Ford can reach 60 mph in 8 seconds. Assume the Ford can maintain this level of acceleration for the entire race. First express the bike speed and the car acceleration in SI (metric) units, then determine who would win such a race and where the loser would be when the winner crosses the finish line.

3.8 Space travellers would not be comfortable travelling in a space ship which accelerated at a rate much greater than g ($9.8\ m/s^2$). If you accelerated at this rate, how long would it take to reach the orbit of Mars, starting from earth, a distance of about 7.8×10^{10} m? How fast would you be going when you got there? How long would it take to reach the orbit of Jupiter, about 6.3×10^{11} m from earth? How fast would you be travelling when you reached there? Do you think the rocket passengers would feel any different when they passed Jupiter as compared to when they passed

Mars at lower speed? Note that the theory we have developed holds only as long as the speeds involved are small compared to the velocity of light, 3×10^8 m/s. At speeds near this value one must use the special theory of relativity, which we will study later. Note also that if we call the distance from the sun to the earth 1 astronomical unit (1 a.u.), then the distance from the sun to Mars is about 1.5 a.u. and from the sun to Jupiter about 5.2 a.u.

3.9 You probably heard about the time the mayor of New York fell off the top of the Empire State Building. As he passed the 57th floor they heard him muttering, "So far, so good. So far, so good." As he passed the 23rd floor he was still mumbling, "So far, so good. So far, so good." That's sometimes the way it is with physics students. The drop date passes, and then mid-term, and then dead week, and then the final exam is approaching. So far, so good; so far, so good. Then, kerpowie! Of course, there is always the chance of a savior. In the mayor's case, Superman sprang downward to the rescue 2 seconds after our hero fell. Assuming they both started 313.6 meters above the ground, with what downward speed would Superman have had to jump in order to catch the mayor before he splattered on the pavement? Do you think Superman would be capable of springing forth with such speed?

3.10 A model rocket is fired straight up with an initial velocity of 60 m/s. (a) How high will it go, and how long will this take? (b) At how many seconds after blast-off will it be 100 meters above the ground, and what will its velocity be then?

3.11 Researchers at the Rand Corporation have designed a "tubecraft" system called the VHST (Very High Speed Transit) capable of carrying passengers through an airless underground tunnel at speeds in excess of 14,000 miles per hour. They envision a 21 minute non-stop trip between Los Angeles and New York, with optional stops at Amarillo and Chicago (which would increase the time to 37 minutes). Let us check these numbers out in order to see what accelerations are required. As a first approximation, suppose that the train accelerates uniformly for the first half of the trip ($10\frac{1}{2}$ min) and then decelerates at the same rate for the second half of the trip. The approximate distance from L.A. to New York is 3200 miles. What is the acceleration required, expressed in m/s^2? What is the ratio of this acceleration to g, the acceleration of gravity? Do you think this acceleration would be too uncomfortable? (Note: A fighter pilot in a pressurized suit can withstand an acceleration of about 7 g's before passing out.)

3.12 Suppose that the VHST system described in the preceding problem operated like this: Accelerate uniformly at 9 m/s^2 for the 2000 km trip from Los Angeles to Amarillo, continue at constant speed form Amarillo to Chicago (1800 km) and then decelerate uniformly at 12 m/s^2 for the 1500 km leg from Chicago to New York. How long would such a trip require, and what would be the maximum speed attained?

3.13 Suppose that a woman driving a Mercedes zooms out of a darkened tunnel at 120 km/hr (about 72 mph). She is momentarily blinded by the light, and when she recovers she sees that she is fast overtaking a truck ahead in her lane moving at a moderate 50 km/hr. She hits her brakes as fast as she can (her reaction time is 0.8 seconds, about average). If she can decelerate at 2 m/s^2, what is the minimum distance between the driver and the truck when she first sees it if they are not to collide?

3.14 It is doubtful if any major league baseball outfielder can throw a ball faster than about 80 mph. What would be the maximum range of such a throw, neglecting air resistance? (Note: Outfielders don't try for maximum range, rather they try to get the ball back quickly, and to do so they use a low trajectory and let the ball skip once or twice.)

3.15 When a baseball outfielder throws a ball in to a base his aim is not to achieve maximum range, but rather to get the ball there as quickly as possible. He can sometimes accomplish this by skipping the ball in on one bounce. To see how this might work, consider the following simplified example.

(a) Suppose the outfielder throws the ball with a speed of 30 m/s from ground level (not a realistic assumption, but it makes the math easier). How far away will the ball strike the ground if the ball is thrown for maximum range, i.e. at 45°? above horizontal? Suppose home plate is at the point where the ball lands. How long does it take the ball to reach the plate with this high trajectory?

(b) Now consider the case where the ball is thrown at a lower angle, say 30° above horizontal. Assume that when the ball bounces on artificial turf it comes off at the same angle at which it hits and that its speed is reduced by 10 percent. How many times will the ball bounce before reaching home plate when thrown at 30°? How long will it take the ball to reach the plate in this case? How far from the plate will the ball first bounce?

(c) The ball comes off artificial turf with greater speed than would be the case for a similar trajectory on natural grass. Would this lead you to aim a little higher or a little lower when throwing on artificial turf, as compared to grass? Where would you aim if there were no friction on the bounce at all? Remember that the aim is to get the ball to the target as quickly as possible.

3.16 During World War II allied planes used a technique called "skip bombing" in an attempt to blow up some dams in Holland. The idea was something like this (I don't remember if it worked or not). A fast plane approaches the dam in level flight at low altitude h_1 with speed v. A bomb is released which skips off the water and bounces up against the dam, striking it at a height h_2 above the water. Given the speed of the plane, and assuming the bomb loses no speed when it bounces off the water (not very realistic, but a useful first approximation), and that the bomb bounces off the water at the same angle at which it hits, how far from the dam should the bomb be released? Solve for this distance x in terms of v, h_1, h_2 and g, and evaluate the answer for the case where v = 500 km/hr, h_1 = 80 meters and h_2 = 20 meters.

• Problem Solutions

3.1 $d = Vt$, so $t = \frac{d}{V}$ IN STILL WATER $V = 0.4\ m/s = V_1$

SO $t_1 = \frac{d}{V_1} = \frac{100\ m}{0.4\ m/s} = 250\ s$

GOING AGAINST 0.2 m/s CURRENT HER SPEED WITH RESPECT TO EARTH IS $V_2 = 0.4\ m/s - 0.2\ m/s = 0.2\ m/s$, SO TIME REQUIRED IS

$$t_2 = \frac{d}{V_2} = \frac{100\ m}{0.2\ m/s} = 500\ s$$

GOING WITH CURRENT HER SPEED IS $V_3 = 0.2\ m/s + 0.4\ m/s = 0.6\ m/s$

THUS $t_3 = \frac{d}{V_3} = \frac{100\ m}{0.6\ m/s} = 167\ s$

3.2 DRAW VELOCITY VECTORS, LABELLING TIP OF ARROW WITH MOVING OBJECT AND TAIL OF ARROW WITH REFERENCE FRAME. DRAW ARROWS TO SCALE ACCORDING TO APPROPRIATE SPEEDS.

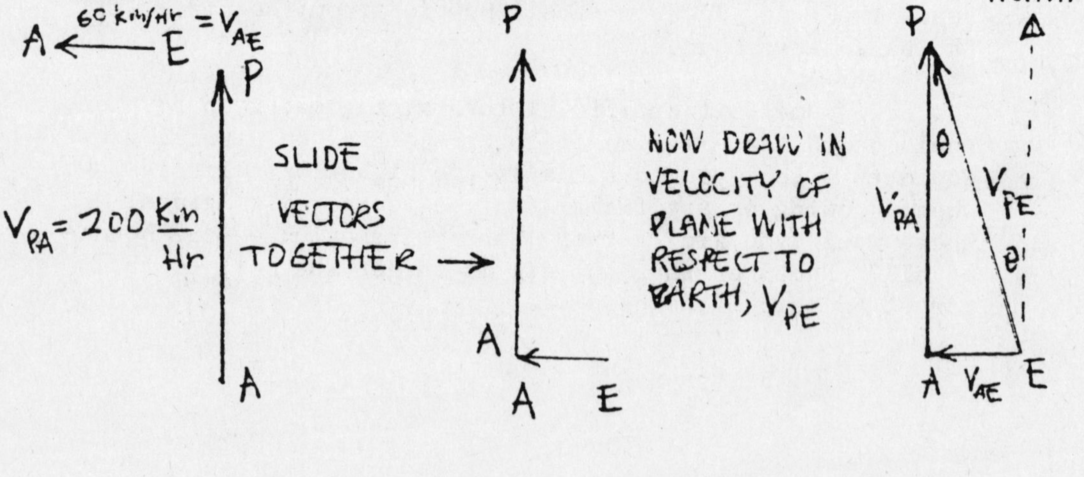

A = AIR, P = PLANE, E = EARTH

$V_{PA} = 200$ km/hr = "AIRSPEED", $V_{PE} = ?$ = "GROUNDSPEED"

FROM THE RIGHT TRIANGLE WE FIND $V_{PE}^2 = V_{AE}^2 + V_{PA}^2 = (60\text{ km/hr})^2 + (200\text{ km/hr})^2$

$$V_{PE} = \underline{\underline{208.8\text{ km/hr}}}$$

THE "GROUND VELOCITY" VECTOR IS POINTING AT AN ANGLE θ WEST OF NORTH, WHERE $\tan\theta = \frac{60}{200}$, SO $\underline{\underline{\theta = 16.7^\circ}}$ WEST OF NORTH.

IN 2 HRS PLANE WILL TRAVEL NORTH A DISTANCE $d_1 = V_{PA}t$

$$= (200\text{ km/hr})(2\text{ hr})$$
$$d_1 = 400\text{ km}$$

IN 2 HRS WIND WILL BLOW PLANE A DISTANCE d_2 TO THE WEST,

$$d_2 = V_{AE}t = (60\text{ km/hr})(2\text{ hr}) = \underline{\underline{120\text{ km WEST OF ST. LOUIS}}}$$

THUS PLANE WILL END UP 120 KM DUE WEST OF ST. LOUIS.

3.3 LET $V_{BW} = 20$ km/hr = SPEED OF BOAT WITH RESPECT TO WATER.

$V_{WE} = 6$ km/hr = SPEED OF WATER WITH RESPECT TO EARTH.

$V_{BE} = ?$ = VELOCITY OF BOAT WITH RESPECT TO EARTH, DIRECTED DUE NORTH WITH UNKNOWN MAGNITUDE. USE METHOD EXPLAINED IN PRECEEDING PROBLEMS.

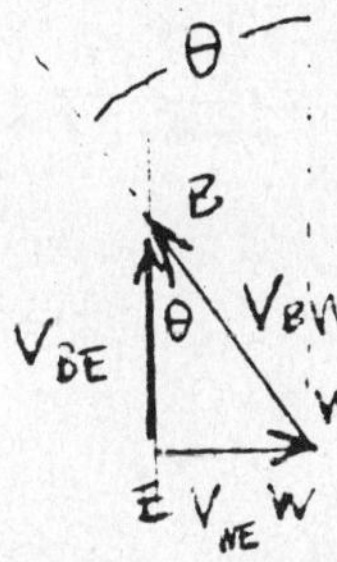

THUS $V_{BE}^2 = V_{BW}^2 - V_{WE}^2$

$= (20)^2 - (6)^2$

$V_{BE} = 19.1$ km/hr.

$t = \frac{d}{V_{BE}} = \frac{10\text{ km}}{19.1\text{ km/hr}} = 0.52$ hr

$\underline{\underline{t = 31\text{ min}}}$

$\tan\theta = \frac{6}{19.1}$, $\theta = 17^\circ$

HEAD 17° WEST OF NORTH

3.4 LET t = TIME FOR CUTTER TO REACH SMUGGLERS. IN THIS TIME CUTTER TRAVELS A DISTANCE d_c AND SMUGGLERS TRAVEL DISTANCE d_s, WHERE $d_c = v_c t = 75t$ AND $d_s = v_s t = 25t$, SO $d_c = 3d_s$.

THE DISPLACEMENT VECTORS INVOLVED ARE DRAWN HERE:

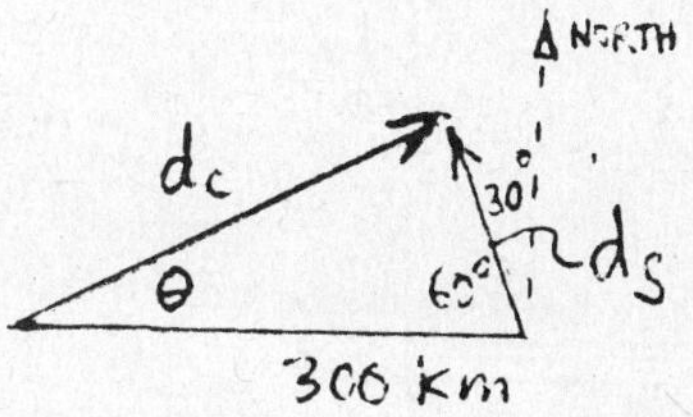

USE LAW OF COSINES:

$d_c^2 = d_s^2 + (300)^2 - 2(d_s)(300)\cos 60^\circ$

$\cos 60^\circ = 1/2$, $d_c = 3d_s$

$(3d_s)^2 = d_s^2 + (300)^2 - 2d_s(300)(1/2)$

$9d_s^2 - d_s^2 + 300d_s - (300)^2 = 0 \Longrightarrow 8d_s^2 + 300d_s - (300)^2 = 0$

USE QUADRATIC FORMULA TO SOLVE FOR d_s:

$$d_s = \frac{-300 \pm \sqrt{(300)^2 + (4)(8)(300)^2}}{(2)(8)} = \frac{-300}{16} \pm \frac{300}{16}\sqrt{33} = \frac{300}{16}\left[-1 \pm \sqrt{33}\right]$$

CHOOSE POSITIVE SIGN SINCE d_s IS POSITIVE AND $d_s = \frac{300}{16}\left[-1+\sqrt{33}\right]$

$d_s = 89$ km

$d_s = v_s t$, $t = \frac{d_s}{v_s} = \frac{89 \text{ km}}{25 \text{ km/hr}} = 3.56$ HRS $= \underline{\underline{3 \text{ HRS } 34 \text{ MIN}}}$

APPLY LAW OF SINES TO FIND θ : $\frac{d_s}{\sin\theta} = \frac{d_c}{\sin 60}$, $\frac{d_s}{\sin\theta} = \frac{3d_s}{\sin 60}$

$\sin\theta = 1/3 \sin 60°$, $\theta = 16.8°$ NORTH OF EAST

THE CUTTER SHOULD STEER 16.8°N OF E OR ON A BEARING OF $90 - 16.8 = \underline{73.2° \text{ EAST OF NORTH}}$ TO INTERCEPT THE SMUGGLERS.

3.5 FROM EQ. (2-21), $2gs = v_o^2$, $s = 80$ m $v_o^2 = (2)(9.8 \text{ m/s}^2)(80 \text{ m})$

NOLAN HAS THROWN THE BALL IN EXCESS OF 100 MPH, SO THIS IS PHYSICALLY POSSIBLE

$v_o = \underline{\underline{39.6 \text{ m/s}}}$

$= (39.6)(2.24 \text{ mi/hr}) = \underline{\underline{88.7 \frac{\text{mi}}{\text{hr}}}}$

3.6 AN OBJECT FALLING FROM REST WILL DROP A DISTANCE y IN TIME t, WHERE $y = \frac{1}{2}gt^2$ $(v_o = 0)$

FOR FIRST DIVER, $t_1 = 5$ secs, $y_1 = \frac{1}{2}(9.8 \text{ m/s}^2)(5 \text{ s})^2 = \underline{\underline{122.5 \text{ m}}}$

FOR SECOND DIVER $t_2 = 4.5$ secs $y_2 = \frac{1}{2}(9.8 \text{ m/s}^2)(4.5 \text{ s})^2 = \underline{\underline{99.2 \text{ m}}}$

SEPARATION IS $\Delta y = y_1 - y_2 = 122.5 \text{ m} - 99.2 \text{ m} = \underline{\underline{23.3 \text{ m}}}$

3.7 BIKE SPEED $V_B = 30 \text{ mi/hr} = 30(0.447) \text{ m/s} = 13.4 \text{ m/s}$

CAR ACCELERATION $a_c = \frac{60 \text{ mi/hr}}{8 \text{ sec}} = 7.5 \text{ mi/s-hr} = 7.5 \frac{(1608 \text{ m})}{(s)(3600 s)} = 3.35 \text{ m/s}^2$

SUPPOSE CAR WINS RACE. THIS WOULD REQUIRE AN ACCELERATION TIME t_c, WHERE $x = 0 + \frac{1}{2}a_c t_c^2$, $t_c = \sqrt{\frac{2x}{a_c}} = \sqrt{\frac{(2)(100 \text{ m})}{3.35 \text{ m/s}^2}} = 7.7$ s

IN 7.7 s BIKE WOULD TRAVEL A DISTANCE $x_B = v_B t_c = (13.4 \text{ m/s})(7.7 \text{ s})$

$x_B = 103.5$ m

BUT THIS MEANS BIKE REACHED FINISH LINE FIRST.

TIME FOR BIKE TO GO 100 m IS t_B , $t_B = x/v_B = 100 \text{ m}/13.4 \text{ m/s}$

$= 7.46$ s

IN THIS TIME CAR TRAVELS A DISTANCE x_c,

$x_c = \frac{1}{2}a_c t_B^2 = (\frac{1}{2})(3.35 \text{ m/s}^2)(7.46 \text{ s})^2 = \underline{\underline{93.2 \text{ m}}}$

THUS $\underline{\underline{\text{BIKE BEATS THE CAR BY ABOUT 7 METERS IN THIS RACE}}}$.

3.8 FROM EQ. (2.10) $x = \frac{1}{2}gt^2$ $(x_o = 0, v_o = 0, a = g)$

$t = \sqrt{\frac{2x}{g}}$

FOR MARS $x_M = 7.8 \times 10^{10}$ m, $t_M = \left[\frac{(2)(7.8)(10^{10})m}{9.8 m/s^2}\right]^{1/2} = \underline{\underline{1.26 \times 10^5 \text{ sec}}}$.

NOTE THAT 1 YEAR = (365)(24)(3600) SECS = 3.15×10^7 SEC
1 DAY = (24)(3600) SEC = 0.86×10^5 SEC.

THUS $t_M = 1.26 \times 10^5 (1/0.86 \times 10^5 \text{ DAY}) = \underline{\underline{1.46 \text{ DAYS}}}$

FOR JUPITER $x_J = 6.3 \times 10^{11}$ m, $t_J = \left[\frac{(2)(6.3)(10^{11})}{9.8}\right]^{1/2} = \underline{\underline{3.59 \times 10^5 \text{ sec}}}$

THUS $t_J = 3.59 \times 10^5 (1/0.86 \times 10^5 \text{ DAY}) = \underline{\underline{4.17 \text{ DAYS}}}$

THE SPEEDS ACQUIRED WOULD BE $v_M = g t_M = (9.8 m/s^2)(1.26 \times 10^5 s)$

$v_M = \underline{\underline{1.2 \times 10^6 \text{ m/s}}}$ AT MARS

$v_J = g t_J = (9.8 m/s^2)(3.59 \times 10^5 s)$

$v_J = \underline{\underline{3.5 \times 10^6 \text{ m/s}}}$ AT JUPITER

THE PASSENGERS WOULD HAVE NO SENSATION OF THE SPEED, JUST AS WE HAVE NO SENSE OF THE HIGH SPEED WITH WHICH WE ARE MOVING AROUND THE SUN.

3.9 DISTANCE FALLEN BY SUPERMAN IS $y_s = v_0 t_s + \frac{1}{2} g t_s^2$
(TAKING y POSITIVE DOWNWARDS)

DISTANCE FALLEN BY THE MAYOR IS $y_F = 0 + \frac{1}{2} g t_F^2$
(THE MAYOR STARTS AT REST)

IF SUPERMAN CATCHES THE MAYOR, $y_s = y_F$ ALSO $t_F = t_s + 2$

THUS $v_0 t_s + \frac{1}{2} g t_s^2 = \frac{1}{2} g (t_s + 2)^2 = y_F$

THE LAST CHANCE TO CATCH THE MAYOR IS JUST BEFORE HE HITS THE GROUND, i.e. WHEN $y_F = 313.6$ m

SOLVE FOR t_F: $\frac{1}{2} g t_F^2 = y_F$, $t_F = \sqrt{\frac{2y_F}{g}} = \sqrt{\frac{(2)(313.6 m)}{9.8 m/s^2}} = 8$ SECS.

$t_s = t_F - 2 = 8s - 2s = 6s$

$y_s = v_0 t_s + \frac{1}{2} g t_s^2 \Longrightarrow 313.6 m = (6s)v_0 + \frac{1}{2}(9.8 m/s^2)(6s)^2$

$313.6 m = (6s) v_0 + 176.4 m$

$(6s) v_0 = 313.6 m - 176.4 m = 137.2 m$; $v_0 = \frac{137.2}{6} \frac{m}{s} = \underline{\underline{22.9 \frac{m}{s}}}$

CURRENT DATA SUGGESTS THAT EVEN SUPERMAN COULD NOT SPRING DOWNWARD WITH SUCH A HIGH VELOCITY SO IT APPEARS THE MAYOR IS OUT OF LUCK. C'EST LA VIE.

V_0 = INITIAL SPEED OF SUPERMAN

3.10 (a) FROM EQ. (2.13) $2g y_{max} = v_0^2$, $y_{max} = \frac{v_0^2}{2g} = \frac{(60 m/s)^2}{(2)(9.8 m/s^2)} = \underline{\underline{184 \text{ m}}}$ HIGH

$v = v_0 - gt = 0$ AT TOP, SO $t = \frac{v_0}{g} = \frac{60 m/s}{9.8 m/s^2} = \underline{\underline{6.1 s}}$

(b) FROM EQ (2.16). $y = v_0 t - \frac{1}{2} g t^2$

$100 m = 60t - (\frac{1}{2})(9.8 m/s^2) t^2 \longrightarrow (4.9 m/s^2) t^2 - (60 m/s) t + 100 m = 0$

USING QUADRATIC FORMULA, $t = \frac{60 \pm \sqrt{(60)^2 - (4)(100)(4.9)}}{(2)(4.9)}$ s

$t_2 = \underline{\underline{10.3\text{ s}}}$ $\qquad t = \underline{\underline{1.99\text{ s}}}$

$v_1 = v_0 - gt_1$

$v_1 = 60\text{ m/s} - (9.8\text{ m/s}^2)(1.99\text{ s}) = \underline{\underline{40.5\text{ m/s}}}$ MOVING UPWARD

$v_2 = 60\text{ m/s} - (9.8\text{ m/s}^2)(10.3\text{ s}) = \underline{\underline{-40.9\text{ m/s}}}$ MOVING DOWNWARD

WHEN NUMERICAL VALUES ARE NOT ROUNDED OFF, ONE FINDS $v_1 = -v_2$ EXACTLY

3.11 FROM EQ. (2.10), $x = \frac{1}{2}at^2$ WHILE SPEEDING UP.

$t = 10.5\text{ min} = (10.5)(60)\text{ s} = 630\text{ s}$

$$a = \frac{2x}{t^2} = \frac{(2)(1600\text{ mi})}{(630\text{ s})^2} = \frac{(2)(1600\text{ mi})(1609\text{ m})}{(630\text{ s})^2} = 13\text{ m/s}^2$$

$\frac{a}{g} = \frac{13\text{ m/s}^2}{9.8\text{ m/s}^2} = 1.3$ THIS WOULD BE A LITTLE UNCOMFORTABLE, BUT NOT UNBEARABLE.

3.12 LET S_1 = DISTANCE FROM LA TO AMARILLO

FROM EQ. (2.12), $2aS_1 = v^2$, SO $v^2 = (2)(9\text{ m/s}^2)(2000 \times 10^3\text{ m})$

$v = \underline{\underline{6000\text{ m/s}}}$

OR $v = \underline{\underline{13{,}424\text{ mph}}}$

THIS WOULD REQUIRE TIME t_1,

$$S_1 = \frac{1}{2}at_1, \quad t_1 = \sqrt{\frac{2S_1}{a}} = \sqrt{\frac{(2)(2000 \times 10^3\text{ m})}{9\text{ m/s}^2}} = 667\text{ s}$$

TRIP FROM AMARILLO TO CHICAGO REQUIRES TIME t_2.

$$S_2 = vt_2, \quad t_2 = \frac{S_2}{v} = \frac{(1800)(1000)\text{ m}}{6000\text{ m/s}} = 300\text{ s}$$

CHICAGO TO NY REQUIRES TIME t_3:

$$S_3 = v_0t_3 - \frac{1}{2}at_3^2$$

$$1500 \times 10^3 = (6000\text{ m/s})t_3 - \frac{1}{2}(12\text{ m/s}^2)(t_3^2)$$

$$(6\text{ m/s}^2)t_3^2 - (6000\text{ m/s})t_3 + 1500 \times 10^3 = 0$$

$$t_3^2 - (1000\text{ m/s})t_3 + 250 \times 10^3 = 0$$

$$t_3 = \frac{1000 \pm \sqrt{(1000)^2 - 4(1)(250 \times 10^3)}}{2}\text{ s} = 500\text{ s}$$

TOTAL TIME IS $t = t_1 + t_2 + t_3 = 667\text{ s} + 300\text{ s} + 500\text{ s} = 1467\text{ s}$

$= 1467(1/60)\text{ min} = \underline{\underline{24.5\text{ min}}}$

3.13 CAR MUST SLOW TO 50 km/hr BEFORE REACHING TRUCK IF COLLISION IS TO BE AVOIDED. LET t_B = BRAKING TIME DURING WHICH CAR VELOCITY DROPS FROM 120 km/hr TO 50 km/hr.

$V_0 = 120 \text{ km/hr} = 120\left(\frac{1000\text{m}}{3600\text{s}}\right) = 33.3 \text{ m/s}, \; V_F = 50 \text{ km/hr} = 50\left(\frac{1000\text{m}}{3600\text{s}}\right) = 13.9 \text{ m/s}$

$V_F = V_0 - at_B \qquad t_B = \frac{V_0 - V_F}{a} = \frac{33.3\text{m/s} - 13.9\text{m/s}}{2\text{m/s}^2} = 9.7\text{s}$

DISTANCE MOVED BY CAR IN SLOWING TO 50 km/hr IS THUS

$$X_c = \underbrace{V_0 t_R}_{\text{REACTION}} + \underbrace{\left(V_0 t_B - \tfrac{1}{2} a t_B^2\right)}_{\text{BRAKING}}$$

$$= (33.3 \text{ m/s})(0.8\text{s}) + (33.3 \text{ m/s})(9.7\text{s}) - \tfrac{1}{2}(2 \text{ m/s}^2)(9.7\text{s})^2$$

$X_c = 256 \text{ m}$

DURING THIS TIME TRUCK MOVES $X_T = V_T(t_R + t_B)$

$= (13.9 \text{ m/s})(0.8\text{s} + 9.7\text{s})$

$X_T = 146 \text{ m}$

THUS INITIAL SEPARATION MUST BE $256\text{m} - 146\text{m} = \underline{\underline{110 \text{ m}}}$ IF A COLLISION IS TO BE AVOIDED.

3.14 FOR A PROJECTILE $V_y = V_{y_0} - gt_1$, $V_{y_0} = V_0 \sin\theta$ = INITIAL VERTICAL VELOCITY

AT TOP OF FLIGHT $V_y = 0 = V_{y_0} - gt_1$, SO $t = V_{y_0}/g$

TOTAL TIME IN AIR IS $T = 2t_1 = 2\frac{V_{y_0}}{g} = \frac{2V_0 \sin\theta}{g}$

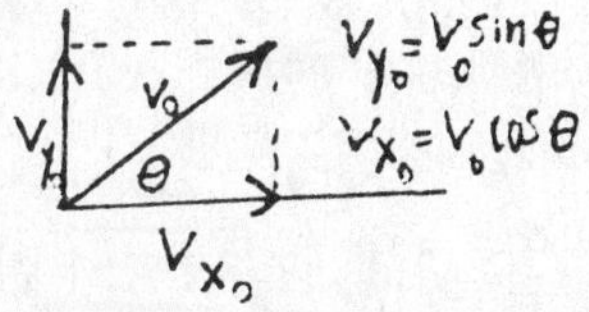

HORIZONTAL RANGE IS $X = V_x T = (V_0 \cos\theta)\left(\frac{2V_0 \sin\theta}{g}\right)$

$= \frac{2V_0^2}{g} \sin\theta \cos\theta$

MAXIMUM RANGE OCCURS FOR $\theta = 45°$. HERE $V_0 = 80 \text{ mi/hr} = (80)(.447 \frac{\text{m}}{\text{s}})$

$= 35.8 \text{ m/s}$

$$X = \frac{(2)(35.8\text{m/s})^2(\sin 45°)(\cos 45°)}{9.8 \text{m/s}^2} = \underline{\underline{131 \text{ m}}} \text{ NEGLECTING AIR}$$

3.15 (a) RANGE IS $X = \frac{2V_0^2}{g} \sin\theta \cos\theta$, EQ. (3.10)

$X = \frac{(2)(30 \text{ m/s})^2}{9.8 \text{m/s}^2} \sin 45° \cos 45° = \underline{\underline{91.8 \text{ m}}}$

TIME OF FLIGHT IS $t = \frac{x}{V_x} = \frac{x}{V_0 \cos\theta} = \frac{91.8\text{m}}{(30 \text{m/s})\cos 45°} = \underline{\underline{4.33 \text{ secs}}}$.

(b) IF BALL IS THROWN AT 30°, $X_1 = \frac{2V_0^2 \sin\theta\cos\theta}{g} = \frac{(2)(30)^2 \sin 30° \cos 30°}{9.8}$

$= 79.5 \text{ m}$

THUS BALL WILL FIRST BOUNCE A DISTANCE d FROM HOME PLATE, WHERE $d = 91.8\text{m} - 79.5\text{m} = \underline{12.3 \text{ m}}$

BALL WILL BOUNCE UP WITH SPEED 27 m/s (i.e. 90% OF 30 m/s) AT ANGLE OF 30°. RANGE FOR THIS SECOND TRAJECTORY WILL BE

$$x_2 = \frac{2(v_0')^2 \sin\theta\cos\theta}{g} = \frac{(2)(27\ m/s)^2 \sin 30^\circ \cos 30^\circ}{9.8\ m/s^2} = 64.4\ m$$

THUS BALL WILL EASILY MAKE IT TO HOME PLATE ON ONE HOP. TOTAL TIME IN AIR WILL NOW BE $t = t_1 + t_2$

$$t = \frac{x_1}{v_{x_1}} + \frac{d}{v_{x_2}} = \frac{79.5\ m}{(30\frac{m}{s})\cos 30^\circ} + \frac{12.3\ m}{(27\frac{m}{s})\cos 30^\circ} = 3.06\ s + 0.53\ s = 3.59\ s$$

THUS WE SEE THAT BALL WILL ARRIVE FASTER IF THROWN AT AN ANGLE LOWER THAN 45°. EXACT OPTIMUM ANGLE DEPENDS ON FRICTIONAL LOSS OF SPEED ON THE BOUNCE.

(C) WITH LESS FRICTION (ARTIFICIAL TURF) AIM LOWER. WITH ZERO FRICTION THE BALL WILL GET THERE FASTEST IF THROWN HORIZONTAL (i.e. RIGHT ALONG THE GROUND IN THIS EXAMPLE)

3.16 PATH OF BOMB IS SKETCHED HERE.

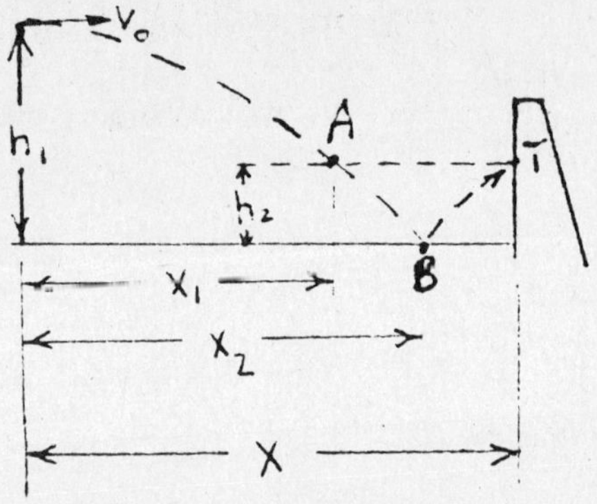

TIME TO FALL DISTANCE h_1 IS t_1,

$$h_1 = \frac{1}{2}gt_1^2, \quad t_1 = \sqrt{\frac{2h_1}{g}}$$

IN TIME t_1 BOMB TRAVELS A HORIZONTAL DISTANCE x_2,

$x_2 = v_x t_1 = v_0 t_1$ (v_x IS CONSTANT)

TIME TO FALL TO POINT A, A DROP OF $h_1 - h_2$ IS t_A, $h_1 - h_2 = \frac{1}{2}gt_A^2$,

$$t_A = \sqrt{\frac{2(h_1 - h_2)}{g}}$$

IN TIME t_A BOMB TRAVELS HORIZONTAL DISTANCE x_1, $x_1 = v_0 t_A$

BECAUSE THE BOMB TRAJECTORY IS SYMMETRIC ABOUT THE BOUNCE POINT B, THE HORIZONTAL DISTANCE FROM A TO B IS THE SAME AS FROM B TO THE TARGET T.

THUS $X = x_2 + \overline{BT}_{HORIZ} = x_2 + \overline{AB}_{HORIZ} = x_2 + (x_2 - x_1)$

OR $X = 2x_2 - x_1 = 2v_0 t_1 - v_0 t_A = 2v_0\sqrt{\frac{2h_1}{g}} - v_0\sqrt{\frac{2(h_1 - h_2)}{g}}$

$$X = v_0\sqrt{\frac{2}{g}}\left\{2\sqrt{h_1} - \sqrt{h_1 - h_2}\right\}$$

FOR $v_0 = 500\ km/hr = 500(1/3.6\ m/s) = 138.9\ m/s$

$h_1 = 80\ m$, $h_2 = 20\ m$

$$X = (138.9\ m/s)\sqrt{\frac{2}{9.8\ m/s^2}}\left\{2\sqrt{80\ m} - \sqrt{80\ m - 20\ m}\right\} = 636\ m$$

Dynamics

4

• Summary of Important Ideas, Principles, and Equations

1. Newton's First Law states that a body subject to no net force will remain at rest or move in a straight line with constant speed.

2. Newton's Second Law states that a body of mass m subject to a net force F will experience an acceleration a in the direction of F given by

   $$\vec{F} = m\vec{a}$$

   m is in kilograms, a in m/s^2 and force in newtons. 1 n = 0.225 lb.

3. Newton's Third Law states that forces always occur in pairs as the result of the interaction between two bodies. The force on body A due to B is of the same strength but opposite direction to the force on B due to A.

4. The weight of an object on earth is equal to the gravitational force on the object. The magnitude of the weight, W, may be written

   $$W = mg$$ Here m is the mass, in kg, of an object of weight W.

   g is determined experimentally to have the value 9.8 m/s^2 at the earth's surface.

   An accelerating object will have an apparent weight different from mg. The apparent weight is what a bathroom scale would read if the object were placed on it. Thus a freely falling object (such as an astronaut in a satellite) has zero apparent weight.

   Weight (a force in newtons) and mass (in kilograms) are proportional, but they are distinct different concepts. An object will always have mass, but in a region where there is no gravitational force it would have no weight.

5. In a closed room you cannot tell if your weight is due to a gravity force (because of interaction with the earth, say) or because you are in a giant accelerating space ship. The effects of acceleration and of gravity are indistinguishable in a uniform gravitational field. This is called the principle of equivalence.

6. The frictional force on an object supported by a surface is proportional to the normal reaction force R exerted by the surface.

$$\boxed{\vec{F} = \mu\vec{R}}$$

μ is the coefficient of friction. The coefficient of kinetic friction, μ_k, for a sliding object is usually less than μ_s, the coefficient of static friction for the object when stationary.

7. To solve a dynamics problem proceed as follows:

First, try to envision in your mind what is happening. Draw a small pictorial sketch if needed. Identify what constitutes the "system".

Second, draw a careful free body diagram which shows all forces acting on the system. Draw the force vectors as arrows with their tails all together.

Determine whether or not the system is in equilibrium. A system in equilibrium is one which is not accelerating and for which, therefore, the total force acting on it is zero. A body which is at rest or which has constant velocity (i.e. moves with constant speed in a straight line) is in equilibrium.

If the system is in equilibrium apply the conditions

$$\Sigma F_x = 0, \text{ i.e. } F_{left} = F_{right}$$

$$\Sigma F_y = 0, \text{ i.e. } F_{up} = F_{down}$$

$$\Sigma F_z = 0, \text{ i.e. } F_{out} = F_{in}$$

Solve these equations algebraically to obtain the desired results.

If the system is not in equilibrium (i.e. the total force is not zero and the system is accelerating) apply F = ma. Note that this relation may be applied to the entire composite system or to individual bodies within the system, as long as the correct forces are identified.

When several parts of the system are connected together rigidly they all have the same acceleration.

• Qualitative Questions

4M.1 In everyday terms we would say that a force is the same as

A. pressure.
B. a push or a pull.
C. strain.
D. the energy of something.
E. a dynamical effect.

4M.2 Two tug-of-war teams pull on the opposite ends of a rope. Each team exerts a horizontal force of 4000 n. The tension in the rope is thus

A. zero.
B. 2000 n
C. 4000 n
D. 8000 n
E. not possible to determine without knowing the length of the rope.

4M.3 A 10 kg fish is weighed with two scales of negligible weight, as sketched.

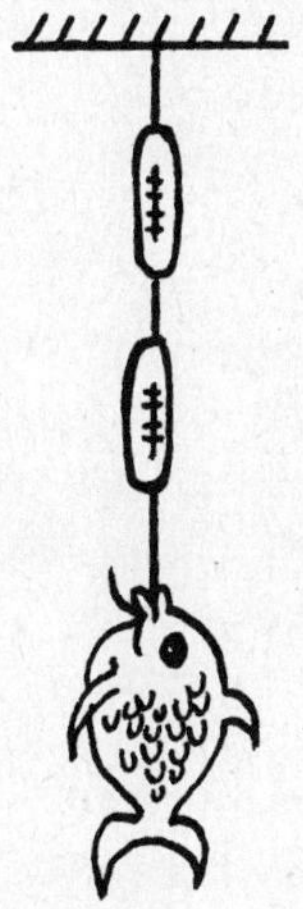

A. Each scale will read 5 kg.
B. Each scale will read 10 kg.
C. The top scale will read 10 kg, and the bottom one will read zero.
D. The top scale will read zero, and the bottom one will read 10 kg.
E. Each scale may show a different reading, but the sum of the two values will be 10 kg.

4M.4 The force required to maintain a rocket ship moving at constant velocity in free space, far from all stellar objects, is

A. equal to the force required to stop it.
B. equal to its weight.
C. zero.
D. dependent on how fast the rocket is moving.
E. dependent on the particular inertial reference frame to which the force is referred.

4M.5 If an object is acted on by a net external force

A. it must be accelerating.
B. it must be moving.
C. its speed will increase.
D. its velocity will increase.
E. its acceleration will increase.

4M.6 Which of the following is a true statement?

A. Mass and weight are the same thing measured in different units.
B. When an object is falling freely it is massless.
C. If the mass of an object is doubled, its weight will not necessarily be doubled also.
D. The mass of an object is independent of the force of gravity acting on the object.
E. An object which has mass does not necessarily have inertia.

4M.7 Crack-the-whip is a game which can be played on skates. A group of kids hold hands and skate in a long line following the leader. As the leader zigzags back and forth the kids at the end are "cracked" off. The object is to hang on as long as possible. Suppose that in this game Charlie is holding the hand of his little sister. He gives her a hard pull of 200 n. In this case

A. the force she exerts on him will be less than 200 n, since she weighs less.
B. the force she exerts on him will be exactly 200 n, no matter what the other circumstances.
C. the force she exerts on him will depend on whether or not she is holding on to someone else with her other hand.
D. the force that she exerts on him will depend on whether or not they are accelerating.
E. there are to be so many complicating variables (speed, number of kids in the line, position in the line) that no definitive statement of the kind above can be made.

4M.8 Consider a shipping crate which is sliding down a loading ramp at constant speed. In this case

A. the crate is not in equilibrium since it is moving.
B. the net force acting on the crate is zero.
C. gravity is the only force acting on the crate.
D. we see that this could not happen in real life, since in fact a crate must continually gain speed as it slides down.
E. friction is the only force acting on the crate.

4M.9 Suppose that a piece of string is cut into two pieces and one piece is used to suspend a lead weight. The second piece is tied to the bottom of the lead weight. You now pull down on the lower string. Assuming you pull hard enough,

A. both strings will always break at the same time (at least in principle).
B. the upper string will always break first.
C. the lower string will always break first.
D. the shortest piece will always break first.
E. the longest piece will always break first.
F. whether the upper or lower piece breaks first depends on whether you pull slowly or give a sudden jerk.

4M.10 Which of the following has a meaning closest to that of "mass"?

A. weight.
B. force.
C. inertia.
D. volume.
E. acceleration.

4M.11 A boy is pulling a sled across a level field where the coefficient of friction between the sled and the snow is μ. He pulls with force F on a rope inclined at angle θ with the horizontal. The mass of the sled is m and the mass of the boy is M. If the sled is moving with constant velocity the force F is thus

A. $\dfrac{\mu mg}{\cos\theta}$

B. $\dfrac{\mu mg}{\cos\theta + \mu\sin\theta}$

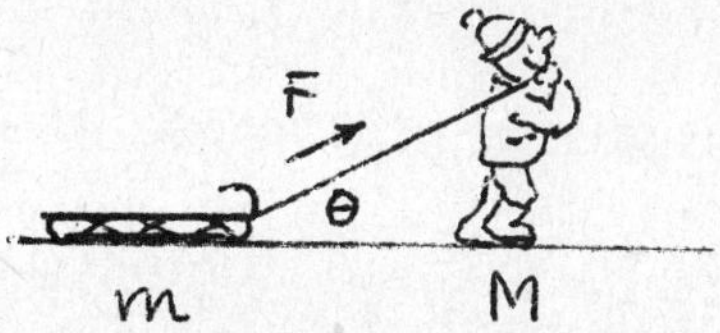

C. $\dfrac{\mu(m + M)g}{\cos\theta + \mu\sin\theta}$

D. μmg

E. $\dfrac{mg}{1 + \mu\sin\theta}$

4M.12 A friend of mine told me that he saw something like the following happen. A painter was trying to get up the side of a building to do some work. First he tied a rope to the top of the building and started to climb up the rope. The rope broke right away. It wouldn't support his weight. Then he rigged a pulley and put the rope over it. He attached a sling to one end and was then able to pull himself up. In order to free his hands he then tied the end of the rope to the building. Immediately it broke and down he tumbled. Why do you think this happened? (Don't say you never learned anything useful in physics!)

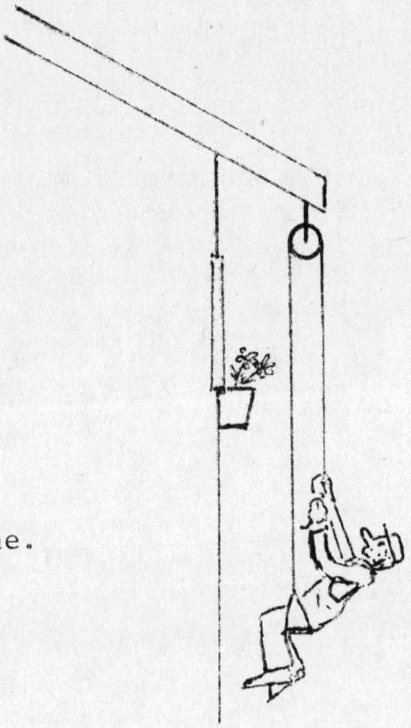

A. The rope came untied as soon as the painter tied it.
B. While the painter was pulling himself up his full weight was not on the rope, but as soon as he stopped the tension increased, causing the rope to break.
C. Tying the rope to the building doubled the tension in the rope.
D. The rope was steadily stretching all the time he was going up, and when he stopped it finally reached the breaking point.
E. The tension in a moving rope is not as great as in a stationary one.
F. The tension in a short rope is greater than for a longer one.

4M.13 A worker weighing 500 n stands on a platform weighing 250 n. The platform is suspended by means of some pulleys, as shown here. The worker supports some of her weight by holding on to the end of the rope. What is the tension in the rope she is holding?

A. 250 n
B. 333 n
C. 375 n
D. 458 n
E. 500 n
F. 750 n

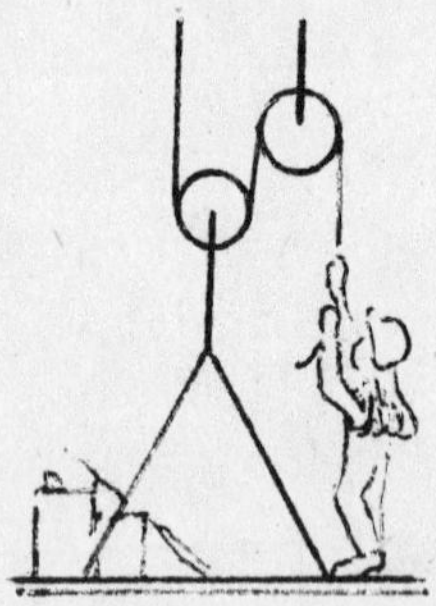

4M.14 Suppose that a person weighing 150 lbs stood on a bathroom scale in an elevator which was accelerating upward with an acceleration of $\frac{1}{2}$g (g = acceleration due to gravity = 32.2 ft/s^2). What would the scale read? Note that this is not just an academic question, since an astronaut in an accelerating rocket ship experiences the same effect, and it can have important consequences for his well-being.

A. 75 lbs
B. 100 lbs
C. 150 lbs
D. 225 lbs
E. 300 lbs

4M.15 I worry a lot about how to escape from burning hotels when I have a room on the top floor. The halls are always full of smoke and flames, and so the window is the only way out. I plan to tear my bedsheets into strips, knot them together and climb down the side of the building. The problem is, I don't think a strip of bed-sheet will support me. I can still use the sheet to get down, though, provided I don't hang on tight and let the sheet slide through my hand while gripping it lightly. Of course, I will still accelerate toward the ground, but not nearly as fast as if I just jumped for it. For example, suppose that the sheet rope could support half my weight. What would be my minimum downward acceleration (as a multiple of g, the acceleration due to gravity) if the rope is not to break?

A. $\frac{1}{4}$ g
B. $\frac{1}{2}$ g
C. 0.75 g
D. 0.8 g
E. g

• Multiple Choice: Answers and Comments

4M.1 B

4M.2 C Imagine a 4000 n weight suspended from a rope attached to the ceiling. You can see that the tension in such a rope is 4000 n. The weight pulls down with 4000 n and the ceiling pulls up with 4000 n. Now turn the picture sideways and replace the weight with one team and the ceiling with the other. Each team exerts a force of 4000 n and, as before, the tension in the rope is 4000 n also.

4M.3 B The tension everywhere in a given rope is the same. We can imagine reading the tension at various points in the rope suspending the fish by putting little spring scales all along the rope. Each will read the same, namely 10 kg in this example. (Note: Remember that 10 kg is a mass, but since mass and weight are proportional, people often measure weight and call it mass.)

4M.4 C From $F = ma$ we see that if $a = 0$, $F = 0$. A rocketship moving with constant velocity has zero acceleration, thus no net external force acts on it. It is just "coasting" and will continue to do so until acted on by a force.

4M.5 A A is correct since $F = ma$, and if $F \neq 0$, $a \neq 0$.
B is false. A ball thrown straight up is always accelerating while in the air, since it is acted on by the force of gravity but it is not moving at the top of its trajectory.
C is false, since it may accelerate by changing direction only and not speed.
D is false. The velocity could, for example, decrease.
E is false since, if the force acting is constant, the acceleration will be constant also, as seen from $a = \frac{F}{m}$.

4M.6 D A is false, since mass and weight are different concepts.
B is false. A falling object has zero apparent weight, but still has mass.
C is false, since weight is proportional to mass, and doubling one will double the other.

4M.7 B Newton's Third Law tells us that forces always occur in pairs of equal magnitude. Thus if Charlie exerts a force of 200 n on his sister, she will automatically exert this same force on him, independent of all other considerations. To see this, imagine there was some kind of spring scale between them. If he pushes on one end of the spring and compresses it, she must be pushing back with the same force (otherwise he couldn't compress the spring).

4M.8 B B is correct. The velocity of the crate is constant, which means the acceleration is zero. From $F = ma$ this means $F = 0$ also.

4M.9 F Try this for yourself. When string is pulled slowly, the upper string must support the weight plus the force you apply. The lower string doesn't have to support weight, so the top string breaks first in this case. If you give a sudden jerk you are trying to cause the weight to have a large acceleration. This can only be accomplished if the lower string exerts a large force. If the force needed exceeds the strength of the string, it breaks.

4M.10 C

4M.11 B $F_F = \mu R$

$F_F = F\cos\theta$

$R + F\sin\theta = mg$

Solve, $F\cos\theta = \mu R$

$= \mu(mg - F\sin\theta)$

$F(\cos\theta + \mu\sin\theta) = \mu mg$

$$F = \frac{\mu mg}{\cos\theta + \mu\sin\theta}$$

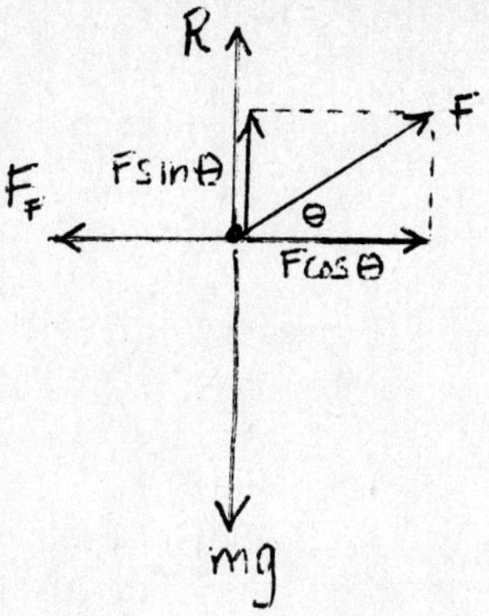

Note that $R \neq mg$ here, so $F_F = \mu R \neq \mu mg$.

This is because the rope lifts up on the sled, thereby reducing R and thus F_F.

4M.12 C Observe that while the painter is holding the end of the rope, two strands of rope support him. As soon as he ties the rope to the building he is supported by only one strand of rope, and it breaks because the tension has doubled.

4M.13 A Note that the tension in a given rope, T, is the same everywhere in the rope. Draw a dashed line around the platform and the painter. This is our "system". Three strands of rope (all part of the rope she is holding) act upon the system. The force of gravity (the total weight of the system) acts downward. The system is in equilibrium, so 3T = W, or T = (250 n + 500 n)/3 = 250 n.

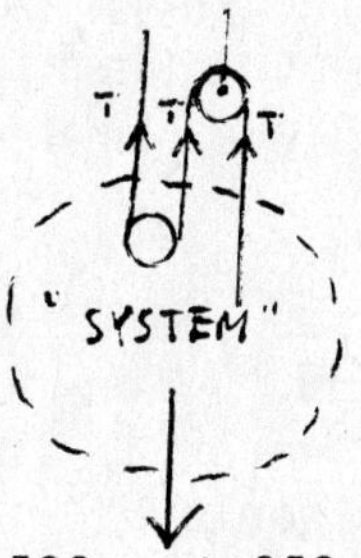

W = 500 n + 250 n

4M.14 D

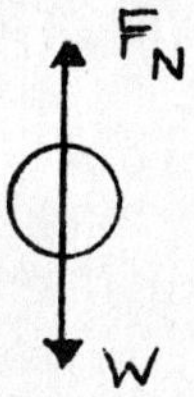

$F_{Net} = F_N - W = ma$

$a = \frac{1}{2}g$, $F_N = ma + W = \frac{1}{2}mg + W$

$= \frac{1}{2}W + W = 3/2\ W = 3/2(150 \text{ lbs}) = 225 \text{ lbs}$

4M.15 B $F_{Net} = mg - \frac{1}{2}mg = ma \rightarrow a = \frac{1}{2}g$

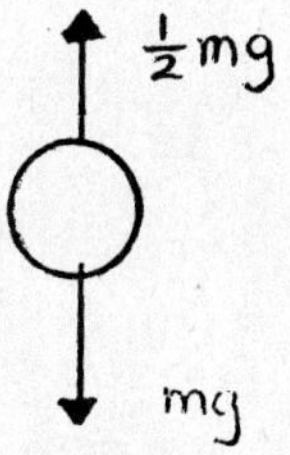

• Problems

4.1 How much would a 150 lb astronaut weigh on the moon where the acceleration due to gravity is 1.62 m/s^2? This value is about 1/6 the acceleration due to gravity on the surface of the earth.

4.2 A girl of 40 kg mass when waterskiing accelerates uniformly from rest to 12 m/s (about 23.4 mph) in 2 seconds. What average force must be exerted on her by the tow rope?

4.3 A locomotive of mass M pulls a train consisting of three identical cars, each of mass m. Friction is negligible. What force is exerted by the coupling between each car if the locomotive exerts a force F?

4.4 When I was a kid we built a tree house elevator which worked like this: A fifty gallon oil drum was attached to a box (in which we rode) by a wire cable which passed over a pulley made from an old wheel. A garden hose, connected to a faucet, was tied high in the tree by the treehouse. When someone wanted to go up she would climb in the box while the operator turned on the water and filled the oil drum. When there was enough water in the drum, up you would go. After you got to the top and climbed out, the water was allowed to drain out and the drum went back to the top again. It worked pretty well, but my mother didn't approve of flooding the yard with 10 gallons of water every trip. Robert Goddard probably encountered the same kind of opposition when he was building the first rockets for space travel.

See if you can calculate the tension in the cable and the acceleration of the elevator given the following design parameters:

Mass of box plus child = 40 kg

Mass of oil drum plus water = 42 kg

Neglect friction and assume the wheel has negligible mass.

4.5 Five hundred kg of sand are placed in a dump truck. The coefficient of static friction between the sand and the bed of the truck is 0.5. At what angle above horizontal must the bed of the truck be tipped before the sand will start to slide out?

4.6 A 500 kg load of steel plate is carried in a flatbed truck travelling 60 km/hr. Posts keep the steel from sliding off the sides of the truck, but only friction prevents it from sliding forward or back. If the coefficient of static friction between the steel and the truck bed is 0.4, what is the shortest distance in which the truck can stop without having the steel slide forward into the cab?

4.7 A 4 kg flowerpot is hanging in my office from the contraption sketched here. The strut is a very light aluminum rod. What is the compressive force exerted by the rod and what is the tension in the string?

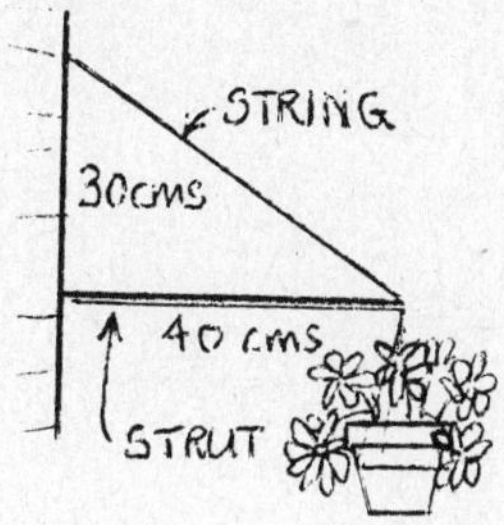

4.8 One time when I was working on my VW I was faced with the following problem. I had taken the engine, with a mass of 150 kg, out of the car and managed to place it on some blocks 30 cms high. I now wanted to raise the engine onto the bed of my pick-up truck, a distance 50 cms above the floor. Here is how I did it. I attached a chain from the engine to the ceiling of the garage, 3 meters above the engine. Then I pulled sideways on the engine until it was high enough off the floor for my friend to back the truck under it. Worked like a charm. What horizontal force did I have to exert? What force would I have had to exert if I had lifted the engine straight up? Where did the "extra" force come from which helped me?

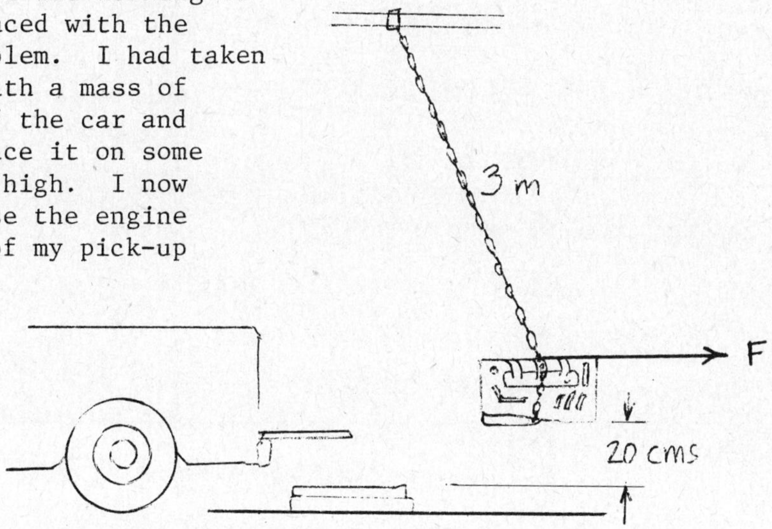

• Problem Solutions

4.1 $W_M = mg_M$ $\quad W_E = mg_E$ $\quad \dfrac{W_M}{W_E} = \dfrac{mg_M}{mg_E}$ $\quad \dfrac{W_M}{150\text{ lbs}} = \dfrac{1.62\text{ m/s}^2}{9.8\text{ m/s}^2}$

$$W = \left(\frac{1.62}{9.8}\right)(150\text{ lbs}) = \underline{\underline{24.8\text{ lbs}}}.$$

4.2 $a = \dfrac{V_2 - V_1}{t} = \dfrac{12\text{ m/s} - 0}{2\text{ s}} = 6\text{ m/s}^2$

$$F = ma = (40\text{ kg})(6\text{ m/s}^2) = \underline{\underline{240\text{ n}}}$$

4.3

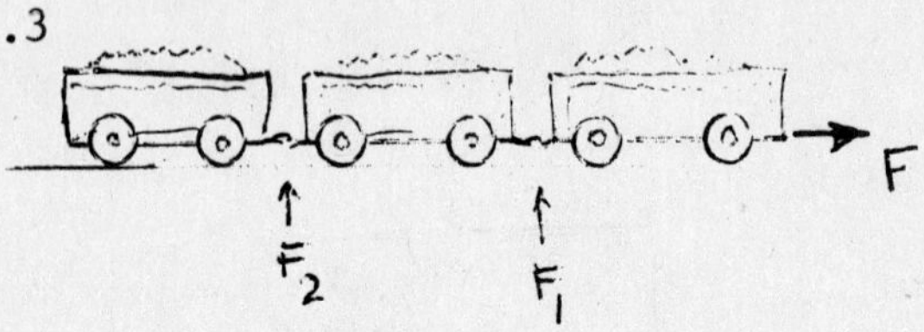

FIRST CONSIDER ALL THREE CARS AS A SINGLE SYSTEM OF MASS 3m SUBJECT TO AN EXTERNAL FORCE F.

$$F = (3m)a\,,\ \text{SO}\ a = \frac{F}{3m}$$

NEXT CONSIDER JUST THE FIRST CAR, SUBJECT TO FORCES F AND F_1.

$$F_{NET} = F - F_1 = ma = m\left(\frac{F}{3m}\right) = \frac{F}{3}$$

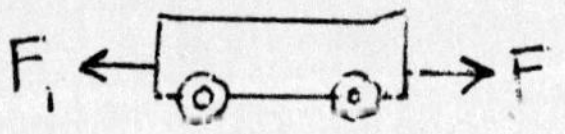

THUS $\underline{\underline{F_1 = \frac{2}{3}F}}$

SIMILARLY FOR THE SECOND CAR, $F_1 - F_2 = ma = m\left(\dfrac{F}{3m}\right) = \dfrac{F}{3}$

THUS $\underline{\underline{F_2 = \frac{1}{3}F}}$

THE KEY TO THIS PROBLEM IS THE RECOGNITION THAT EACH CAR HAS THE SAME ACCELERATION AND THAT A GIVEN COUPLING EXERTS A FORCE OF THE SAME MAGNITUDE ON THE CAR AHEAD AND ON THE CAR BEHIND (NEWTON'S 3RD LAW).

4.4

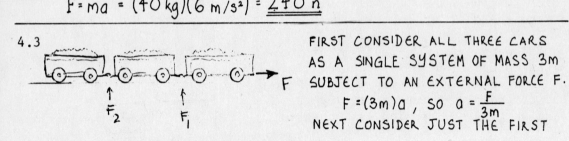

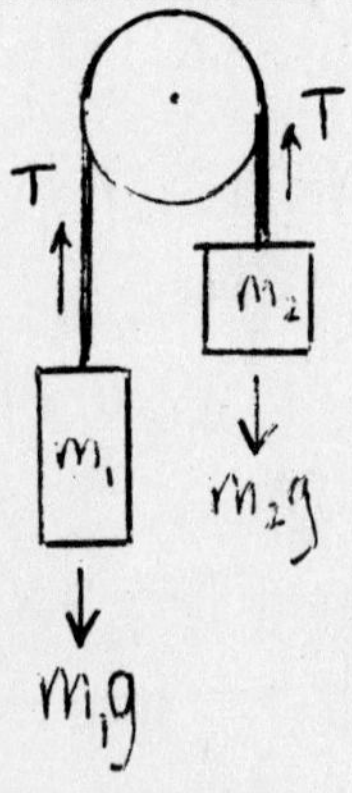

$m_1 = 42$ kg
$m_2 = 40$ kg

THE NET FORCE ACCELERATING THE SYSTEM DOWNWARD IS $\quad F_{NET} = m_1g - m_2g = (m_1 + m_2)a$

HERE m + m IS THE TOTAL MASS OF THE SYSTEM ACCELERATING.

THUS $\quad a = \dfrac{(m_1 - m_2)}{(m_1 + m_2)}g = \left(\dfrac{42 - 40}{42 + 40}\right)(9.8\text{ m/s}^2)$

$$a = 0.24\text{ m/s}^2$$

TO DETERMINE THE TENSION LOOK AT JUST ONE OF THE MASSES, SAY m_1. THE NET FORCE ACCELERATING m_1 IS $m_1g - T$, SINCE m_1 IS

ACCELERATING DOWNWARD. APPLY $F_{1NET} = m_1 a$.

$$m_1 g - T = m_1 a$$

$$T = m_1 g - m_1 a = (g - a)m_1 = (9.8\,m/s^2 - 0.24\,m/s^2)(42\,kg)$$

$$T = 402\,n$$

WE COULD EQUALLY WELL HAVE APPLIED THIS ANALYSIS TO MASS m_2, WHICH IS ACCELERATING UPWARD.

$$T - m_2 g = m_2 a \;,\quad T = m\,(a + g) = 40\,kg(0.24 + 9.8)\,m/s^2 = 402\,n$$

4.5

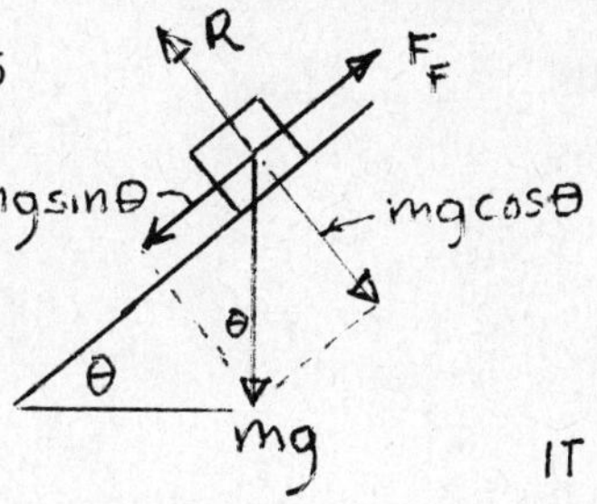

THE FORCES ACTING ON THE SAND ARE THE FORCE OF GRAVITY mg, THE NORMAL REACTION FORCE OF THE TRUCKBED R, AND THE FORCE OF FRICTION F_F.

JUST BEFORE THE SAND STARTS TO SLIP IT IS IN EQUILIBRIUM. FIRST REPLACE mg BY ITS COMPONENTS $mg\cos\theta$ AND $mg\sin\theta$.

$$R = mg\cos\theta$$

$$F_F = mg\sin\theta$$

ALSO, $F_F = \mu R$

THUS $\mu R = mg\sin\theta$

$$R = mg\cos\theta$$

DIVIDE THESE EQUATIONS: $\dfrac{\mu R}{R} = \dfrac{mg\sin\theta}{mg\cos\theta} \longrightarrow \mu = \tan\theta$

GIVEN $\mu = 0.5$, $\theta = \tan^{-1} 0.5$

$$\theta = 26.6^\circ$$

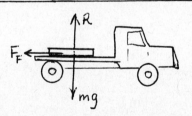

4.6 THE FORCE OF FRICTION ACTING ON THE LOAD IS $F_F = \mu R = \mu mg$. THIS CAUSES THE LOAD TO DECELERATE WITH ACCELERATION a,

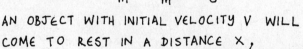

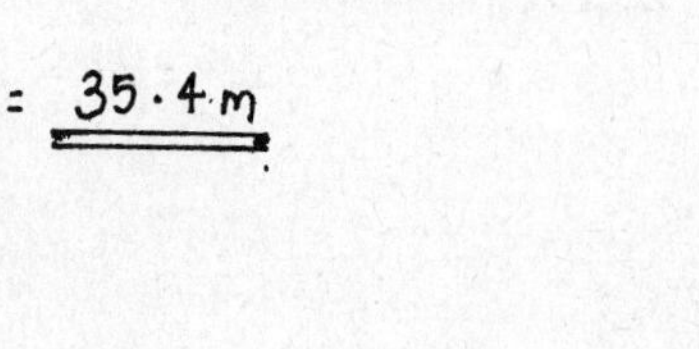

$$a = \frac{F_F}{m} = \frac{\mu mg}{m} = \mu g$$

AN OBJECT WITH INITIAL VELOCITY V WILL COME TO REST IN A DISTANCE X,

$$2ax = V^2$$

SO $$x = \frac{V^2}{2a} = \frac{V^2}{2\mu g} = \frac{\left(60 \times \frac{1000\,m}{3600\,s}\right)^2}{(2)(0.4)(9.8\,m/s^2)} = 35.4\,m$$

4.7 THE FLOWERPOT IS IN EQUILIBRIUM, SO THE TOTAL FORCE ACTING IS ZERO. LOOK AT THE FORCES ACTING AT THE POINT WHERE THE CABLE IS ATTACHED TO THE STRUT. THESE ARE SHOWN IN THE FREE BODY DIAGRAM DRAWN HERE. FIRST RESOLVE THE TENSION T INTO ITS COMPONENTS.

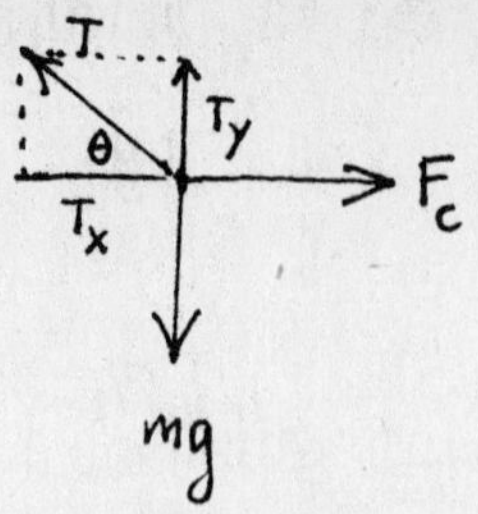

$$T_x = T\cos\theta$$

$$T_y = T\sin\theta$$

WHERE $\tan\theta = \dfrac{30\text{ cms}}{40\text{ cms}} = 0.75 \quad \theta = 36.9°$

NOTE: STORE θ IN YOUR CALCULATOR FOR USE LATER.

NEXT APPLY CONDITION FOR EQUILIBRIUM:

$$F_{UP} = F_{DOWN}$$

$$F_{LEFT} = F_{RIGHT}$$

THUS $T\sin\theta = mg$ T = TENSION IN CABLE

$T\cos\theta = F_c$ F_c = COMPRESSION IN STRUT

TO SOLVE DIVIDE THESE EQUATIONS.

$$\frac{T\sin\theta}{T\cos\theta} = \frac{mg}{F_c}$$

$$\tan\theta = \frac{mg}{F_c}$$

$$F_c = \frac{mg}{\tan\theta} = \frac{(4\text{ kg})(9.8\text{ m/s}^2)}{0.75} = \underline{\underline{52\text{ n}}}$$

SUBSTITUTE THIS VALUE IN $T\cos\theta = F_c$

$$T = \frac{F_c}{\cos\theta} = \frac{52}{\cos 36.9°} = \underline{\underline{65\text{ n}}}$$

4.8 THE ENGINE IS IN EQUILIBRIUM.

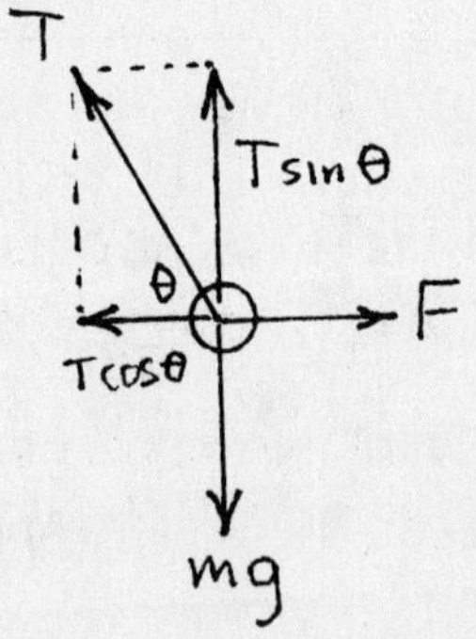

$$F_{LEFT} = F_{RIGHT}$$

$$F_{UP} = F_{DOWN}$$

$$T\cos\theta = F$$

$$T\sin\theta = mg$$

DIVIDE THESE EQNS: $\dfrac{T\sin\theta}{T\cos\theta} = \dfrac{mg}{F}$

$$\tan\theta = \frac{mg}{F}$$

$$F = \frac{mg}{\tan\theta}$$

$$\theta = \frac{L-h}{L} = \frac{3-.2}{3} = 0.93$$

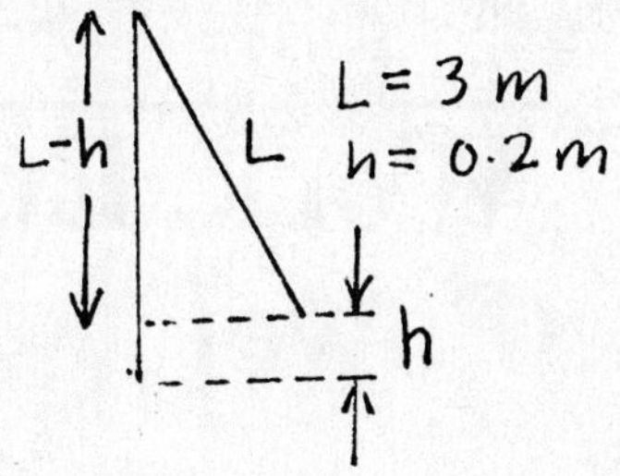

$\theta = 69°$, $\tan\theta = 2.6$

$$F = \frac{mg}{\tan\theta} = \frac{(150kg)(9.8m/s^2)}{(2.6)} = \underline{\underline{565\ n}}$$

TO LIFT THE ENGINE STRAIGHT UP WOULD REQUIRE A FORCE mg.

$$mg = (150kg)(9.8\,m/s^2) = 1470\ n$$

THUS A MUCH SMALLER FORCE IS NEEDED WHEN THE OBJECT IS PULLED SIDEWAYS. NOTICE, HOWEVER, THAT THE ENGINE HAD TO BE PULLED SIDEWAYS A GREATER DISTANCE THAN 20 cms, NAMELY A DISTANCE $L\sin\theta \doteq 2.8\,m$. THE EXTRA UPWARD FORCE IS PROVIDED BY THE CEILING WHICH PULLS UP ON THE CHAIN.

5 Work, Energy, and Power

• Summary of Important Ideas, Principles, and Equations

The work done by a constant force F is

$$\boxed{W = F\Delta s\ \cos\theta = F_s \Delta s} \qquad (5.1)$$

Here $\Delta\vec{s}$ is the displacement of the point of application of the force $\vec{F}$ and θ is the angle between $\vec{F}$ and $\Delta\vec{s}$. F_s is the component of $\vec{F}$ in the direction of $\Delta\vec{s}$, and it is only this component of $\vec{F}$ that does work.

Work is a scalar measured in joules, where 1 newton-meter = 1 joule.

The dimension of work is ML^2/T^2.

Definition: The energy of a system is a measure of its ability to do work.

Energy and work are both measured in joules. Other energy units are calories, kilocalories, ft-lbs, BTU (British Thermal Units), and kw-hrs.

When work is done in moving an object against a force such as gravity or a static electric force the object gains potential energy (PE). The energy used to do this work is not lost, but is stored and can be used later to do work on another body, hence it is called potential energy. For example, when work is done in lifting some water to a higher elevation, the water can later be allowed to fall down and do work on the turbines of a hydroelectric generator. A force which gives rise to a potential energy is called a conservative force (because it conserves energy). Conservative forces are ones which "spring back", as opposed to a force like friction, which is not conservative. When work is done against a friction force energy is not stored to be recovered later.

Definition: The potential energy of a body is its capacity to do work by virtue of its position or configuration.

The gravitational PE of an object of mass m at a positio
arbitrary reference level h = 0 is

$$\boxed{PE = mgh}$$

Only changes in PE are significant, so the choice of th
is arbitrary, much as we can measure elevation with res
as h = 0 or to the floor of the room as h = 0. PE is a
in joules, as is work.

A moving body can collide with another object and move
the process. The energy a body has by virtue of its m
kinetic energy ("kinetic" means "motion").

$$\boxed{KE = \tfrac{1}{2}mv^2} \qquad (5.3)$$

v is the speed of the body of mass m. KE is a scalar measured in joules.

Definition: The total mechanical energy of a system is

$$\boxed{E = PE + KE} \qquad (5.4)$$

Energy takes many different forms, such as chemical, thermal, nuclear, light, sound, electrical, magnetic energy and mass itself. These are rough classifications and overlap somewhat. They may involve both kinetic and potential energy. Natural processes all involve the transformation of energy from one form to another. Energy cannot be created nor destroyed, but it may be transformed into a form, such as thermal energy, which is not useful to us, and in this case we may refer to the energy as "lost". Total energy is a scalar measured in joules.

The Law of Conservation of Energy states that the total energy of a closed system is constant. Thus

$$\boxed{KE_i + PE_i = KE_f + PE_f + W} \qquad (5.5)$$

i and f refer to initial and final states of the system, and W is work done by the system against non-conservative forces such as friction. In this sense W is an "energy loss".

Definition: Power is the rate at which work is done or at which energy is transferred.

If work ΔW is done in time Δt, the power is $P = \frac{\Delta W}{\Delta t}$

If the rate of doing work is constant then

$$\boxed{P = \frac{W}{t}} \qquad (5.6)$$

ar with dimension ML^2/T^3 measured in joules/sec = watts.
w = 1 kilowatt = 10^3 W and 1 mw = 1 megawatt = 10^6 watts.

rce F causes an object to undergo a displacement Δx in a
, the work done is $F_x \Delta x$ and the rate of doing work by the
F is

$$P = \frac{\Delta W}{\Delta t} = \frac{F_x \Delta x}{\Delta t} = F_x\left(\frac{\Delta x}{\Delta t}\right) = F_x v_x$$

$$P = F_x v_x \qquad (5.7)$$

where v_x is the velocity of the object and F_x is the component of the force F in the direction of displacement x.

• Qualitative Questions

5M.1 When you slide an object of mass m a distance d across a level floor the work that you do against gravity is

A. proportional to the weight of the object.
B. dependent on how rough the floor is.
C. dependent on how fast the object is moved.
D. proportional to the distance the object is moved.
E. zero.
F. More than one of the above are true.

5M.2 Which of the following is a true statement?

A. Work and energy are both measured in units of joules.
B. The total energy of a system can never be negative.
C. Kinetic energy can be negative or positive.
D. The work done by a force is never negative.
E. It is possible to define a potential energy associated with friction forces much as is done for the gravitational force.

5M.3 Any quantity in an equation which has the dimension ML^2/T^2 can be identified as

A. a force.
B. a weight.
C. an acceleration.
D. an energy.
E. a velocity.

5M.4 Which of the following best explains the meaning of the term "energy"?

A. Energy is a measure of how fast an object is moving.
B. Energy is the ability to do work.
C. Energy is a measure of the amount of work done per second.
D. Energy is essentially the same as force.
E. Energy is a material substance (which used to be called "caloric" long ago) which can flow from one object to another. A given amount of energy has a definite mass and volume, much as a glass of water does.
F. Energy is the same as power.

5M.5 From the equation PE = mgh it appears that the gravitational potential energy of an object depends on our choice of the point at which h = 0. Which of the following is a valid observation concerning this question?

A. The choice of the place at which h = 0 is arbitrary as long as the PE is always positive.
B. This equation is true only if we choose h = 0 at the surface of the earth.
C. This equation is true only if we choose h = 0 at the center of the earth.
D. Since only changes of potential energy are of significance in physical processes, the point at which we choose to set PE = 0 is arbitrary and can be chosen in any convenient way.
E. If we change the location of the point at which h = 0 the resulting change in PE would be compensated for by a corresponding change in kinetic energy.

5M.6 Which of the following has a meaning closest to that of "kinetic energy"?

A. Heat energy.
B. Stored energy.
C. Potential energy.
D. Motion energy.
E. Metabolic energy.

5M.7 If when driving you double the speed of your car, the kinetic energy of the car will

A. be unchanged.
B. also double.
C. increase by a factor of four.
D. increase by a factor of eight.
E. not change, since energy is always conserved.
F. change by a factor which depends on whether or not friction is present.

5M.8 One frequently reads in the newspapers about the "energy crisis" and the "energy shortage". What are they talking about?

A. Energy is constantly being used up, thus the total supply of energy in the universe is steadily decreasing.
B. Energy is steadily being transported out of our "closed system", consisting of the earth and its atmosphere, leaving less energy behind for humans to use.
C. The supply of forms of energy which are cheap and relatively easy to use is decreasing here on earth.
D. We are using energy faster than we are making it, hence this has resulted in an ecological imbalance.
E. It now appears that the so-called "law of conservation of energy" is only a theoretical idea which is not in fact valid in the real world, so energy is not in fact "conserved" as was once thought.

5M.9 Pregnant women are usually encouraged to follow a diet which will minimize their weight gain. At one time people thought a pregnant woman should eat heartily since she was "eating for two". (Maybe some still think this.) Suppose you estimate how much added food a woman should eat each day during pregnancy to provide for the growth of the fetus. Assume the fetus grows at a constant rate (not quite true; growth is most rapid toward the end of gestation). Some useful information is given here:

1) 4 kcal will generate 1 gram of protein (a fetus is mostly protein).
2) Gestation period: 270 days (approximate).
3) 25 percent of human tissue is organic material (mostly protein); remainder is water and inorganic minerals.
4) Average baby has mass of 3 kg at birth.

Which of the following would come closest to fulfilling the daily caloric requirements for the growth of a fetus during pregnancy?

A. 2 hamburgers
B. 1 cup of milk
C. 1 egg
D. 1 slice of rye bread
E. 1 cup of carrots
F. 1 large bite of a doughnut
G. 1 deep breath of Los Angeles air.

Energy Content of Some Common Foods in kcal

Whole milk, 1 quart	660	1 baked potato	100
1 egg	75	1 apple	70
1 hamburger	245	1 slice rye bread	55
1 cup carrots	45	1 doughnut	135

5M.10 A roller coaster car subject to no friction is released from rest at point X. For which of the paths shown will it have the greatest speed at point Y?

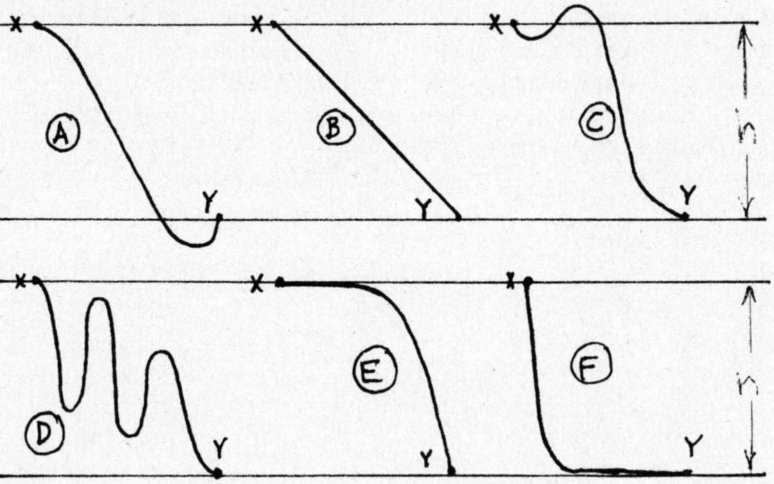

A. A
B. B
C. C
D. D
E. E
F. F
G. More than one of the above.

5M.11 Suppose that the outlet from a water tank is directed upward to make a fountain, as sketched here. If friction is negligible, how high will the fountain rise?

A. To a height which depends on the size of the outlet nozzle.
B. To a height greater than that of the water level in the tank.
C. To the same height as the water level in the tank.
D. To a lower height than the water level in the tank.

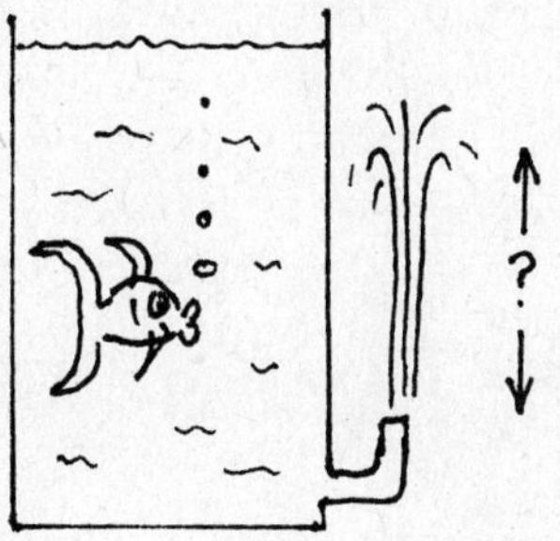

5M.12 Suppose that in accelerating your car from rest to speed v you use an amount of fuel Q. If we neglect friction, we would then expect that the additional amount of fuel needed to increase the speed from v to 2v would be

A. ½ Q
B. Q
C. 2 Q
D. 3 Q
E. 4 Q

5M.13 The oxygen you breathe reacts with fats, carbohydrates and proteins in your body and releases energy which then heats your body or enables you to do work. 2×10^4 J of energy are released for each liter of oxygen. Suppose that a runner of mass 80 kg uses oxygen at the rate of 3 liters/minute while running 10 mph. A milkshake (12 oz.) yields about 420 kcal when eaten (1 kcal = 4190 J). How far would the runner have to run to burn up the energy intake from one milkshake?

A. 0.5 mi.
B. 1.5 mi.
C. 1.8 mi.
D. 3.0 mi.
E. 4.9 mi.
F. 22 mi.

5M.14 The following is a practical scheme for conserving energy. When an electric streetcar goes downhill in a hilly place like San Francisco it is braked by using its electric motor as an electric generator. The electricity generated in this way can be fed back into the power grid and used to run other streetcars going uphill. Suppose that a car of mass 5,000 kg maintained constant speed of 8 m/s while going down the hill shown here. Assuming 100 percent efficiency, at what rate could it generate electric power?

A. 24 kw
B. 68 kw
C. 116 kw
D. 235 kw
E. 300 kw

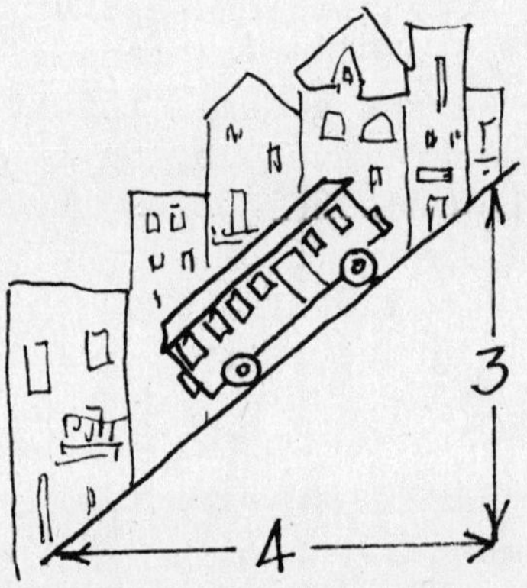

5M.15 A tractor of mass 1000 kg pulls a plow across a field at a steady speed of 4 m/s by exerting a horizontal force of 5000 n. At what power level is the tractor doing work?

A. 4.1 kw
B. 8 kw
C. 12 kw
D. 20 kw
E. It is not possible to determine the answer without knowing the mass of the plow.

• Multiple Choice: Answers and Comments

5M.1 E One does work only when there is a component of force in the direction of motion. Here the force of gravity is vertical and the displacement is horizontal, so no work against gravity is done.

5M.2 A B is false since the choice of a zero point of potential energy is arbitrary. Depending on where we choose the zero of potential energy, the potential energy (and hence the total energy also) may be positive or negative. This is analagous to choosing the zero of elevation at, say, sea level. With this choice our elevation is negative at Death Valley.
C is false since KE is always positive. This is so because m and v^2 are always positive.
D is false because if the force and the displacement are, for example, antiparallel, then the angle between them is 180° and $\cos 180^\circ = -1$, so $W = F\ d \cos\theta = -F\ d$. If the work done by a force is negative, this means the system is doing work <u>against</u> the force. E is false because friction is a non-conservative force. Work done against it is lost. This force does not "spring back" as does gravity or a spring force.

5M.3 D Observe that L/T is the dimension of a velocity, so ML^2/T^2 has the same dimension as $\frac{1}{2}mv^2$, the kinetic energy.

5M.4 B A is false; velocity is a measure of how fast an object is moving.
C is false; power is the amount of work done per second.
D is false; energy and force are different concepts.

5M.5 D

5M.6 D

5M.7 C KE is proportional to v^2, thus if $KE_1 = \frac{1}{2}mv_1^2$ and $KE_2 = \frac{1}{2}mv_2^2$

$$\frac{KE_2}{KE_1} = \frac{\frac{1}{2}mv_2^2}{\frac{1}{2}mv_1^2} = \frac{\frac{1}{2}m(2v_1)^2}{\frac{1}{2}mv_1^2} = 4$$

5M.8 C Energy is never created or destroyed in the universe as far as we know. It is just changed from one form to another, and some forms (such as thermal energy in the atmosphere) may not be useful to us. We have a shortage of cheap, convenient forms of energy such as oil, but there is no shortage of more expensive forms of energy such as sunlight. When I say more expensive, I mean more expensive to use to run cars, light buildings, etc.

5M.9 F The daily growth of the fetus is $\frac{3 \text{ kg}}{270 \text{ days}} = 0.011$ kg/day. Twenty-five percent of this tissue is protein, so the amount of protein added per day is (0.25)(0.011 kg) = 0.0027 kg/day = 2.7 gm/day.
Four kcal will generate about 1 gm of protein, so the energy intake needed each day to generate 2.7 gms of protein is

$$(4 \text{ kcal/gm})(2.7 \text{ gm}) = 11.1 \text{ kcal/day}$$

This is not very much energy, and so one bite of a doughnut probably provides enough energy for the baby's daily growth. Note, however, that a doughnut does not provide all of the needed nutrients, and energy intake is just one small aspect of nutrition. Needless to say, it would be easy for a pregnant woman to gain too much weight if she were to eat extra amounts for the baby.

5M.10 G The speed at Y depends only on the difference in height between points X and Y, assuming it is released from rest at X. This follows from the conservation of energy. Measuring h = 0 at the level of Y,

$$KE_X + PE_X = KE_Y + PE_Y$$

$$0 + mgh = \tfrac{1}{2}mv^2 + 0 \text{ thus } v = \sqrt{2gh}$$

5M.11 C Consider a tiny piece of water which flies up in the fountain. The water in the tank drops a little when this piece of water is shot upward, so it is as if the piece of water went from the top level of the tank (where its kinetic energy is zero) to the top of the fountain (where its kinetic energy is again zero, since it stops at the top of the fountain). Since total energy is conserved, the potential energy, mgh, must be the same for the piece of water inside and outside the tank. Notice that this is not quite the same as making a fountain with a garden hose, where a smaller nozzle will shoot higher. In this case the nozzle is not connected directly to a tank, as is the case here, but rather it is connected to a tank by means of a long water pipe which leads to the nozzle. There are frictional forces on the water which increase with increasing water flow in the pipes. Thus if you pinch off the nozzle at the outlet there is not so much energy lost to friction and the small stream of water will squirt higher.

5M.12 D Given $Q = \tfrac{1}{2}mv^2$,
The KE of a car with speed 2v will be

$$KE = \tfrac{1}{2}m(2v)^2 = 4(\tfrac{1}{2}mv^2) = 4Q$$

Thus the added fuel needed to accelerate from v to 2v is 4Q - Q = 3Q.

5M.13 E In one minute, the runner uses 3 liters of oxygen, or

$$(3)(2 \times 10^4 \frac{J}{min}) = 6 \times 10^4 \frac{J}{min}$$

Convert this to kcal:

$$E = (6 \times 10^4)(\frac{1}{4190} \frac{kcal}{min}) = 14.3 \text{ kcal/min}$$

To use up 420 kcal a runner would have to run for

$$\frac{420}{14.3} \frac{kcal}{kcal/min} = 29.3 \text{ min}$$

At 10 mph the distance covered would be

$$x = (10 \text{ mi/hr})(29.3/60)\text{hr} = 4.9 \text{ miles}$$

5M.14 D The speed of the car is constant, so the kinetic energy does not change. Thus the decrease in potential energy is all converted to electric energy, assuming 100 percent efficiency. In one second the car travels 8 meters forward and drops vertically a distance h. From the drawing we see that a right triangle with sides 3 and 4 has a hypotenuse

$$x = \sqrt{3^2 + 4^2} = 5$$

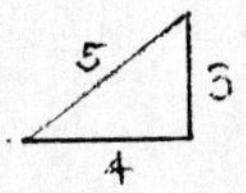

Thus $\frac{h}{3} = \frac{8}{5}$ using the properties of similar triangles. Thus, h = 4.8 m. The loss in PE per second is the power generated.

$$P = \frac{mgh}{t} = \frac{(5000 \text{ kg})(9.8 \text{ m/s}^2)(4.8\text{m})}{(15)s} = 235 \times 10^3 \text{ W}$$

P = 235 kw

5M.15 D P = Fv = (5000 n)(4 m/s) = 20,000 w = 20 kw.

• Problems

5.1 A person slides a packing crate a distance of 4 meters along the floor by pushing on it with a force of 200 n directed at an elevation of 30° above horizontal. How much work does she do?

5.2 In doing "chin-ups" a person lifts himself 30 cms each time. If his mass is 70 kg, how much work does he do against gravity in doing 100 chin-ups? Express the result in joules and in kcals. Note that one slice of bread yields about 50 kcal.

5.3 What is the kinetic energy of a bullet of mass 10 grams (about 150 grains) and velocity 1000 m/s. Express your answer in joules and in kilocalories.

5.4 A child of mass 20 kg is lifted a distance of 1 meter in a swing and given an initial speed of 5 m/s. How high will she go? What maximum speed will she attain? Neglect friction.

5.5 A motorist finds that if he switches off his engine while travelling 40 km/hr he will coast to a stop in 150 meters. In what distance would he come to a stop if he had been travelling up a 7 percent grade (a hill which rises 7 meters for every 100 meters along the road)?

5.6 A kid on a bobsled pushes off with a speed of 4 m/s at the top of a small hill 2 meters high. He slides down a 30° slope and then runs out on a horizontal field. Assuming the coefficient of kinetic friction is 0.1, how far from the base of the hill will he stop?

5.7 Suppose you tried to determine the average frictional force acting on an arrow by the following means. Shoot the arrow straight up. With a high speed movie camera take a picture of the arrow at a fixed point (say 2 meters above the ground) and from these pictures one can determine the speed on the way up, v_1, and the speed on the way down, v_2. One could also use a strobe light and a regular camera to obtain such information. Given v_1, v_2 and the mass, m, of the arrow, determine the average frictional force and the height to which the arrow rose.

5.8 A 70 kg person (154 lbs) expends on the average about 2500 kcal per day. What is his metabolic rate, i.e. at what level of power is he using energy?

5.9 A lamp rated at 80w provides 8 w of light power. What is its efficiency?

5.10 Two men, each of whom can generate 0.2 HP, lift a 500 lb piano to an apartment 30 meters above the street. Assuming the pulley system they use has an efficiency of 70 percent, how long would it take them to lift the piano if they worked at maximum output?

5.11 A bicyclist whose metabolic rate is 120 w while standing still finds that her rate rises to 600 w when riding on level ground at 10 mph. (a) What is the average frictional force acting on her? (b) Suppose that she could reduce the frictional force by 20 percent by bending low over the handlebars. What increase in speed would this provide, assuming her power output was constant?

5.12 A typical small car of mass 1200 kg (wt. 2640 lbs) may get 30 miles per gallon of gas while travelling at a steady speed of 45 mph. The energy content of 1 gallon of gas is about 1.6×10^{8} J, and a car engine has an efficiency of perhaps 20 percent. (a) From the above information, deduce the horsepower generated by the car engine when cruising at 45 mph. (b) What is the average frictional force acting on the car at this speed? (c) If you went up a 7 percent grade hill at 45 mph, what would you expect your miles per gallon to be?

• Problem Solutions

5.1 $W = Fd\cos\theta = (200\,n)(4\,m)(\cos 30^\circ) = \underline{\underline{693\,J}}$

5.2 $W = 100\,mgh = (100)(70\,kg)(9.8\,m/s^2)(0.3\,m) = \underline{\underline{2.1 \times 10^4\,J}}$

$1\,kcal = 4190\,J$, so $W = 2.1 \times 10^4 (1/4190\ kcal) = \underline{\underline{5.0\ kcal}}$

5.3 $K.E. = \frac{1}{2}mv^2 = \frac{1}{2}(0.01\,kg)(1000\,m/s)^2 = \underline{\underline{5000\,J}}$

$K.E. = \left(\frac{5000}{4190}\right)kcal = \underline{\underline{1.2\ kal}}$

5.4 TOTAL ENERGY IS CONSERVED, SO $KE_1 + PE_1 = KE_2 + PE_2$.

AT MAXIMUM HEIGHT $KE_2 = 0$, SO

$$KE_1 + PE_1 = 0 + PE_2$$

$$\frac{1}{2}mv_1^2 + mgh_1 = 0 + mgh_2$$

$$h_2 = \frac{v_1^2}{2g} + h_1$$

LET $h = 0$ BE LOWEST POINT SWING REACHES, SO $h_1 = 1\,m$

$$h_2 = \frac{(5\,m/s)^2}{(2)(9.8\,m/s^2)} + 1 = \underline{\underline{2.3\,m}} = \text{MAXIMUM HEIGHT}$$

MAXIMUM SPEED OCCURS WHEN $P.E = 0$, SO

$$\frac{1}{2}mv^2 + mgh_1 = \frac{1}{2}mv_{MAX}^2 + 0$$

$$V_{MAX} = \sqrt{v_1^2 + 2gh_1} = \sqrt{(5\,m/s)^2 + (2)(9.8\,m/s^2)(1\,m)}$$

$$V_{MAX} = \underline{\underline{6.7\,m/s}} = \text{MAXIMUM SPEED}$$

5.5 ON LEVEL GROUND LOSS IN K.E. = WORK DONE AGAINST FRICTION

SO $\frac{1}{2}mv_1^2 = F_F d_1$

$$F_F = \frac{mv_1^2}{2d_1}$$

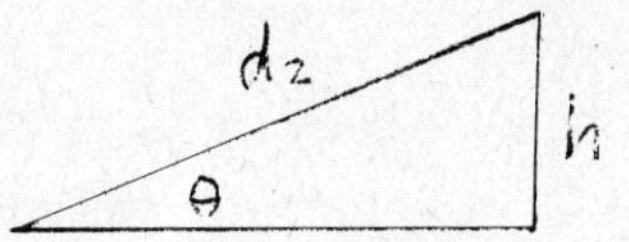

WHEN COASTING UPHILL,

$$KE_1 + PE_1 = KE_2 + PE_2 + \text{ENERGY LOSS}$$

$$\frac{1}{2}mv_1^2 + 0 = 0 + mgh + F_F d_2$$

(first 0:) PE = 0 AT BOTTOM OF HILL

(second 0:) KE = 0 WHEN FINALLY STOPPED.

$$h = d_2 \sin\theta$$

THUS $\frac{1}{2}mv_1^2 = mgd_2\sin\theta + \left(\frac{mv_1^2}{2d_1}\right)d_2$

$$v_1^2 = (2g\sin\theta + v_1^2/d_1)\,d_2$$

$$v^2 = \left(\frac{2gd_1\sin\theta + v_1^2}{d_1}\right)d_2$$

$$d_2 = \frac{d_1 v_1^2}{2gd_1\sin\theta + v_1^2}$$

$$V_1 = 40\ km/hr = (40)\left(\frac{1000\,m}{3600\,s}\right) = 11.1\ m/s$$

$$\sin\theta = 7/100 = 0.07$$

$$d_1 = 150\ m$$

$$d_2 = \frac{(150m)(11.1 m/s)^2}{(2)(9.8 m/s^2)(150m)(0.07) + (11.1 m/s)^2} = \underline{\underline{56 m}}$$

THUS CAR WILL STOP IN 56 m GOING UPHILL.

5.6 $KE_1 + PE_1 = KE_2 + PE_2 +$ WORK DONE AGAINST FRICTION

LET d = DISTANCE SLID ON HILL WHERE $h = d\sin\theta$

x = DISTANCE SLID ON LEVEL

ON THE HILL $R_1 = mg\cos\theta$

$F_{F_1} = \mu R_1 = \mu mg\cos\theta$

ON THE LEVEL $R_2 = mg$

$F_{F_2} = \mu R_2 = \mu mg$

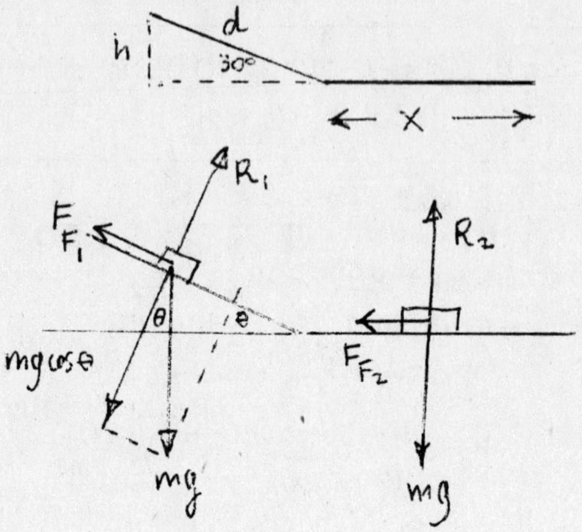

THUS

$\frac{1}{2}mv_1^2 + mgh = 0 + 0 + F_{F_1}d + F_{F_2}x$

$\frac{1}{2}mv_1^2 + mgh = \mu mg\cos\theta\, d + \mu mgx$

$x = \frac{v_1^2}{2g\mu} + \frac{h}{\mu} - d\cos\theta$

$= \frac{v_1^2}{2g\mu} + \frac{h}{\mu} - \left(\frac{h}{\sin\theta}\right)\cos\theta$

$= \frac{(4 m/s)^2}{(2)(9.8 m/s^2)(.1)} + \frac{(2m)}{(.1)} - (2m)\frac{\cos 30°}{\sin 30°}$

$x = 8.2 m + 20 m - 3.5 m = \underline{\underline{24.7 m}}$

$\theta = 31°$ $\mu = 0.1$
$h = 2 m$ $v_1 = 4 m/s$

5.7 LET E_1 = INITIAL TOTAL ENERGY.

E_2 = ENERGY OF ARROW AT TOP OF TRAJECTORY.

E_3 = ENERGY OF ARROW AFTER IT FALLS BACK TO GROUND.

Fh = WORK DONE AGAINST FRICTION IN GOING UP OR DOWN A DISTANCE h.

$E_1 = E_2 + Fh$ (GOING UP)

$E_2 = E_3 + Fh$ (COMING DOWN)

OR $KE_1 + PE_1 = KE_2 + PE_2 + Fh \longrightarrow \frac{1}{2}mv_1^2 + 0 = 0 + mgh + Fh$ (1)

AND $KE_2 + PE_2 = KE_3 + PE_3 + Fh \longrightarrow 0 + mgh = \frac{1}{2}mv_3^2 + 0 + Fh$ (2)

SUBTRACT EQN(2) FROM EQN(1):

$\frac{1}{2}mv_1^2 - mgh = mgh - \frac{1}{2}mv_3^2$

SOLVE, $2mgh = \frac{1}{2}mv_1^2 + \frac{1}{2}mv_3^2$, $h = \underline{\underline{\frac{v_1^2 + v_2^2}{4g}}}$

SUBSTITUTE THIS VALUE BACK IN EQN (1) AND SOLVE FOR F:

$$\frac{1}{2}mv_1^2 = mg\left(\frac{v_1^2 + v_2^2}{4g}\right) + \left(\frac{v_1^2 + v_2^2}{4g}\right)F$$

$$\left(\frac{v_1^2 + v_2^2}{4g}\right)F = \frac{mv_1^2}{2} - \frac{mv_1^2}{4} - \frac{mv_2^2}{4} = \frac{m}{4}(v_1^2 - v_2^2)$$

$$F = \underline{\underline{\left(\frac{v_1^2 - v_2^2}{v_1^2 + v_2^2}\right)mg}}$$

5.8 $P = \frac{\text{ENERGY}}{\text{TIME}} = \frac{2500 \text{ kcal}}{24 \text{ Hrs}} = 104$ kcal/Hr

OR $P = \frac{2500}{24}\frac{(4190 J)}{(3600 s)} = 121$ J/s $= \underline{\underline{121 \text{ WATTS}}}$.

5.9 EFFICIENCY = $\dfrac{\text{LIGHT POWER OUT}}{\text{TOTAL POWER IN}} = \dfrac{8\,W}{80\,W} = 0.1 = 10\%$

5.10 WORK DONE TO LIFT PIANO IS $W = mgh$

IF W = WORK DONE BY MEN, $W = 0.7W_M$ (70% EFFICIENCY)

$P = \dfrac{W}{t} = \dfrac{0.7W_M}{t} = 0.7P_M$

So $t = \dfrac{W}{0.7P_M} = \dfrac{mgh}{0.7P_M}$

$mg = 500\,\text{lbs} = (500)(4.45\,n) = 2225\,n$

$P_M = (2)(.2\,Hp) = (2)(.2)(746\,W) = 298.4\,W$

$t = \dfrac{mgh}{0.7P_M} = \dfrac{(2225\,n)(30\,m)}{(0.7)(298.4\,W)} = 320\,s$

$t = 5\,\text{min}\ 20\,\text{sec}$ (1 min = 60 sec)

5.11 EXTRA POWER EXPENDED IN RIDING IS $600\,W - 120\,W = 480\,W$

$P = F_1V_1$, SO $F_1 = \dfrac{P}{V_1} = \dfrac{480\,W}{10\,\text{mi/Hr}} = \dfrac{480\,W}{(10)\left(\dfrac{1609\,m}{3600\,s}\right)}$

$F_1 = 107\,n$ (ABOUT 24 lbs)

IF $F_2 = 0.8F_1$, $P = F_1V_1 = F_2V_2$

$V_2 = \dfrac{F_1}{F_2}V_1 = \dfrac{F_1}{0.8F_1}V_1 = 1.25\,V_1$

$V = (1.25)(10\,\text{mph}) = 12.5\,\text{mph}$

5.12(a) IN ONE HOUR CAR TRAVELS 45 MILES AND AT 30 MILES/GAL THIS REQUIRES 45 MI/(30 MI/GAL) = 1.5 GALS. THE ENERGY YIELD FROM THIS MUCH GAS IS $Q = (1.5\,\text{GAL})(1.6\times10^8\,J/\text{GAL}) = 2.4\times10^8\,J$.
THE ENGINE HAS 20% EFFICIENCY, SO THE WORK DONE WITH THIS MUCH FUEL IS $W = (0.2)(2.4\times10^8\,J) = 4.8\times10^7\,J$.
THIS WORK IS DONE IN ONE HOUR, OR 3600 s, SO POWER IS

$P = \dfrac{W}{t} = \dfrac{4.8\times10^7\,J}{3600\ s} = 1.33\times10^4\,W$

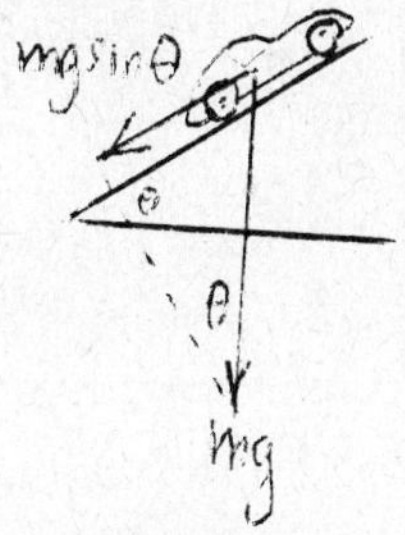

1 Hp = 746 W, so $P = 1.33\times10^4\left(\dfrac{1}{746}\,Hp\right) = 17.9\,Hp$

(b) $P = Fv$, $F = \dfrac{P}{v} = \dfrac{1.33\times10^4\,W}{(45)\left(\dfrac{1609\,m}{3600\,s}\right)} = 663\,n$

(c) $v = \dfrac{P_1}{F_1} = \dfrac{P_2}{F_2}$

$F_1 = 663\,n$ = FORCE OF FRICTION $\qquad F_2 = F_1 + mg\sin\theta$

GOING UPHILL ENGINE MUST PUSH AGAINST BOTH FRICTION F_1 AND COMPONENT OF WEIGHT, $mg\sin\theta$, PARALLEL TO GROUND.

THUS $P_2 = (F_2/F_1)P_1 = [(F_1 + mg\sin\theta)/F_1]P_1 = \dfrac{663\,n + (1200\,kg)(9.8\,m/s^2)(0.07)}{663\,n}(1.33\times10^4\,W)$

$P_2 = 2.98\times10^4\,W$

THE MILES PER GALLON (MPG) FOR GIVEN CAR SPEED IS PROPORTIONAL TO THE POWER GENERATED.

$\dfrac{(MPG)_2}{(MPG)_1} = \dfrac{P_2}{P_1}$, $(MPG)_2 = \dfrac{P_2}{P_1}(MPG)_1 = \left(\dfrac{1.33\times10^4\,W}{2.98\times10^4\,W}\right)(30\,\text{mi/gal})$

$(MPG)_2 = 13.4\ \text{mi/gal}.$

6 Impulse and Momentum

• Summary of Important Ideas, Principles, and Equations

1. Definition: The linear momentum of a particle of mass m and velocity v is

$$\vec{p} = m\vec{v} \tag{6.1}$$

For a system of particles with masses m_1, m_2, ... and velocities $\vec{v}_1$, $\vec{v}_2$, ... the total linear momentum is

$$\vec{p} = m_1\vec{v}_1 + m_2\vec{v}_2 + \ldots = \Sigma m_i\vec{v}_i \tag{6.2}$$

2. Newton's second law can be written in terms of momentum as

$$\vec{F} = \frac{\Delta\vec{p}}{\Delta t} \text{ or } F_x = \frac{\Delta p_x}{\Delta t}, \quad F_y = \frac{\Delta p_y}{\Delta t}, \quad F_z = \frac{\Delta p_z}{\Delta t} \tag{6.3}$$

3. The change in momentum when an average force $\langle\vec{F}\rangle$ acts for a time Δt is called the impulse of the force $\vec{F}$.

$$\Delta\vec{p} = \langle\vec{F}\rangle\Delta t \tag{6.4}$$

Impulse and momentum both have dimension ML/T and are measured in kg-m/s.

4. Definition: An isolated system is one on which no external forces act.

The Principle of Conservation of Momentum states that the momentum of an isolated system of interacting bodies is constant.

5. A collision in which the kinetic energy of the system is unchanged before and after the collision is an elastic collision. Collisions for which this is not the case are inelastic.

Momentum is conserved, i.e. the same before and after, for all collisions, elastic and inelastic.

6. The center of mass of a system is located at a point $\vec{r}_{CM}$ defined by

$$\boxed{M\vec{r}_{CM} = \Sigma m_i \vec{r}_i} \qquad (6.5)$$

where M is the total mass of the system, $M = \Sigma m_i$ and $\vec{r}_i$ is the position of the piece of mass m_i.

Qualitatively, the center of mass point is like the "balance point" of the system (like the fulcrum of a teeter totter). For a solid object the CM need not be within the object.

The total momentum of a system is

$$\vec{p} = \Sigma m_i \vec{v}_i = M\vec{v}_{CM}$$

where $\vec{v}_{CM}$ is the velocity of the center of mass. For an isolated system, $\vec{p}$ = constant, so $\vec{v}_{CM}$ = constant also. For an isolated system the CM moves in a straight line with constant speed.

When solving collision problems it is often easiest to use a coordinate system with the CM at the origin. In this moving reference frame the CM is always at rest.

• Qualitative Questions

6M.1 We can be certain that a moving (translating) object has

A. positive total energy.
B. linear momentum.
C. potential energy.
D. an impulse.
E. velocity, but not necessarily speed.

6M.2 Which of the following pairs of quantities have the same dimension?

A. Weight and mass.
B. Force and acceleration.
C. Force and energy.
D. Impulse and momentum.
E. Momentum and force.

6M.3 Suppose that a ball flies through the air and strikes horizontally the wall of a building. For a given collision time the force experienced by the wall will be determined primarily by

A. the kinetic energy of the incident ball.
B. the force carried by the ball.
C. the linear momentum of the ball.
D. the mass of the building.
E. the mass of the ball.

6M.4 Suppose that you are faced with the choice of ramming your car head on into an immovable brick wall or making a head on collision with an approaching car of the same mass and speed as your car. In which case would the force of impact be greatest?

A. Colliding with the wall would be worse.
B. Colliding with the other car would be worse.
C. The forces encountered would be the same in both cases.
D. The choice here would depend upon whether your car knocked the other one back or it knocked you back. The car knocked back would experience the greater force.

6M.5 During World War I the Germans used some long range cannons called Big Berthas with which they lobbed artillery shells for distances up to 20 miles. Such long range cannons have very long barrels. The reason such long barrels are used is

A. to increase the force which acts on the cannonball.
B. to increase the time during which the force acts on the cannonball.
C. to reduce friction on the cannonball.
D. to improve the ability to aim a high speed projectile.
E. to increase the momentum of the cannonball without having to increase its kinetic energy.

6M.6 In an auto safety test a car is driven into a wall at high speed with a dummy in the front seat. A high speed camera records the time of impact during which the head of the dummy is in contact with the dashboard of the car. Which of the following would be the most useful additional piece of information if you wished to estimate the average force exerted on the dummy's head during the crash?

A. The mass of the head.
B. The kinetic energy of the head just before impact.
C. The linear momentum of the head just before impact.
D. The velocity of the head just before impact.
E. The acceleration of the head just before impact.

6M.7 A boy throws a golf ball against a brick wall. It bounces back as fast as it hit. If the ball was thrown horizontally from a distance of 4 meters and its speed when it hit the wall is 20 m/s, what is the average force exerted on the wall by the ball? The mass of the boy is 50 kg and the mass of the ball is 100 grams.

A. 2 N
B. 4 N
C. 8 N
D. 11.2 N
E. The force cannot be determined without more information.

6M.8 When a bullet is fired from a rifle, the bullet and the rifle acquire

A. velocities of equal magnitude but opposite direction.
B. momenta of equal magnitude but opposite direction.
C. equal amounts of kinetic energy.
D. More than one of the above are true.
E. None of the above is true.

6M.9 A 100 kg linebacker moving 2 m/sec tackles head-on an 80 kg halfback running 3 m/sec. Neglecting any effects due to digging in cleats,

A. the linebacker will drive the halfback backwards.
B. the halfback will drive the linebacker backwards.
C. Neither player will drive the other back.
D. This is an example of an elastic collision.
E. More than one of the above are true.
F. None of the above is true.

6M.10 A fire hose is turned on the door of a burning building in order to knock the door down. This requires a force of 1000 N. If the hose delivers 40 liters per second, what is the minimum velocity of the stream needed, assuming the water doesn't bounce back?

A. 10 m/sec
B. 12.5 m/sec
C. 25 m/sec
D. 40 m/sec
E. 50 m/sec

6M.11 A 200 lb logger stands on a 400 lb log floating in a pond. The logger walks 24 meters along the log in 8 seconds. At what speed does the log move through the water, assuming negligible friction?

A. 1 m/s
B. 1.5 m/s
C. 2 m/s
D. 2.4 m/s
E. 4 m/s
F. 8 m/s
G. 12 m/s

6M.12 Water runs out of a horizontal drain pipe at the rate of 1200 liters/minute. It falls 3.2 meters to the ground. Assuming the water doesn't splash up when it hits, what average force does it exert on the ground?

A. 20 n
B. 64 n
C. 96 n
D. 160 n
E. 320 n
F. 640 n

6M.13 A bullet of mass m and velocity v strikes a large wooden block of mass M suspended from a string of length L. The block is initially at rest. The bullet sticks in the block. Using the conservation of momentum and conservation of energy, we deduce that the block will swing up a height

A. $\frac{v^2}{2g}(\frac{m}{m+M})$

B. $\frac{v^2}{2g}\sqrt{\frac{m}{m+M}}$

C. $\frac{v^2}{2g}(\frac{m}{m+M})^2$

D. $\frac{v^2}{2g}$

E. $\frac{v^2}{2g}(\frac{m}{M})^2$

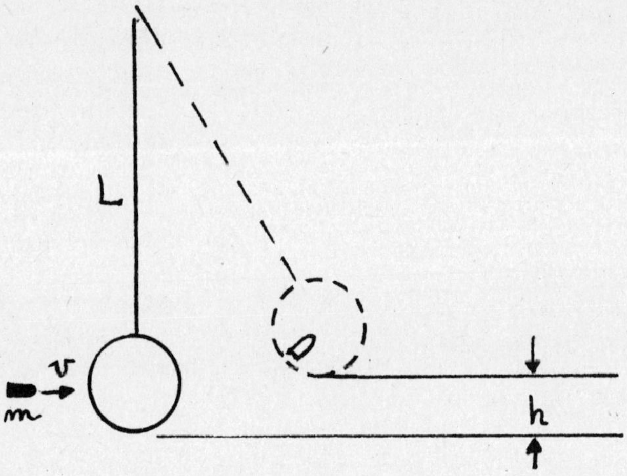

6M.14 A ball is dropped from rest and allowed to bounce on a concrete floor. A graph of its position is sketched here. From this we can see that

A. momentum of the earth-ball system was not constant during the first four collisions.
B. the impulse for each successive collision increased.
C. the collisions with the floor were inelastic.
D. the total mechanical energy of the ball remained constant as a function of time.
E. this graph describes a process which is not physically possible because the time between bounces is not shown as constant.

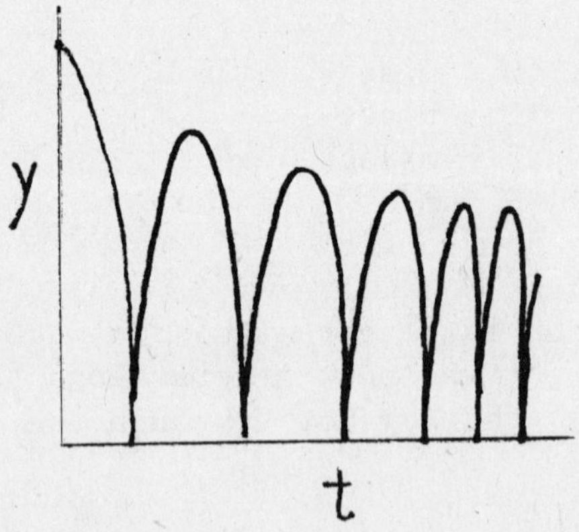

• Multiple Choice: Answers and Comments

6M.1 B A is false, since total energy is the sum of KE plus PE, and is defined only to within an additive constant, so total energy can be negative even though KE is always positive. C is false; a moving object has KE, but not necessarily PE. D and E don't make any sense.

6M.2 D

6M.3 C B is false, since a particle does not "carry" a force. The force exerted depends on the initial momentum, the momentum after the collision (i.e. on the change in momentum) and on the collision time. You know this from practical experience. If you jump from a table down on to the floor and land stiff-legged (so that the collision time is very short) the force on your legs is great and could break them. However, if you bend your knees and thereby lengthen the collision time, the force exerted is much less. In each case the impulse of the force is the same, but the force itself is different.

6M.4 C In either case your car comes to rest after the collision. Thus, since the time of collision and the change in momentum are the same in each case, the force exerted on your car is the same in both cases.

6M.5 B One wants to give the cannonball the maximum speed, and hence maximum momentum. The impulse (the change in momentum) is equal to $F\Delta t$, so by increasing Δt with a long barrel, Δp is made larger.

6M.6 C $F = \Delta p/\Delta t$, so to determine F we need to know Δt (given) and Δp. Right after the crash $p = 0$, so if we know the initial momentum we can calculate F.

6M.7 E To calculate $F = \Delta p/\Delta t$ we need to know Δp and Δt. We can calculate Δp from the data given, but we are not given Δt, so we can't calculate F. Note that most of the other information, including the page number, is irrelevant to the solution of the problem. You must learn to recognize what data is relevant and what is not. In real life they do not tell you what chapter you are in, or even what book or course you are in. You may or may not have all the information you need, or you may have way too much. This is something you have to learn to figure out for yourself.

6M.8 B Momentum is conserved, and the initial momentum of the system before firing is zero, so the total momentum after firing is also zero.

6M.9 B Momentum is conserved, so $M_{LB}v_{LB} + M_{HB}v_{HB} = (M_{LB} + M_{HB})\ v_{fin}$

$$v_{fin} = \frac{M_{LB}v_{LB} + M_{HB}v_{HB}}{(M_{LB} + M_{HB})}$$

$$= \frac{(100\ kg)(2m/s) + (80\ kg)(-3\ m/s)}{100\ kg + 80\ kg}$$

$$v_{fin} = -0.22\ m/s$$

D is false. Since players stick together, collision is inelastic.

6M.10 C $F = \frac{\Delta p}{\Delta t} = \frac{\Delta (mv)}{\Delta t} = \frac{(40\ kg)(v)}{1\ sec}$

Given F = 1000 n, $v = \frac{1000}{40}$ m/s = 25 m/s

6M.11 A Let m_1 = mass of logger and m_2 = mass of log.

v_1 = velocity of logger with respect to log = 24 m/8s = 3 m/s.

v_2 = velocity of log with respect to earth.

$v_1 - v_2$ = velocity of logger with respect to earth.

The momentum of the system is conserved, so

$$m_1(v_1 - v_2) - m_2v_2 = 0$$

$$m_1v_1 - (m_1 + m_2)v_2 = 0$$

$$v_2 = \frac{m_1v_1}{m_1 + m_2} = -\left(\frac{m_1g}{m_1g + m_2g}\right)v_1 = -\frac{w_1}{w_1 + w_2}\ v_1$$

$$= -\left(\frac{200\ lbs}{200\ lbs + 400\ lbs}\right)(3\ m/s) = -1\ m/s$$

Note that it is O.K. to use the British units "lbs" here since we are just calculating a ratio and the units cancel out.

6M.12 D $F = \frac{\Delta p}{\Delta t} = \frac{mv_i - mv_f}{\Delta t}$

$v_f = \sqrt{2gh}$, $v_i = 0$

1200 liters of water has a mass of 1200 kg.

$t = 1$ minute = 60 secs.

Thus $F = \frac{(1200 \text{ kg})(\sqrt{(2)(9.8 \text{ m/s}^2)(3.2 \text{ m})})}{60 \text{ s}} = 160$ n

6M.13 C When bullet strikes the block momentum is conserved, but KE is not conserved because collision is inelastic.

Thus $mv = (m + M)V$

$V = \frac{m}{m + M} v$

When block swings up to height h energy is conserved, so

$$(PE)_1 + (KE)_1 = (PE)_2 + (KE)_2$$

$$0 + \tfrac{1}{2}(m + M)V^2 = (m + M)gh + 0$$

$$h = \frac{V^2}{2g} = \frac{1}{2g}\left(\frac{m}{m + M}\right)^2 v^2$$

6M.14 C Momentum is conserved for each collision, whether or not the collisions are elastic.
B is false, since the impulse Δp gets smaller and smaller, as p itself gets smaller each time. Eventually $p = 0$ after many bounces (losing energy each time) so of course $\Delta p = 0$ also.
C is true, since the rebound height gets smaller and smaller, showing that energy is lost in each collision.
D is false; the total energy is decreasing, as indicated by the decreasing maximum height of each bounce.
E is false, since lower bounces require less time, and hence the time between bounces should decrease, as sketched.

• Problems

6.1 What is the momentum of a 10 gram bullet which has a KE of 3200 J?

6.2 A baseball of mass 150 g is thrown with a speed of 20 m/s to a batter who strikes it and sends a line shot straight back at the pitcher with a speed of 60 m/s. The ball is in contact with the bat for 25 msec. What average force is exerted on the ball by the bat? The mass of the bat is 1.02 kg.

6.3 Consider three identical freight cars, each of mass 10,000 kg. One is sitting at rest and the other two are moving toward it, each with velocity of 6 m/s. When a car collides with another car it couples to it. What will be the final velocity of the car which is initially at rest?

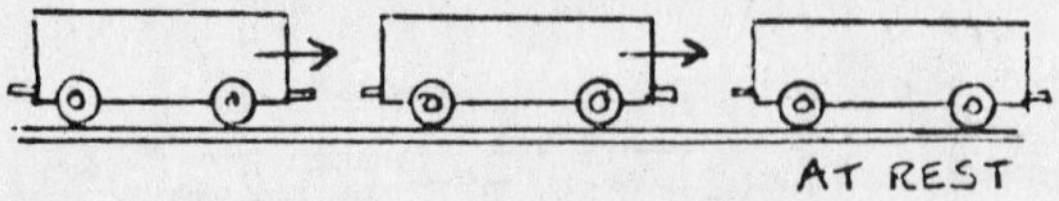

6.4 An astronaut in a small one woman rocket ship is drifting directly toward a distant star with a speed of 1200 m/s. She wants to change her course, but since she is low on fuel she decides not to fire her retrorockets. Instead, she fires her port lugwumper hurling a 500 kg projectile at 800 m/s (with respect to the ship) at 90^0 to her direction of travel. Before the firing the mass of the ship and passenger was 2500 kg. Through what angle was she able to change her direction of travel by this manuever?

6.5 A remarkable demonstration of momentum conservation is the following: Place a small ball of mass m (a marble or a pingpong ball work well) on top of a rubber superball of mass M (a ball about the size of a golf ball which is very elastic). Drop them from rest together from shoulder height, a distance h above the floor. How high will the small ball bounce in the limit $m << M$. This has to be seen to be believed. Try it, but be sure there is no skylight or glass light fixture overhead. I put a lecture room out of commission showing this when using a steel ball bearing! (Hint: Imagine that first the superball makes an eleastic collision with the floor, and then on its way up it makes an elastic collision with the little ball which is coming down.)

• Problem Solutions

6.1 $KE = \frac{1}{2}mv^2 = \frac{1}{2m}(mv)^2 = \frac{p^2}{2m}$

$p^2 = 2m(KE)$

$p = \sqrt{2m\,KE} = \sqrt{(2)(0.01\,kg)(3200\,J)} = \underline{\underline{8\ kg\,m/s}}$

6.2 $\Delta p = \langle F\rangle \Delta t$ FROM EQN. (5.4)

$\langle F\rangle = \frac{\Delta p}{\Delta t} = \frac{p_f - p_i}{\Delta t} = \frac{mv_f - mv_i}{\Delta t}$

$= \frac{(0.15kg)(60m/s) - (0.15kg)(-20m/s)}{0.025s}$

$\underline{\underline{\langle F\rangle = 480\ n}}$

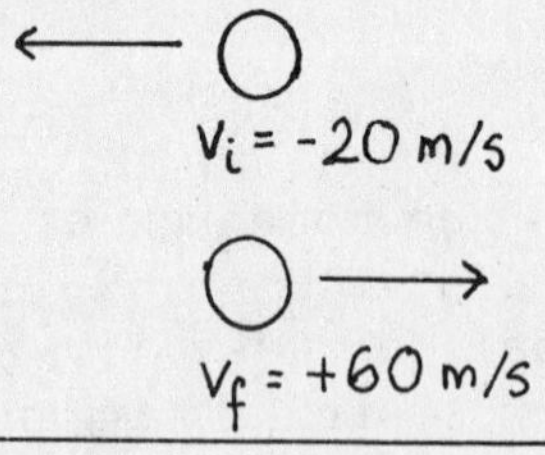

6.3 TOTAL MOMENTUM IS CONSERVED, SO

$mv_1 + mv_1 + 0 = (3m)V_2$

$2mv = 3mv_2$, $V_2 = \frac{2}{3}V_1 = \left(\frac{2}{3}\right)(6\,m/s) = \underline{\underline{4\,m/s}}$

6.4 NO EXTERNAL FORCES ACT, SO THE TOTAL LINEAR MOMENTUM OF THE SYSTEM IS CONSTANT. IN PARTICULAR, THE CENTRE OF MASS (CM) MOVES IN THE SAME STRAIGHT LINE WITH CONSTANT VELOCITY $V_{cM} = 1200$ m/s DIRECTED TOWARD THE STAR. IN THE CM FRAME (i.e. TO A PERSON INITIALLY RIDING IN THE SHIP). THE TOTAL MOMENTUM IS ZERO HENCE IT REMAINS ZERO AFTER THE LUGWUMPER IS FIRED.

LET $\vec{P}_{cM}$ = MOMENTUM OF CM IN REFERENCE FRAME ATTACHED TO STAR.

$\vec{P}_{LW}$ = MOMENTUM OF LUGWUMPER IN CM REFERENCE FRAME.

$\vec{P}_R$ = MOMENTUM OF ROCKET IN CM FRAME AFTER LUGWUMPER HAS BEEN FIRED.

WE ARE TOLD THAT $\vec{P}_{LW}$ IS PERPENDICULAR (FIRED AT 90°) TO $\vec{P}_{CM}$, SO THESE VECTORS ARE RELATED LIKE THIS:

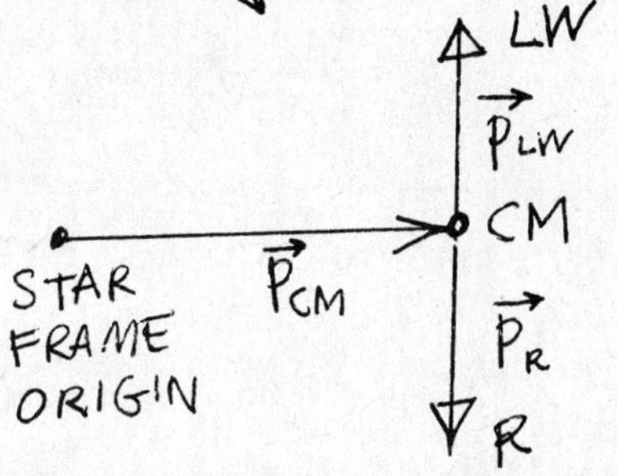

NOTE THAT $\vec{P}_{LW} + \vec{P}_R = 0$

THE MOMENTUM OF THE ROCKET WITH RESPECT TO THE STAR REFERENCE FRAME IS OBTAINED FROM "STAR" TO "R", LIKE THIS:

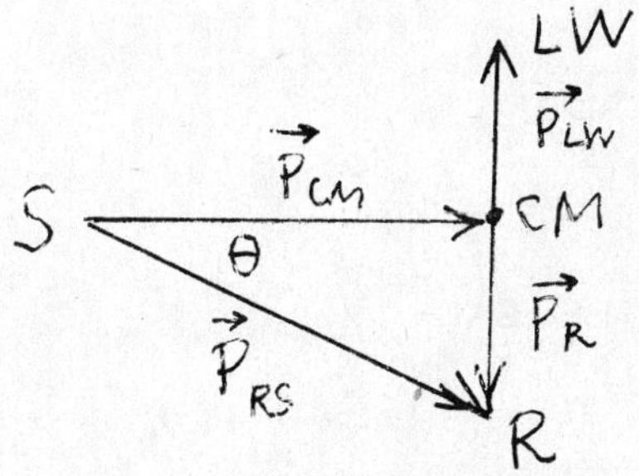

THUS THE ROCKET IS NOW HEADING OFF AT ANGLE θ WITH RESPECT TO THE ORIGINAL DIRECTION, AND

$$\tan\theta = \frac{P_R}{P_{CM}}$$

SINCE $P_R = P_{LW}$ $\tan\theta = \frac{P_{LW}}{P_{CM}}$

$P_{LW} = m_{LW}V_{LW} = (500\text{ kg})(800\text{ m/s}) = 4 \times 10^5$ kg·m/s

$P_{CM} = m_{SHIP}V_{CM} = (2500\text{ kg})(1200\text{ m/s}) = 3 \times 10^6$ kg-m/s

$$\tan\theta = \frac{4 \times 10^5 \text{ kg-m/s}}{3 \times 10^6 \text{ kg-m/s}} = 0.133 \qquad \underline{\underline{\theta = 7.6°}}$$

6.5 LARGE BALL FALLS FROM HEIGHT h AND AQUIRES SPEED V JUST BEFORE IT HITS FLOOR. ENERGY IS CONSERVED, SO

$$KE_1 + PE_1 = KE_2 + PE_2$$

$$0 + Mgh = \tfrac{1}{2}MV^2 + 0$$

$$V = \sqrt{2gh} \qquad (1)$$

LARGE BALL MAKES ELASTIC COLLISION WITH FLOOR AND THUS BOUNCES UPWARD WITH SPEED V. NOW IT COLLIDES HEAD ON WITH SMALL BALL COMING DOWN WITH SPEED V. AGAIN MOMENTUM AND KE ARE CONSERVED, SO

$$MV - mV = MV_1 + mv_1 \qquad (2)$$

$$\frac{1}{2}MV^2 + \frac{1}{2}mV^2 = \frac{1}{2}MV_1^2 + \frac{1}{2}mv_1^2 \qquad (3)$$

(2) AND (3) ARE TWO EQUATIONS FOR THE TWO UNKNOWNS v_1 AND V_1.

v_1 = VELOCITY OF m JUST AFTER COLLISION.

V_1 = VELOCITY OF M JUST AFTER COLLISION.

SOLVE (2) FOR V_1: $V_1 = \left(1 - \frac{m}{M}\right)V - \frac{m}{M}v_1$

SUBSTITUTE THIS IN (3):

$$\frac{1}{2}(M+m)V^2 = \frac{1}{2}M\left[\left(1-\frac{m}{M}\right)V - \frac{m}{M}v_1\right]^2 + \frac{1}{2}mv_1^2$$

MULTIPLY BY 2 AND DIVIDE THROUGH BY M AND LET $\frac{m}{M} \equiv \alpha$

$$(1+\alpha)V^2 = \left[(1-\alpha)V - \alpha v_1\right]^2 + \alpha v_1^2$$

$$(1+\alpha)V^2 = (1-\alpha)^2V^2 - 2\alpha(1-\alpha)Vv_1 + \alpha^2 v_1^2 + \alpha v_1^2$$

SIMPLIFY AND COLLECT TERMS:

$$(3\alpha - \alpha^2)V^2 = -2\alpha Vv_1 + 2\alpha^2 Vv_1 + (\alpha + \alpha^2)v_1^2$$

IN THE LIMIT $\alpha = \frac{m}{M} << 1$, $\alpha^2 << \alpha$

THUS NEGLECT TERMS IN α^2 AS A FIRST APPROXIMATION.

$$3\alpha V^2 = -2\alpha Vv_1 + \alpha v_1^2$$

$$3V^2 = -2Vv_1 + v_1^2$$

$$v_1^2 - 2Vv_1 - 3V^2 = 0$$

FACTOR, $(v_1 - 3V)(v_1 + V) = 0$

SO $v_1 = 3V$ OR $-V$

THE POSITIVE SOLUTION CORRESPONDS TO THE LITTLE BALL BOUNCING UPWARD. IT WILL RISE TO A HEIGHT h_F WHERE FROM ENERGY CONSERVATION

$$\frac{1}{2}mv^2 + 0 = mgh_F + 0$$

$$h_F = \frac{v^2}{2g} = \frac{(3V)^2}{2g} = 9\left(\frac{V^2}{2g}\right)$$

BUT ORIGINAL HEIGHT h FROM WHICH BALLS WERE DROPPED IS

$$h = \frac{V^2}{2g}$$

THUS LITTLE BALL RISES TO A FINAL HEIGHT h_F,

$$h_F = 9h$$

TRY THIS FOR YOURSELF. IT IS AMAZING!

Circular Motion

7

• Summary of Important Ideas, Principles, and Equations

1. Consider an object which moves in a circle of radius r. It travels a distance s along the arc in time Δt, sweeping out an angle θ in so doing. θ is the angular displacement.

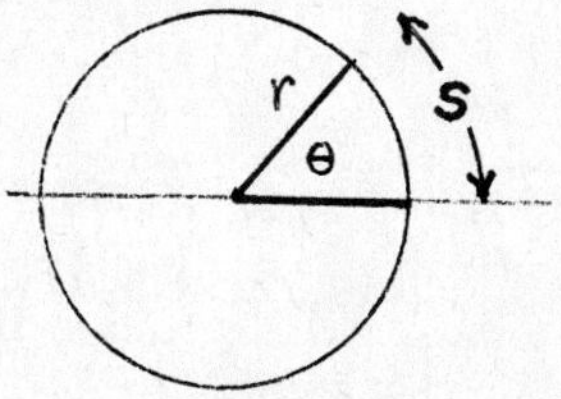

$$s = r\theta \qquad (7.1)$$

Here θ is measured in radians,

1 radian = 57.3°

2π radians = 360°

The average angular velocity $\langle\omega\rangle$ is defined as

$$\langle\omega\rangle = \frac{\theta_f - \theta_o}{t_f - t_o} = \frac{\Delta\theta}{\Delta t} \qquad (7.2)$$

The instantaneous angular velocity is defined as

$$\omega = \lim_{\Delta t \to 0} \frac{\Delta\theta}{\Delta t} \qquad (7.3)$$

The angular velocity is also called the angular frequency. It is measured in radians/second.

The number of revolutions made per second is the frequency, f. Since one revolution is 2 π radians,

$$\omega = 2\pi f \qquad (7.4)$$

The average and instantaneous angular accelerations, measured in rad/s^2, are

$$\langle \alpha \rangle = \frac{\omega_f - \omega_o}{t_f - t_o} = \frac{\Delta\omega}{\Delta t}, \quad \alpha = \lim_{\Delta t \to 0} \frac{\Delta\omega}{\Delta t} \qquad (7.5)$$

The linear displacement, s, the linear velocity, v, and the linear acceleration a are related to the corresponding angular quantities θ, ω and α as follows:

$$s = r\theta$$

$$v = r\omega$$

$$a = r\alpha \qquad (7.6)$$

2. The kinematic equations for angular motion and the corresponding ones for linear motion are as follows:

Angular Equations	Linear Equations
$\theta = \omega_o t + \frac{1}{2}\alpha t^2$	$x = v_o t + \frac{1}{2}at^2$
$\omega = \omega_o + \alpha t$	$v = v_o + at$
$\omega^2 = \omega_o^2 + 2\alpha\theta$	$v^2 = v_o^2 + 2as$
(choosing $\theta_o = 0$ and $x_o = 0$)	

(7.7)

3. An object of mass m moving with linear speed v in a circle of radius r is accelerating toward the center of the circle. This type of acceleration (in which the speed is constant but the direction of the velocity is changing) is called centripetal acceleration. ("Centripetal" means "toward the center".)

$$a_c = \frac{v^2}{r} = r\omega^2 \qquad (7.8)$$

To cause this acceleration it is necessary that a force act on the particle. The force needed to cause this kind of motion is called centripetal force.

$$F_c = ma_c = \frac{mv^2}{r} = mr\omega^2 \qquad (7.9)$$

Note that centripetal force is not a kind of force, like gravity or friction. Rather, it is a way of using a force. Thus when you draw a force diagram showing the forces acting on an object, do not draw in F_c as one of these forces. The forces which act will be mg, the force of friction, a normal reaction force, a tension force in a rope, etc. One of these, or a combination of them, may constitute the "centripetal force", but you would err if you drew in yet an additional force, F_c.

• Qualitative Questions

7M.1 Which of the following is a FALSE statement concerning a point on the rim of a rotating wheel.

A. If the wheel is rotating with constant angular velocity, the point has a linear velocity.
B. If the wheel rotates with constant angular velocity, the point has a linear acceleration.
C. If the wheel rotates with constant angular velocity, the point has centripetal acceleration.
D. If the wheel rotates with constant angular acceleration, the point has a linear acceleration.
E. If the wheel rotates with constant angular acceleration, the point has centripetal acceleration.

7M.2 The flywheel on a large machine is rotating with an angular velocity of 126 rad/sec. Through what angle will the wheel be displaced after 6 seconds?

A. 12°
B. 23°
C. 60°
D. 79°
E. 116°
F. 233°

7M.3 The speedometer on your car is calibrated so that it gives an accurate speed reading when your car is equipped with regular highway tires. What would happen if you replace the regular tires with thick tread snow tires?

A. This would cause the speedometer to give a reading which is too low.
B. This would cause the speedometer to give a reading which is too high.
C. This would have no effect on the accuracy of the readings, since the speedometer is calibrated for given angular velocity of the wheel, not for linear velocity.
D. This would have no effect on the accuracy of the readings, since the larger tire radius would result in a reduced angular velocity and hence no change in speed.
E. This would have no effect on the accuracy of the readings, since by definition a speedometer measures speed, no matter what kind of tires are used.

7M.4 The wheels of a car have a diameter of 60 cms. With what angular velocity are they rotating when the car is moving 20 m/s?

A. 10.6 rad/sec
B. 14.0 rad/sec
C. 66.7 rad/sec
D. 133.3 rad/sec
E. 179 rad/sec

7M.5 Potlatch, Idaho is located at a latitude of about 47°N. The radius of the earth is 6380 km. How fast is Potlatch moving due to the rotation of the earth?

A. 162 m/sec
B. 316 m/sec
C. 464 m/s
D. 7600 m/s
E. 9000 m/s

7M.6 In the game of tetherball a ball is attached to a long string, one end of which is tied to the top of a pole fixed in the ground. If one hits the ball, it whirls around and around the pole, and the string winds itself up on the pole, drawing the ball in to a smaller and smaller radius. If r is the distance of the ball from the pole and ω is the angular velocity of the ball as it rotates around the pole, it is possible to show that $mr^2\omega$ = constant, where m = mass of the ball. This means that as r gets smaller, ω gets larger. You may have observed this effect yourself. The ball whirls faster and faster as it moves in. What can you say about the centripetal acceleration of the ball as it winds in?

A. The centripetal acceleration remains constant.
B. the centripetal acceleration decreases as the ball winds in.
C. The centripetal acceleration increases as the ball winds in.
D. The centripetal acceleration may either increase or decrease depending on how fast the ball is going.
E. The centripetal acceleration is zero here, since the speed of the ball is changing.

7M.7 A record player rotating 33.3 times per minute slows to a stop in 16 seconds. What is the magnitude of the angular acceleration, which is assumed to be constant?

A. Zero
B. 0.03 rad/s^2
C. 0.22 rad/s^2
D. 2.1 rad/s^2
E. 13 rad/s^2

7M.8 A boy swings a bucket of water in a vertical circle of radius 50 cms. At the top of the arc the bucket is moving 2.5 m/s. From this information

A. we cannot determine whether or not the water will fall out of the bucket unless we know the mass of the water.
B. we cannot determine whether or not the water will fall out, since this depends on how hard the boy pulls on the bucket at the top of the arc.
C. we can see that the water will not fall out at the top of the arc.
D. we can see that the water will fall out at the top of the arc.
E. we deduce that part, but not necessarily all, of the water will fall out at the top of the arc.

7M.9 You may have noticed in the movies that sometimes when they show a picture of a racing stagecoach it appears that the wheels are not turning. At other times it appears that the wheels are turning backward. The reason for this is that a motion picture camera takes one picture every 1/16th of a second. Thus if there are 12 spokes on the wheel, and the wheel rotates through 1/12th of a revolution, or 2/12ths, or 3/12ths, it appears in the pictures as if the wheel had not rotated at all, since all the spokes look alike. If the wheel rotates just a little less than 1/12th of a revolution it looks (to our brain) as if the wheel rotated backwards a little bit (brains are trained to make certain set deductions, and they would rather think the wheel turned back 5° than ahead 25°, which is pretty stupid considering that the stagecoach is going forward). Now that you know how this works, what is the minimum speed a stagecoach with 1 meter diameter wheels (and 12 spokes) can travel if the wheel is to appear as if it isn't turning in the movie (taken with 16 frames per second)?

A. 0.67 m/s
B. 1.33 m/s
C. 4.2 m/s
D. 6.7 m/s
E. 12 m/s

7M.10 Suppose that you are riding around on a merry-go-round at a distance r from the center. You probably know that you have to hold on to keep from falling off. If you were to move in toward the center of the merry-go-round (which rotates at a constant rate) the centripetal force you would experience would

A. be unchanged.
B. increase.
C. decrease.
D. be zero if you were not holding on.
E. be independent of your mass.

7M.11 An ultracentrifuge used to separate and study protein molecules has an effective radius of 3 cms and rotates at 100,000 RPM. What is the centripetal acceleration of a sample in such an instrument, expressed as a multiple of g ($g = 9.8\ m/s^2$)?

A. 32 g
B. 440 g
C. 1260 g
D. 55600 g
E. 336,000 g

7M.12 Two identical masses are attached to two pieces of string of equal lengths, as shown here, and whirled in a horizontal circle. What is the ratio of the tension in the inner string to the tension in the outer string?

A. $\sqrt{2}$
B. 1.5
C. 2
D. $2\sqrt{2}$
E. 4

7M.13 The friction on the tires of your car provides the centripetal force needed to turn your car. Suppose that when travelling 20 km/hr the minimum radius in which you can turn is 6 meters. If you were going twice as fast the minimum radius in which you could turn would be

A. 3 meters.
B. still 6 meters.
C. 12 meters.
D. 24 meters.
E. 48 meters.

• Multiple Choice: Answers and Comments

7M.1 B A is true, since $v = r\omega$.
B is false, since if ω = constant, $\alpha = 0$ and $a = r\alpha$.
C is true, since $a_c = v^2/r$.
D is true, since $a = r\alpha$.
E is true, since a rotating point has centripetal acceleration whether or not it has angular acceleration.

7M.2 E $\theta = \omega t = (126 \text{ rad/s})(6\text{s}) = 756$ radians
2π radians = 1 revolution, so this angular displacement is

$\frac{756}{2\pi} = 120.32$ revolutions. Thus the net effect is to rotate the wheel through 0.32 revolution, since rotation through an integral number of revolutions doesn't change the position of the wheel.

To express θ in degrees it is easy to use a ratio. 0.32 revolution is to one revolution as θ degrees is to 360°.

$$\frac{0.32}{1} = \frac{\theta}{360^o} \qquad \theta = (0.32)(360^o) = 116^o$$

7M.3 A The speedometer is calibrated to give a certain speed reading corresponding to a given angular velocity of the wheels. Since $v = r\omega$, increasing r (bigger tires) for given ω will make the car go faster. Thus the speedometer will read a lower speed than the car actually has.

7M.4 C $\omega = \frac{v}{r} = \frac{20 \text{ m/s}}{0.3 \text{ m}} = 66.7$ rad/s

7M.5 B $v = r\omega = R_E \cos 47^\circ (2\pi f)$ where $\omega = 2\pi f$

f = frequency of rotation of the earth(s^{-1})

$$= \frac{1 \text{ revolution}}{(24)(3600s)}$$

Thus $v = (6380 \times 10^3 m)(\cos 47^\circ)(2\pi)(\frac{1}{(24)(3600)})$

$v = 316$ m/s

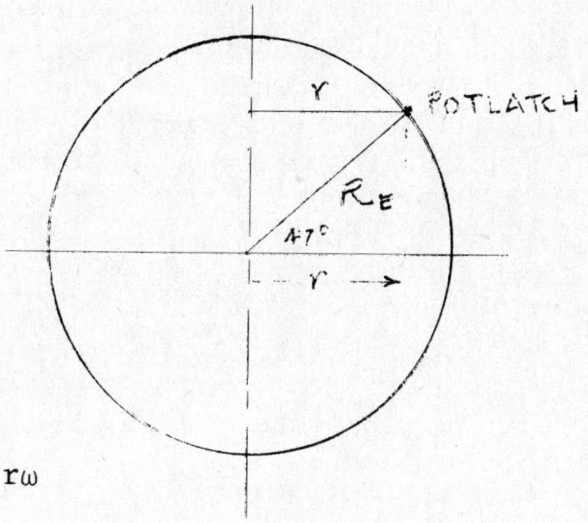

7M.6 C $a_c = \frac{v^2}{r} = r\omega^2$ since $v = r\omega$

If $mr^2\omega$ = constant = c

$$r^2\omega = \frac{c}{m}$$

$$(r^2\omega)^2 = (\frac{c}{m})^2$$

$$a_c = r\omega^2 = (\frac{c}{m})^2 \frac{1}{r^3}$$

Thus if r becomes smaller, a_c increases. Thus the increase in ω outweighs the decrease in r, and the net result is that the centripetal acceleration and the centripetal force, $F_c = ma_c$, both increase as the ball winds in.

7M.7 C $\omega = \omega_o - \alpha t$. I write minus sign because the turntable is decelerating. At $t = 16s$, $\omega = 0$. Thus

$$0 = \omega_o - \alpha t, \ \alpha t = \omega_o, \ \alpha = \frac{\omega_o}{t}$$

$$\alpha = [\frac{(33.3)(2\pi)}{60 \text{ s}}] \frac{1}{16 \text{ s}} = 0.22 \text{ rad/s}^2$$

7M.8 C In order for the water to move in a circle it must experince a force mv^2/r directed toward the center of revolution. The forces acting on the water at the top of the arc are shown here. The force of the bucket, R, and the force of gravity, mg, work together to make the water move in a circle. The smallest value R can have is zero. In this limiting case

$$mg = \frac{mv^2}{r}$$

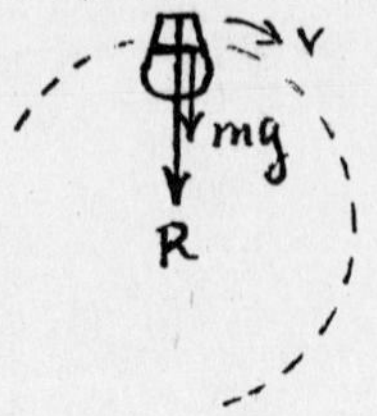

If mg exceeds the value given by this equation, the water, which has a given speed v, will curve in a circle of smaller radius at the top of the arc, i.e. it will fall out of the bucket. Thus we must determine for the case at hand if $mg \geq mv^2/r$.

If $mg > mv^2/r$, the water will fall out.

If $mg < mv^2/r$, the water will stay in and $R + mg = mv^2/r$. For the water to stay in $mg < mv^2/r$. We can divide out m, so the mass of the water does not matter.

Thus $v^2/r = (2.5\ m/s)^2/(0.5m) = 12.5\ m/s^2 < g$.

The water will stay in the bucket.

7M.9 C Angle between spokes is $\frac{2\pi}{12}$ radians.

Wheel moves through this angle in $\frac{1}{16}$ sec.

$$\text{Thus } \omega = \frac{\Delta\theta}{\Delta t} = \frac{(\frac{2\pi}{12})}{1/16}\ rad/s = \frac{(2\pi)(16)}{12}\ (\frac{rad}{s})$$

$$v = r\omega = (0.5\ m)(\frac{(2\pi)(16)}{(12\ s)}) = 4.2\ m/s$$

Notice that "radians" are not a unit in the regular sense.

Thus $(m)(\frac{rad}{s}) = \frac{m}{s}$.

A radian is more of a "reminder" that the relation $s = r\theta$ holds if $\theta = 2\pi$ when s is the arc all the way around a full circle.

7M.10 C $a_c = v^2/r = r\omega^2$. If ω remains unchanged as you move in, then a_c would decrease as r is decreased. Since $F_c = ma_c$, F_c would also decrease.

7M.11 E

$$a_c = \frac{v^2}{r} = r\omega^2 = r(2\pi f)^2$$

$$= (0.03\,m)(2\pi)^2\left(\frac{10^5}{60s}\right)^2$$

$$\frac{a_c}{g} = \frac{(0.03m)(2\pi)^2(10^{10})}{(9.8\,m/s^2)(60s)^2} = 336{,}000$$

7M.12 B

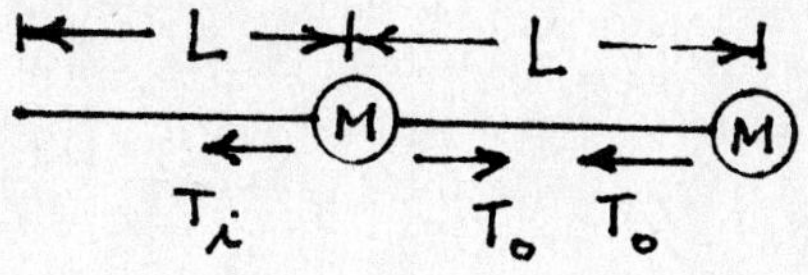

FOR THE OUTER MASS, $F_c = T_o = M(2L)\omega^2$

FOR THE INNER MASS $F_c = T_i - T_o = ML\omega^2$

DIVIDE THESE TWO EQUATIONS :

$$\frac{T_o}{T_i - T_o} = \frac{2ML\omega^2}{ML\omega^2}$$

SO $\frac{T_o}{T_i - T_o} = 2$, $T_o = 2T_i - 2T_o$

$3T_o = 2T_i$ $T_o = 1.5\,T_i$

7M.13 D The available centripetal force due to friction is fixed, thus

$$F_c = \frac{mv_1^2}{R_1} = \frac{mv_2^2}{R_2}$$

$$R_2 = \left(\frac{v_2}{v_1}\right)^2 R_1 = \left(\frac{2v_1}{v_1}\right)^2 R_1 = 4R_1$$

THUS $R_2 = (4)(6\,m) = 24\,m$

• Problems

7.1 Out west in Idaho they have a lot of cowboys and rodeos, and those boys can do some pretty fancy tricks. In one trick a cow-poke rides along on a horse hellbent for leather twirling a rope which he hangs down toward the ground. He keeps a spinning loop parallel to the ground just as easy as can be. I haven't mastered anything that nice yet, but I'm working on one where I tie a pool ball to a string and whirl it around. The string traces out a cone as the hanging ball goes around. So far I've only tried this in poolrooms, but one of these days I'll try it on a horse. Maybe. If the ball is hung on a string 80 cms long and whirled one revolution per second, what angle with the vertical will the string make?

7.2 A 120 lb student riding in a ferris wheel finds that her apparent weight is only 100 lbs at the top of the ride. What is her apparent weight at the lowest point? What would be her apparent weight at the lowest and highest points if the speed of the ferris wheel were doubled?

7.3 An entertaining amusement park ride works like this. You walk into a circular room of diameter 6 meters. The door is closed and everyone stands with his back to the wall, feeling silly and looking at the other sheep doing the same thing. (Reminds me of the drunk who staggered into the crowded elevator and stood there leaning against the closed door. Seeing all of the people staring at him, after a few minutes he speaks out. "I shupose you're all wunderin' why I called this meetin'.") Anyway, after a moment the room starts to spin. Faster and faster it goes. Then, as you stand there pinned against the wall, the floor slowly drops away from beneath your feet. You are left stuck to the wall, like a human fly. My friends always said it was a great way to meet girls, but not for me. Makes me sick at my stomach. Me and Woody Allen. My question is this. If the coefficient of friction between you and the wall is 0.4, how many revolutions per minute does the room have to make in order for you not to slide down? What would then be your linear velocity? Would it be harder or easier for a heavy person to stick to the wall?

• Problem Solutions

7.1 THE FORCES ACTING ON THE BALL ARE mg DUE TO GRAVITY AND T DUE TO THE STRING. NOTE THAT I DO NOT DRAW IN ANY "CENTRIPETAL FORCE" VECTOR. THERE IS NO MOTION IN THE VERTICAL DIRECTION, SO (CONTIN.)

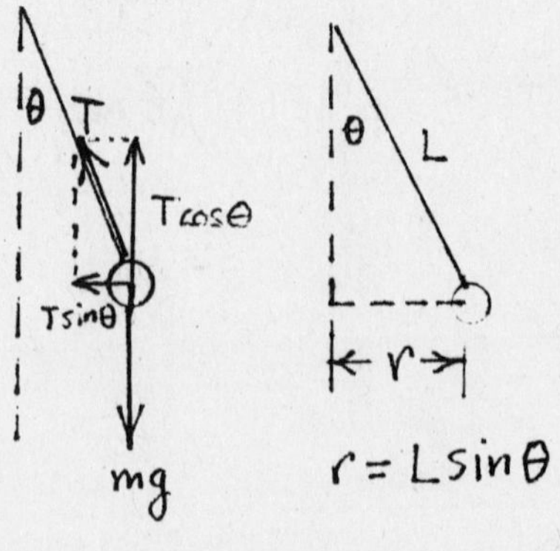

$T\cos\theta = mg \qquad (1)$

THE INWARD COMPONENT OF T PROVIDES THE NEEDED CENTRIPETAL FORCE $mr\omega^2$.

THUS $mr\omega^2 = T\sin\theta \qquad (2)$

SOLVE (1) FOR T AND SUBSTITUTE IN (2).

$$T = \frac{mg}{\cos\theta}, \qquad mr\omega^2 = \frac{mg}{\cos\theta}\sin\theta$$

$$\frac{\sin\theta}{\cos\theta} = \frac{r\omega^2}{g} = \frac{L\sin\theta}{g}\omega^2$$

$$\cos\theta = \frac{g}{L\omega^2} = \frac{g}{L(2\pi f)^2} = \frac{9.8\,m/s^2}{(0.8m)(2\pi)^2(1\,Hz)} = 0.31$$

$$\theta = \underline{\underline{72^\circ}}$$

7.2 AT THE TOP THE NET FORCE TOWARD THE CENTER IS

$mg - R_T = \frac{mv_1^2}{r} \qquad (1)$

AT THE BOTTOM THE NET FORCE TOWARD THE CENTER IS

$R_B - mg = \frac{mv_1^2}{r} \qquad (2)$

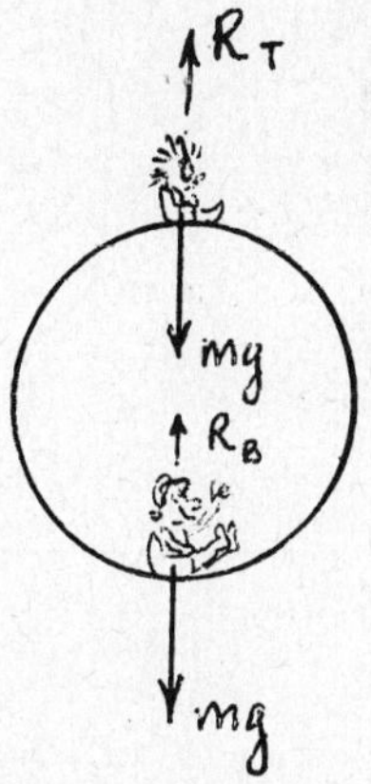

R = "APPARENT WEIGHT"

R_T = FORCE EXERTED BY SEAT AT THE TOP.

R_B = FORCE EXERTED BY SEAT AT THE BOTTOM.

COMBINE THE EQUATIONS.

$\frac{mv_1^2}{r} = mg - R_T = R_B - mg$,

OR $R_B = 2mg - R_T \qquad (3)$

GIVEN $mg = 120$ lbs, $R_T = 100$ lbs.

$R_B = (2)(120\text{ lbs}) - 100\text{ lbs.} = \underline{\underline{140\text{ lbs.}}}$ AT BOTTOM

$\frac{mv_1^2}{r} = mg - R_T = 120\text{ lbs} - 100\text{ lbs} = 20\text{ lbs.}$

IF $V_2 = 2V_1$, $\frac{mV_2^2}{r} = \frac{m(2V_1)^2}{r} = 4\frac{mV_1^2}{r} = 4(20\text{ lbs}) = 80\text{ lbs.}$

THUS $\frac{mV_2^2}{r} = mg - R_{T_2} \Longrightarrow 80\text{ lbs} = 120\text{ lbs} - R_{T_2}$, $\underline{\underline{R_{T_2} = 40\text{ lbs}}}$

USING EQ. (3) $R_{B_2} = 2mg - R_{T_2} = 2(120\text{ lbs}) - 40\text{ lbs}$

$\underline{\underline{R_{B_2} = 200\text{ lbs}}}$

7.3 THE FORCES ACTING ON THE PERSON ARE SHOWN ON THE NEXT PAGE. THE WALL PUSHES IN WITH A NORMAL REACTION FORCE R. FRICTION PUSHES UP WITH FORCE $F_F = \mu R$. GRAVITY PULLS DOWN. R PROVIDES THE NEEDED CENTRIPETAL FORCE.

$F_F = mg$

OR $\mu R = mg \qquad (1)$

$mr\omega^2 = R \qquad (2)$

SUBSTITUTE (2) IN (1):

$$\mu(mr\omega^2) = mg$$

$$\omega = \sqrt{\frac{g}{\mu r}}$$

$$= \sqrt{\frac{9.8\,m/s^2}{(0.4)(3m)}}$$

$$= 2.86 \text{ RAD/SEC}$$

$$f = \frac{\omega}{2\pi} = \frac{2.86}{2\pi}\ \frac{1}{s}$$

$$= 0.45 \text{ REV/S}$$

$$= (60)(.45) \text{ REV/MIN}$$

$$= \underline{\underline{27.3 \text{ R.P.M}}}$$

$$v = r\omega = (3)(2.86)$$

$$= \underline{\underline{8.6\ m/s}}$$

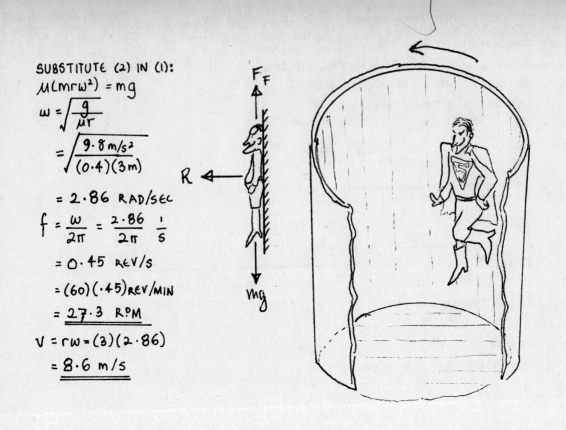

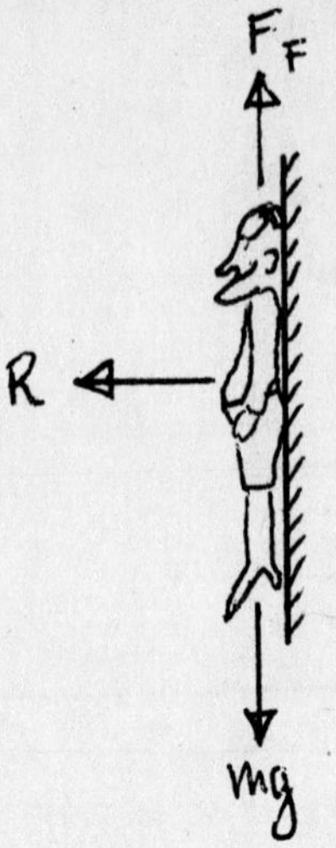

Gravitation

8

• Summary of Important Ideas, Principles, and Equations

The gravitational force between two masses m_1 and m_2 is always attractive and is directed along the line joining the two masses. The magnitude of the force on each object is given by the Law of Universal Gravitational Attraction.

$$F = G\frac{m_1 m_2}{r^2} \qquad (8.1)$$

Here r is the separation of m_1 and m_2 and G is the universal gravitation constant.

$$G = 6.67 \times 10^{-11}\ \mathrm{N \cdot m/kg^2}$$

The gravitational force on a spherically symmetric object of mass M due to another mass m outside M is just that which would result if the mass M were all concentrated at its center.

When dealing with the gravitational attraction between a planet of mass M_p and another object of mass m located on the surface a distance R_p from the center of the planet it is convenient to group the constants G, M_p and R_p^2 together and call them "g_p".

Thus

$$g_p = \frac{GM_p}{R_p^2} \qquad (8.2)$$

The force on m is thus

$$F = \frac{GM_p m}{R_p^2} = mg_p \qquad (8.3)$$

Note that (7.3) correctly gives the force on m only as long as m is on the surface of the planet. If the mass m moves farther away from the center of the planet (as when a satellite is launched high above the earth) the force on m decreases, i.e. the effective g decreases.

If an object of mass m is allowed to fall freely it will accelerate downward with acceleration a given by F = ma, thus

$$F = ma = mg_P , \quad \text{or } a = g_P$$

For this reason the constant g_P is called the <u>acceleration due to gravity</u>.

The value of g_P at the surface of the earth is $g = 9.8\ m/s^2$.

Note that the force on m is mg whether or not it is falling.

Do not be misled by the name, "acceleration due to gravity". m need not be accelerating to be subject to the force of gravity.

• Qualitative Questions

8M.1 If the separation between two small point masses is doubled, the gravitational force acting on one due to the other will

A. double.
B. become smaller by a factor of ½.
C. become smaller by a factor of ¼.
D. decrease, but we cannot say by what factor without knowing whether other masses are nearby and exerting gravitational forces also.
E. not change, since by Newton's third law it is not possible for one of the masses to exert a net force on the other under these circumstances.

8M.2 The radius of the moon, R_M, is much smaller than the radius of the earth, R_E. Further the density ρ_M of the moon is less than the density of the earth, ρ_E. Density is

$$\rho = \frac{\text{mass}}{\text{volume}} .$$

Given that $R_M = 0.27\ R_E$ and $\rho_M = 0.6\rho_E$, how much would a person who weighed 200 lbs. on earth weigh on the moon?

A. 32 lbs.
B. 54 lbs.
C. 90 lbs.
D. 444 lbs.
E. 1646 lbs.

8M.3 Suppose that on a distant planet you could throw a rock nine times as high as here on earth. If air friction effects are neglected, you would deduce that the ratio $g_E:g_P$ was

A. 81:1
B. 9:1
C. 3:1
D. 1:1
E. 1:3
F. 1:9
G. 1:81

8M.4 Observations of Saturn show that it rotates once around the sun every 29.5 years. From this information and the knowledge that the earth rotates around the sun once each year, it is possible to determine the relative distance of Saturn from the sun, as compared to the earth-sun distance (which is called 1 astronomical unit, or 1 a.u.). One thus concludes that the distance from Saturn to the sun is about

A. 2.2 a.u.
B. 6.2 a.u.
C. 9.5 a.u.
D. 30 a.u.
E. 39 a.u.

8M.5 Three identical point masses m are placed at the corners of an equilateral triangle of side a. If $F_o = Gm^2/a^2$, the force acting on one of the masses due to the other will be

A. $\frac{1}{2} F_o$
B. F_o
C. $\frac{\sqrt{3}}{2} F_o$
D. $\sqrt{3}\ F_o$
E. $2\ F_o$

hv

8M.6 Suppose that an unidentified satellite was observed orbiting the earth. Early reports provided only its altitude in a circular orbit and its speed. Could we then determine its mass?

A. Yes, from the altitude alone we could determine the mass of the satellite.
B. Yes, from the speed alone we could determine the mass.
C. Yes, by using both the speed and mass we could determine the altitude.
D. No, we could not determine the mass unless we also knew the period of revolution.
E. No, we could not determine the mass.

• Multiple Choice: Answers and Comments

8M.1 C $F = G\frac{m_1m_2}{r^2}$ If we replace r by 2r the force will become

$$F' = G\frac{m_1m_2}{(2r)^2} = \frac{1}{4}G\frac{m_1m_2}{r^2} = \frac{1}{4}F$$

8M.2 A The mass of a sphere of density ρ is

$$\text{mass} = \left(\frac{\text{mass}}{\text{volume}}\right)(\text{volume}) = (\text{density})(\text{volume})$$

or $M = \rho V$

For a sphere $V = \frac{4}{3}\pi R^3$

Thus $M = \frac{4}{3}\pi R^3 \rho$

A person's weight is the gravitational force acting, so on earth

$$W_E = G\frac{M_E m}{R_E{}^2} \text{ and on the moon } W_m = G\frac{M_m m}{R_m{}^2}$$

$$\frac{W_M}{W_E} = \frac{GM_m m/Rm^2}{GM_E m/RE^2} = \frac{M_m}{M_E}\left(\frac{R_E}{R_m}\right)^2 = \frac{4/3\,\pi R_m^3 \rho_m R_E^2}{4/3\,\pi R_E^2 \rho_E R_m^2}$$

$$\frac{W_M}{W_E} = \frac{R_m}{R_E}\;\frac{\rho m}{\rho E} = (0.27)(0.6) = 0.162$$

So $Wm = 0.162\ W_E = (0.162)(200\text{ lbs}) = 32\text{ lbs.}$

8M.3 B Energy of ball is conserved, so

$$KE_1 + PE_1 = KE_2 + PE_2$$

$$\tfrac{1}{2}mv^2 + 0 = 0 + mgh$$

$$h = \frac{v^2}{2g}$$

$$hg = \frac{v^2}{2}$$

$$h_E g_E = h_\rho g_p = \frac{v^2}{2}$$

$$\frac{h_p}{h_E} = \frac{g_E}{g_p} = 9, \text{ so } g_p = \frac{1}{9}\,g_E$$

8M.4 C From Kepler's third law, $\frac{r^3}{T^2}$ = constant.

Thus $\frac{r_E{}^3}{T_E{}^2} = \frac{r_s{}^3}{T_s{}^2}$, $r_s{}^3 = (\frac{Ts}{TE})^2 \; r_E{}^3$

$$r_s{}^3 = (\frac{29.5}{1})^2 \; r_E{}^3$$

$$r_s = (29.5)^{2/3} \; r_E = 9.5 \; r_E$$

To evaluate $(29.5)^{2/3}$ on your calculator proceed as follows:

Let $x = (29.5)^{2/3}$

$$\ln x = \ln(29.5)^{2/3} = \frac{2}{3}\ln(29.5)$$

$$x = e^{2/3 \ln(29.5)}$$

This looks more complicated than it really is. Just do this.

1. Punch 29.5
2. Punch "Ln" (You can also use "log" just as well.)
3. Punch "x" (multiply)
4. Punch "2"
5. Punch " " (divide)
6. Punch "3"
7. Punch "="
8. Punch "INV" (inverse or antilog)
9. Punch "Ln" (or "log" if you used it first)

Answer will appear in display.

8M.5 D $F = 2\,F_o \cos 30^o$

$= 2\,F_o(\frac{\sqrt{3}}{2})$

$F = \sqrt{3}\,F_o$

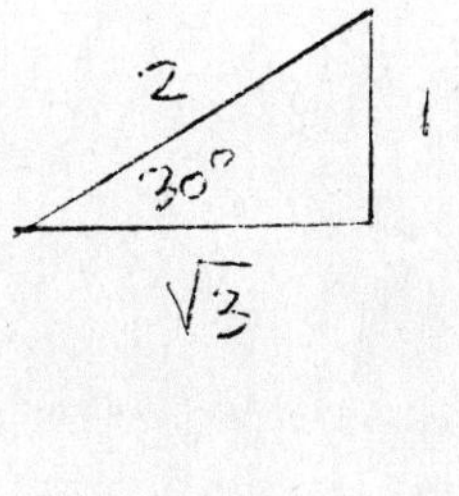

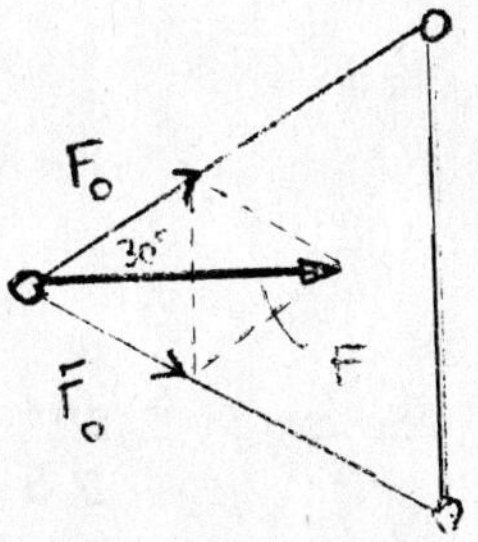

8M.6 E For a satellite gravity provides the needed centripetal force.

$$G\frac{M_E m}{R^2} = \frac{mv^2}{R}$$

$$\text{so } Rv^2 = GM_E.$$

Note that m, the mass of the satellite, drops out of this equation, so it is not possible to determine m given R and v. All objects with particular speed will occupy the same circular orbit independent of their mass. This is why an astronaut, his satellite and his sandwiches all fly around happily together.

D is false. Given v and R we can find the period of revolution, τ from $v\tau = 2\pi R$, but this is of no help in finding the mass.

• Problems

8.1 People who believe in astrology believe that the planets act on each of us to influence our lives. The only interaction observed between two such masses as you and Mars is gravity. Calculate this force of gravity on you as a fraction of your weight, given that the mass of Mars is 6.5×10^{23} kg (about 0.1 times the mass of the earth) and that Mars never comes closer to us than about 5.5×10^{10} meters.

8.2 A space adventurer discovers a small planet while exploring a new star system. He puts his ship in circular orbit around the planet and finds that he travels once around the planet in 10 hours. With his navigational equipment he determines that the radius of his orbit is 7000 km. What is the mass of the planet? We assume G is known.

• Problem Solutions

8.1 $F = G\frac{M_M m}{R^2}$

YOUR WEIGHT IS $W = mg$

THUS $\frac{F}{W} = \frac{GM_M m/R^2}{mg} = \frac{GM_M}{R^2 g} = \frac{(6.67 \times 10^{-11}\ \frac{n\cdot m^2}{kg^2})(6.5 \times 10^{23}\ kg)}{(5.5 \times 10^{10}\ m)^2 (9.8\ m/s^2)}$

$$\frac{F}{W} = 1.5 \times 10^{-9}$$

THIS RESULT CERTAINLY SEEMS TO INDICATE NO GRAVITATIONAL BASIS FOR ASTROLOGY, AND NO OTHER BASIS FOR IT HAS EVER BEEN OBSERVED EITHER. IT IS REMARKABLE THAT IN THIS DAY AND AGE THERE ARE STILL HUNDREDS OF MILLIONS OF PEOPLE WHO STILL BELIEVE THAT "THE STARS CONTROL OUR DESTINY". IT IS

PERHAPS TRUE THAT THEY DO, BUT NOT IN THE WAY ASTROLOGERS BELIEVE.

8.2 FOR A CIRCULAR ORBIT, $F_c = \frac{mv^2}{R} = G\frac{M_p m}{R^2}$

m = MASS OF SHIP

M = MASS OF PLANET

$$M_p = \frac{v^2 R}{G}$$

IN TIME T SHIP TRAVELS DISTANCE $2\pi R$, SO

$$vT = 2\pi R, \quad v = \frac{2\pi R}{T}$$

$$M_p = \left(\frac{2\pi R}{T}\right)^2 \frac{R}{G} = \frac{(2\pi)^2 R^3}{T^2 G} \qquad \text{(KEPLER'S 3RD LAW)}$$

$$M_p = \frac{(2\pi)^2(7\times10^6\,\text{m})^3}{[(10)(3600\,\text{s})]^2\left[6.67\times10^{-11}\,\frac{\text{n.m}^2}{\text{kg}^2}\right]} = \underline{\underline{1.57\times10^{23}\ \text{kg}}}$$

---

9 Rotational Equilibrium; Dynamics of Extended Bodies

• Summary of Important Ideas, Principles, and Equations

1. Torque means twist. The torque about a center O due to a force F is defined as

   Torque = (Force)x(Perpendicular moment arm about O)

$$\tau = F\ell = F\ r\ \sin\theta \qquad (9.1)$$

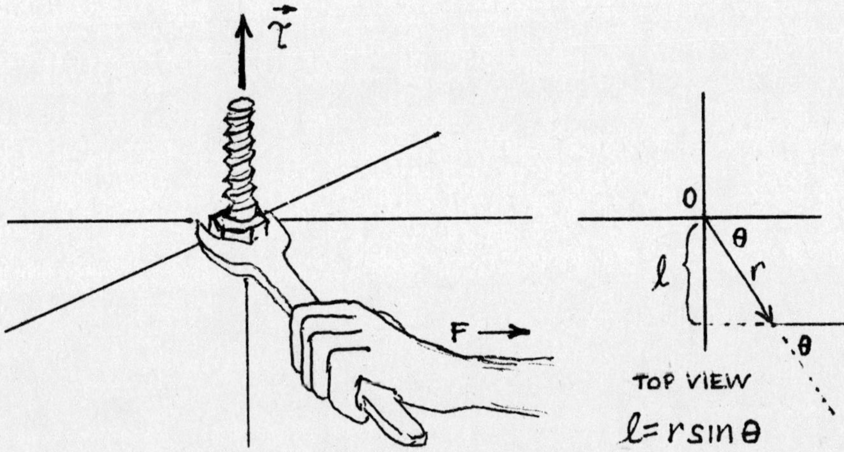

Torque is a vector which points along the axis of twist in the direction a right hand screw would advance. It does not point along $\vec{F}$.

Torque is measured in newton-meters and has the dimension $kg\cdot m^2/s^2$.

For a system to be in equilibrium, $\Sigma\vec{F}_i = 0$

and $\boxed{\Sigma\vec{\tau}_i = 0}$ (9.2)

We will consider only cases where the torque is directed along one axis, in which case the condition for zero total torque in equilibrium may be written

$$\boxed{\tau_{clockwise} = \tau_{counterclockwise}} \quad (9.3)$$

A system is in rotational equilibrium if it is not rotating or if it is rotating about a fixed axis with constant angular velocity, just as a system is in translational equilibrium if it is at rest or moving with constant linear velocity.

2. The <u>center of gravity</u> (CG) and the <u>center of mass</u> (CM) are at the same point for an object in a uniform gravitational field, as we assume is the case for the earth's field at the surface of the earth.

   We can treat an object as if all of its mass is located at the CM as far as gravitational effects are concerned. Thus the torque due to gravity about the CM is zero.

3. Newton's Second Law of motion for rotating objects is

   $\boxed{\tau = I\alpha}$ (like $F = ma$) (9.4)

   Here τ = torque
   α = angular acceleration
   I = moment of inertia or "rotational inertia"

   <u>Definition</u>: $I = mr^2$ for point mass m at distance r from origin
   = <u>moment of inertia</u> of m with respect to origin.

   $I = \Sigma m_i r_i^2$ for a system of masses

   Moment of inertia plays the same role for rotational motion that mass does for translational motion. The moment of inertia for some simple objects rotated about axis XX' is given here.

Moments of Inertia of Simple Regular Bodies

Hoop		MR^2
Disk or solid cylinder	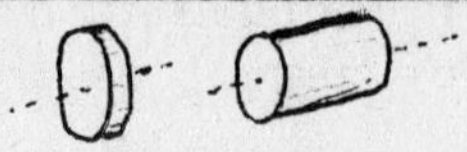	$\frac{1}{2}MR^2$
Sphere		$\frac{2}{5}MR^2$
Rod, axis through center	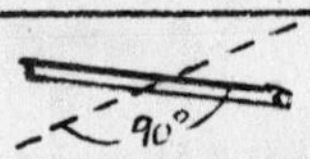	$\frac{1}{12}ML^2$
Rod, axis through end	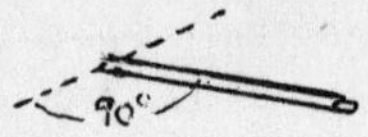	$\frac{1}{3}ML^2$

4. The <u>angular momentum</u> $\vec{L}$ of a body rotating with angular velocity $\vec{\omega}$ is

$$\boxed{\vec{L} = I\vec{\omega}} \qquad (9.5)$$

This is analagous to linear momentum $\vec{p}$, $\vec{p} = m\vec{v}$.

$$\boxed{\vec{\tau} = \lim_{\Delta t \to 0} \frac{\Delta \vec{L}}{\Delta t}} \qquad (9.6)$$

<u>Law of Conservation of Angular Momentum:</u>

$$\boxed{\text{If } \tau = 0,\ L = I\omega = \text{constant.}}$$

Thus if no external torque acts on a system, the angular momentum of the system stays constant. Note that moment of inertia I can change if position of mass in the system is changed, so angular velocity ω can also change even when angular momentum is constant.

• Qualitative Questions

9M.1 Which of the following has a meaning closest to "torque"?

A. Force
B. Push
C. Tension
D. Twist
E. Work
F. Angular displacement

9M.2 A torque acting on an object tends to produce

A. translation.
B. rotation.
C. a center of gravity.
D. equilibrium.
E. a moment arm.
F. an imbalance of forces.

9M.3 If you try to loosen a bolt with a wrench, the torque you exert will increase if you

A. slide your hand out nearer the end of the wrench away from the bolt.
B. increase the force you are exerting.
C. change the angle at which you pull so that you now pull perpendicular to the wrench handle.
D. All of the above are true.
E. None of the above is true.

9M.4 Determination of the position of the center of mass (CM) of the human body is very important in kinesiology, the study of animal motion. Sketched here is an arrangement which allows one to determine this information. A plank of weight 40 N is placed on two scales separated by 1.6 meters. A person of weight 800 N lies on the plank. The left scale is observed to read 500 N. How far is the CM from the left scale?

A. 61 cms
B. 64 cms
C. 80 cms
D. 96 cms
E. 99 cms

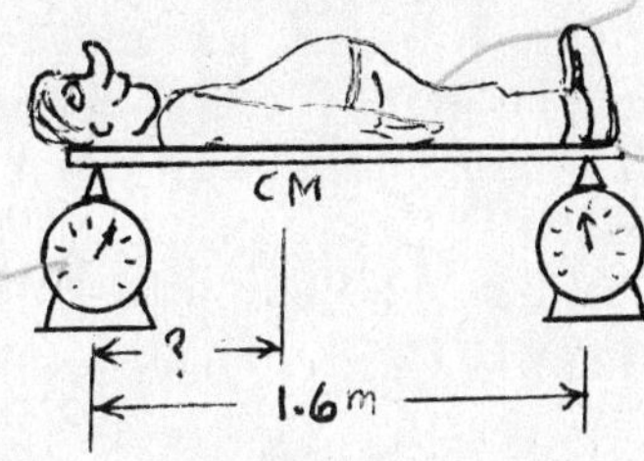

9M.5 Two people are carrying a uniform 120 lb log through the forest, as sketched here. What weight is Sharkey supporting?

A. 30 lbs
B. 40 lbs
C. 60 lbs
D. 75 lbs
E. 90 lbs

9M.6 Two kids are balanced on a teeter-totter they have made from a heavy plank. If each kid now moves halfway in toward the fulcrum

A. the big kid's end will rise.
B. the little kid's end will rise.
C. the teeter-totter will remain in balance.
D. we can't say what will happen without knowing the relative weights of the kids and the plank.
E. the short side of the teeter-totter (if there is one) will rise.

9M.7 If you hold a boomerang at point B and throw it into the air, it will rotate about

A. point A.
B. point B.
C. point C.
D. its center of mass.
E. no particular point.

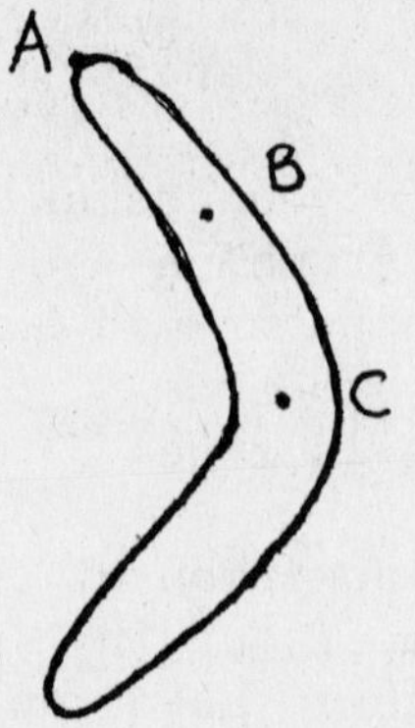

9M.8 It sometimes helps us understand the meaning of "moment of inertia" to consider the analagous quantity for translational motion. This quantity is

A. momentum
B. energy
C. mass
D. force
E. angular velocity

9M.9 Sketched here are several flywheels, each of the same mass and of uniform thickness. Which has the greatest moment of inertia?

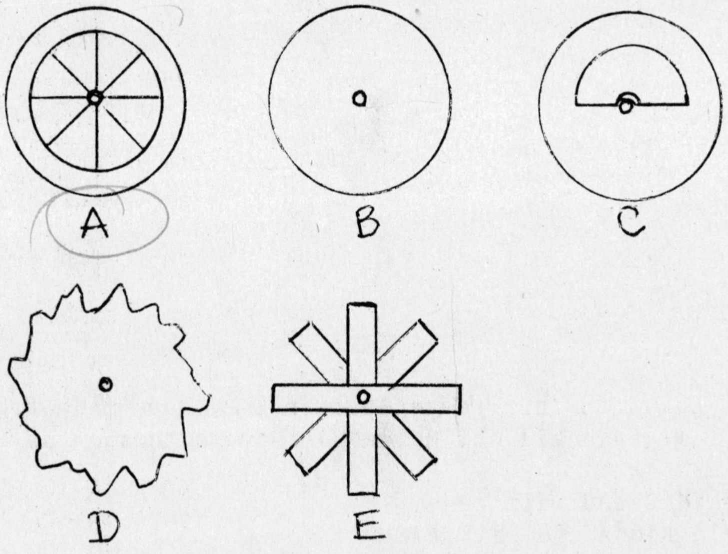

9M.10 The moon causes two tides daily on the earth. This causes water in the oceans to bulge out, as I have sketched here. The earth is rotating, and because of friction the tidal bulges stick to the earth somewhat and are thus not along the earth-moon line. Bulge X is closer to the moon than is bulge Y, so the moon pulls harder on X than on Y. This has some very, very interesting consequences. Among other things, this torque will result in

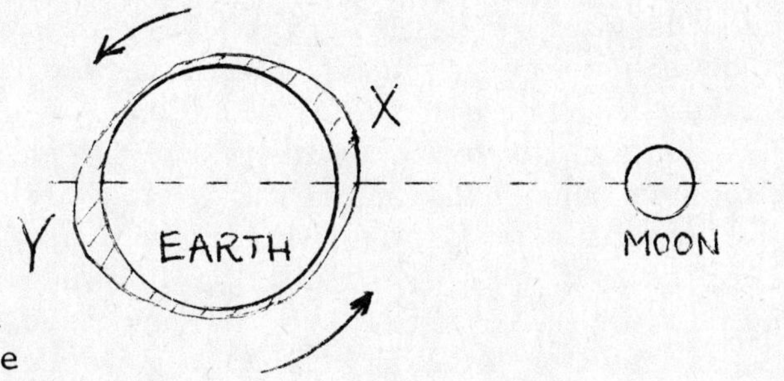

A. a speeding up of the rate of rotation of the earth.
B. bigger and bigger tides each time the earth rotates.
C. a decrease in the spin angular momentum of the earth and hence a gradual lengthening in the period of one revolution (i.e. in the time duration of one day).
D. a decrease in the moment of inertia of the earth.

9M.11 Which of the following is <u>not necessarily true</u> for a system in equilibrium?

A. The total external force acting is zero.
B. The total external torque acting is zero.
C. The linear momentum is constant.
D. The angular momentum is constant.
E. The total energy is constant.

9M.12 We have all seen divers who seem effortlessly to do a double somersault and then gracefully straighten out and enter the water in a perfect vertical position. We can most easily understand this motion by recognizing that while in the air

A. the total energy of the diver is conserved.
B. the linear momentum of the diver is conserved.
C. the angular momentum of the diver is conserved.
D. rotational kinetic energy increases at the expense of translational kinetic energy.
E. the moment of inertia (or rotational inertia) of the diver remains constant.

9M.13 I used to take my two oldest sons to the playground in Golden Gate park while my wife was having her teeth fixed at the U.C. Dental School. There we discovered a madcap game which the kids and I loved and which gave some mothers gray hair. In the park was a small merry-go-round, the kind you push. We would load a bunch of kids on and they would all hang out around the outer edge. Then I would push the merry-go-round as hard as I could, until I had it spinning at top speed. When I then gave the word all the kids would quickly try to pull themselves in to the center of the merry-go-round. When they did, the contraption would take off like a bat out of hell. It would spin so fast it almost came off its axle. Kids would be thrown flying in all directions if they had failed to get in close enough to the center. It was a wondrous thing to see as they tumbled head over teakettle on the grass. If you have never seen this you must definitely make a trip to the park soon to do some extra-curricular physics investigation. If you think about this, you will see that this phenomenon is closely related to the fact that

A. the kinetic energy of the kids stays constant as they move in toward the center of the merry-go-round.
B. the angular momentum of the system stays constant as the kids move in.
C. the angular momentum of the system decreases as the kids move in.
D. the angular momentum of the system increases as the kids move in.
E. The total energy of the system stays constant as the kids move in.

9M.14 Once in Germany I saw an impressive circus act. A juggler had a long pole, at the end of which was a small platform. A little dog sat on the platform while the juggler balanced the pole on his chin, meanwhile twirling hoops on his feet and throwing five frisbees in the air like boomerangs. Amazing. My question is this: Which of the following statements is most accurate?

A. It would have been easier to balance the platform without the dog on it.
B. Having the dog on the platform made it easier to balance.
C. Having the dog on the platform made no difference in the difficulty of balancing the platform.
D. Damn clever, these German dogs!

9M.15 You have probably sometime been out hiking and had to cross a stream by using a fallen log for a bridge. Did you ever notice that when you start to fall you automatically throw your arms out to a wide horizontal position to regain your balance? Why does this work?

A. By raising your arms you convert some of the kinetic energy you gain in falling into gravitational potential energy.
B. This shifts your center of mass back to a position above the log.
C. This increases your moment of inertia, thereby causing you to fall more slowly and thus providing time to shift your center of mass back over the log.
D. This creates a torque which causes you to rotate back up onto the log.
E. Since your angular momentum is constant while falling, lifting your arms up causes your body to compensate by rotating downward.

9M.16 My sons and I used to have lots of fun racing little toy cars in the Cub Scout Pinewood Derby Races. Little cars without motors were allowed to coast down an inclined ramp, and the object was to build the fastest rolling car for a given maximum weight. It would have been nice to let one kid race another, but it always ended up as father versus father as fancier and fancier cars were built. As an example of one design consideration, consider the wheels. For a given total car weight, which of the following is the best option, aside from questions of friction?

A. Make the wheels as heavy as possible and the body as light as possible.
B. Make the wheels as light as possible and the body as heavy as possible.
C. Put as much of the wheel weight as possible near the outside rim.
D. Balance the weight equally between the wheels and the body.
E. Recognize that the relative distribution of weight between wheels and body has no bearing on the final speed, and thus concentrate on other considerations.

9M.17 Suppose that you were along on an archeological dig in Egypt which turned up two beautiful gold cylinders. The project director deduces that these were used in some kind of religious ceremony, and he instructs you to learn as much about them as possible without harming them in any way. First off, you weigh them and find they have equal masses. Someone suggests that one may be a hollow gold shell while the other has thinner walls but is filled with clay to make it as heavy as the pure gold artifact. Can you tell which is which?

A. We cannot answer this question without being told if the cylinders are of the same size and shape.
B. Roll both down an inclined plane. The hollow one will roll most slowly.
C. Roll both down an inclined plane. The hollow one will roll fastest.
D. If both cylinders are identical in size and shape, both will obviously roll at the same rate and so nothing more can be deduced through rolling experiments.
E. If it should happen that the largest cylinder rolls fastest, then it is definitely the hollow one.

9M.18 Suppose two spherical balls are rolled without slipping down a hill. If both are released from rest,

A. the biggest one will roll fastest.
B. the smallest one will roll fastest.
C. the heaviest one will roll fastest.
D. the lightest one will roll fastest.
E. both will roll at the same speed, even if they are of different sizes and weights.

• True–False Questions

9TF.1 The torque vector is directed parallel to the force vector giving rise to the torque.

9TF.2 A force directed radially out from the origin will always exert zero torque about the origin.

9TF.3 A system with a fixed mass thus has a fixed moment of inertia.

9TF.4 The torque exerted on a body by a given force depends on the choice of origin used.

9TF.5 We have seen that the total torque acting on a system in rotational equilibrium iz zero. However, this is true only if we calculate the torque about a particular point, namely the center of mass.

9TF.6 A gymnast tucks her knees in when doing a somersault in order to reduce her moment of inertia.

9TF.7 If no torque acts on a rigid body it will continue to rotate at a constant number of revolutions per second.

9TF.8 All disks roll faster than all hoops when no slipping occurs.

9TF.9 Because angular displacements are measured in radians, whereas linear displacements are measured in meters, rotational kinetic energy is not measured in the same units as is translational kinetic energy.

9TF.10 If you do physics experiments in which weights are connected by strings passed over pulleys, you will find that the weights move more slowly if large pulleys are used instead of small pulleys, all other factors being equal.

• Multiple Choice: Answers and Comments

9M.1 D

9M.2 B

9M.3 D A is true because you increase your moment arm when you slide your hand out.
B is true because torque is proportional to the force creating the torque.
C is true because this increases the perpendicular moment arm.

9M.4 B The man and plank are in equilibrium, so torque clockwise = torque counterclockwise. Calculate torque about point O. CM of plank is at its midpoint.

$$(500N)(1.6m) = (800N)(1.6-x) + (40N)(0.8m)$$

$$800 = 1280 - 800x + 32$$

$$x = 0.64m = 64 \text{ cms}$$

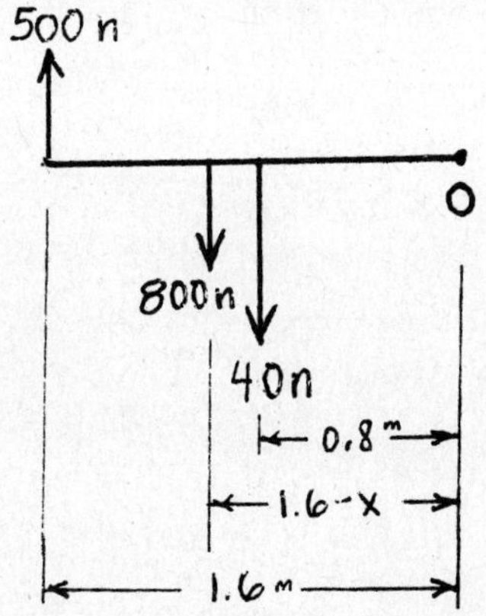

Note: We could have calculated the torque about some other point just as well and found the same result. For example, we could have calculated the torque about the left scale. Then the force exerted by the right scale would come into the calculation, and we would first have to deduce it. Since the total weight

9M.4 of the person plus the plank is 840N, the right scale must read 340N. We could then proceed as above. Some choices of origin lead to simpler calculations than others, but all will yield the correct answer, so don't worry too much about this point. There is more than one correct approach.

9M.5 E Since we do not need to find the force exerted by the person on the right, let us calculate torque about his position (then he has no moment arm and exerts no torque, so his unknown force won't enter the problem). Note that the weight of the log acts downward at its CM at its center. The log is in equilibrium, so torque clockwise = torque counterclockwise.

$$(4\text{m})(F) = (3\text{m})(120 \text{ lbs})$$

$$F = \frac{360}{4} = 90 \text{ lbs.}$$

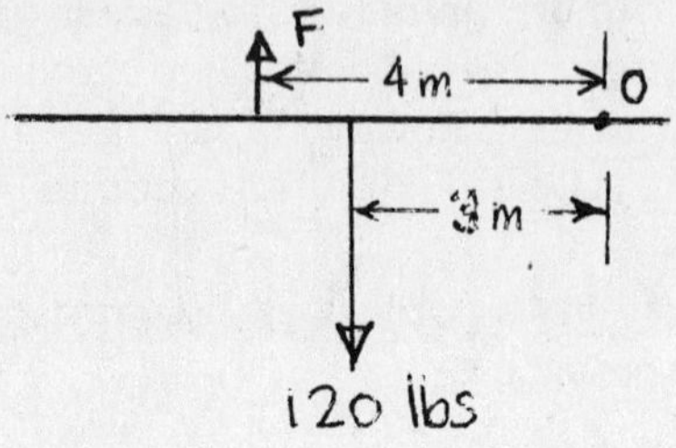

Note that we don't have to worry about changing distances to the British units of "feet", since these units cancel out in the equation anyway.

9M.6 E Let W_1 = weight of one kid and x_1 = his moment arm about fulcrum.

W_2 = weight of second kid, x_2 = his moment arm about fulcrum.

W = weight of plank, x = distance of CM of plank from fulcrum.

Let us suppose the kid with weight W_1 is on the same side of the fulcrum as the CM of the plank.

The condition of zero total torque then becomes

torque clockwise = torque counterclockwise

$$W_2x_2 = W_1x_1 + Wx \qquad (1)$$

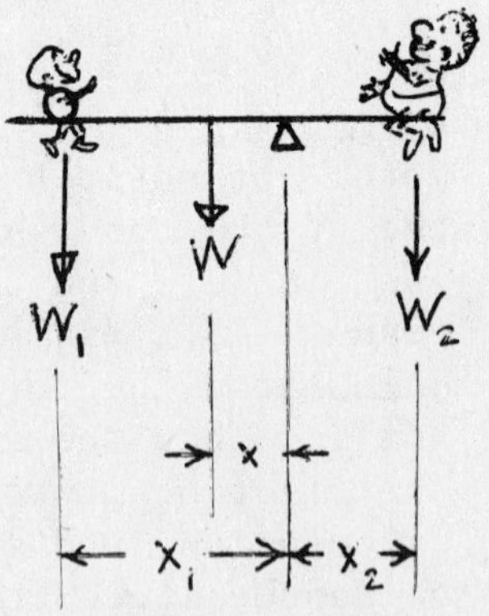

Will the teeter-totter be balanced when kids move in halfway?

$$\text{i.e.} \quad \frac{W_2x_2}{2} \stackrel{?}{=} W_1\frac{x_1}{2} + Wx \qquad (2)$$

From (1) we see

$$\frac{W_2x_2}{2} = W_1\frac{x_1}{2} + W\frac{x}{2}$$

Thus the right side of Eqn (2) is greater than the left side. This means torque due to W_1 plus W is greater, so this side will go down. This is the side where the CM of the plank is, i.e. the long end. If it happened that the CM of the plank was above the fulcrum, then the teeter-totter would stay in balance even after the kids move in.

9M.7 D

9M.8 C

9M.9 A Since $I = \Sigma m_i r_i^2$ the moment of inertia of a rigid body of fixed mass will be greatest when the mass is located as far as possible from the axis of rotation. Thus the spoked wheel, A, which is almost a hoop, will have the greatest moment of inertia of the shapes shown.

9M.10 C From the drawing we see that the moon exerts a net clockwise torque on the earth. Thus the angular momentum of the earth will not be constant. The torque is tending to decrease the angular momentum, and hence the angular velocity. This means that little by little the earth's rotation is slowed down, and the days grow a fraction of a second longer each year. It is believed that long ago the moon was much closer to the earth, and then the tides and the torque due to the moon were much greater. Little by little the rate of revolution of the earth decreased until it reached its present period of about 24 hours.

9M.11 E Consider, for example, a crate sliding down a loading ramp at constant speed. Its velocity is not changing, and thus it is in translational equilibrium. Gravity pulls it down and friction holds it back, and these two forces just cancel to yield zero total force. Note, however, that although the kinetic energy is staying constant, the gravitational potential energy (and hence the total energy E = KE + PE) is decreasing. Energy is being lost to friction, so equilibrium clearly does not mean constant energy in a case like this.

9M.12 C No torque acts on a diver in the air (since the force of gravity acts on the center of mass and exerts no torque about the CM). Thus angular momentum is constant. Since $L = I\omega$, when the diver tucks (thereby reducing I) the angular velocity ω must increase. When she wants to stop rotating she straightens out. This increases I and reduces ω, thereby allowing her to enter the water vertical.

9M.13 B If friction on the axle is negligible, essentially no external torque acts on the system. Thus the angular momentum $L = I\omega$ = constant. When the kids move in they reduce the moment of inertia I, so ω increases to keep L constant. Voila! The merry-go-round is $F_c = mr\omega^2$, so if a child stays at fixed radius r when the angular velocity ω increases drastically, the force needed to stay on goes up dramatically and the kid is thrown off.

9M.14 B Consider what happens when the pole rotates slightly and the platform drops a height h. The loss in PE is equal to the gain in rotational KE for the pole plus platform.

$$\tfrac{1}{2}I\omega^2 = mgh$$

For a given fall h, ω will be less if I is larger. Having the dog on the platform increases the moment of inertia I a lot, which means that ω will be small. But if the platform is slowly rotating down toward the ground, the juggler has ample time to move the point of support for the pole (his chin) over to one side, placing it under the platform. When this is done there is then no torque acting due to gravity and the platform is again in balance. Try this yourself with, say, a pushbroom. It is far easier to balance with the heavy end up (large I) than with the heavy end down.

9M.15 C Same reasoning as for question 8M.14.

9M.16 B Total gain in KE = loss in PE = mgh

$$KE_{tot} = KE_{rot} + KE_{trans}$$

$$KE_{tot} = \tfrac{1}{2}I\omega^2 + \tfrac{1}{2}mv^2$$

$$v = r\omega, \text{ so } KE_{tot} = \tfrac{1}{2}\frac{I}{r^2}v^2 + \tfrac{1}{2}mv^2$$

$$= \tfrac{1}{2}\left(\frac{I}{r^2} + m\right)v^2 = mgh$$

For fixed h, m and r, v will be biggest for smallest moment of inertia, I, so wheels should be as light as possible with these constraints (assuming no friction).

9M.17 B See discussion in text for an explanation of which objects roll fastest. If the moment of inertia is of the form

$I = kMr^2$, objects with larger k will always roll more slowly, no matter what the value of M or r.

9M.18 E See text for explanation, as in 8M.17.

• True–False: Answers and Comments

9TF.1 F The torque vector is along the screw axis, perpendicular to the force vector.

9TF.2 T A radial force has zero perpendicular moment arm and so exerts no torque about the axis.

9TF.3 F Moment of inertia depends on mass and on the position of the mass. Moving mass out away from the axis of rotation increases the moment of inertia. $I = \Sigma m_i r_i^2$.

9TF.4 T Torque = Force x Moment arm, and moment arm depends on the position of origin.

9TF.5 F The total torque on a system in equilibrium is zero, no matter what the choice of origin. Try this on an example for yourself and you will see that this is so.

9TF.6 T

9TF.7 T

9.TF8 T See text for explanation.

9TF.9 F All forms of energy have the same dimension.

9TF.10 T If the pulleys have appreciable moments of inertia, they can acquire appreciable kinetic energy, $\frac{1}{2}I\omega^2$. This means that for a given loss in PE, the gain in translational kinetic energy will be less and the weights will move more slowly.

• Problems

9.1 A simple fishing reel has a spool of diameter 3 cms and a crank handle of length 6 cms. The line gets snagged on the bottom. What is the maximum force which can be applied to the crank if the force on the fishing line is not to exceed 50n? Assume the force on the crank is applied perpendicular to the crank.

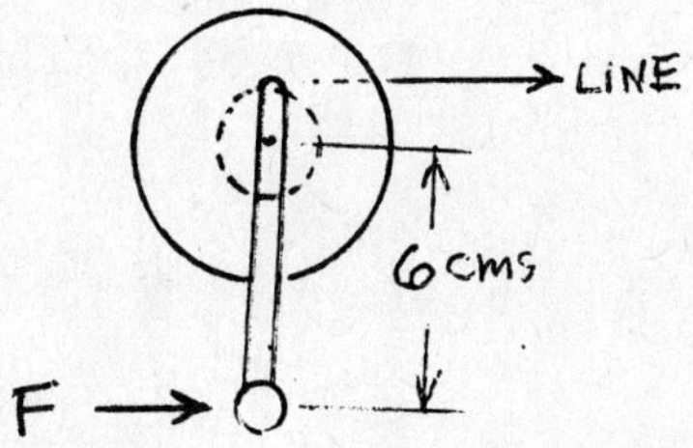

9.2 What is the moment arm of the force F about the point P in each of the following cases?

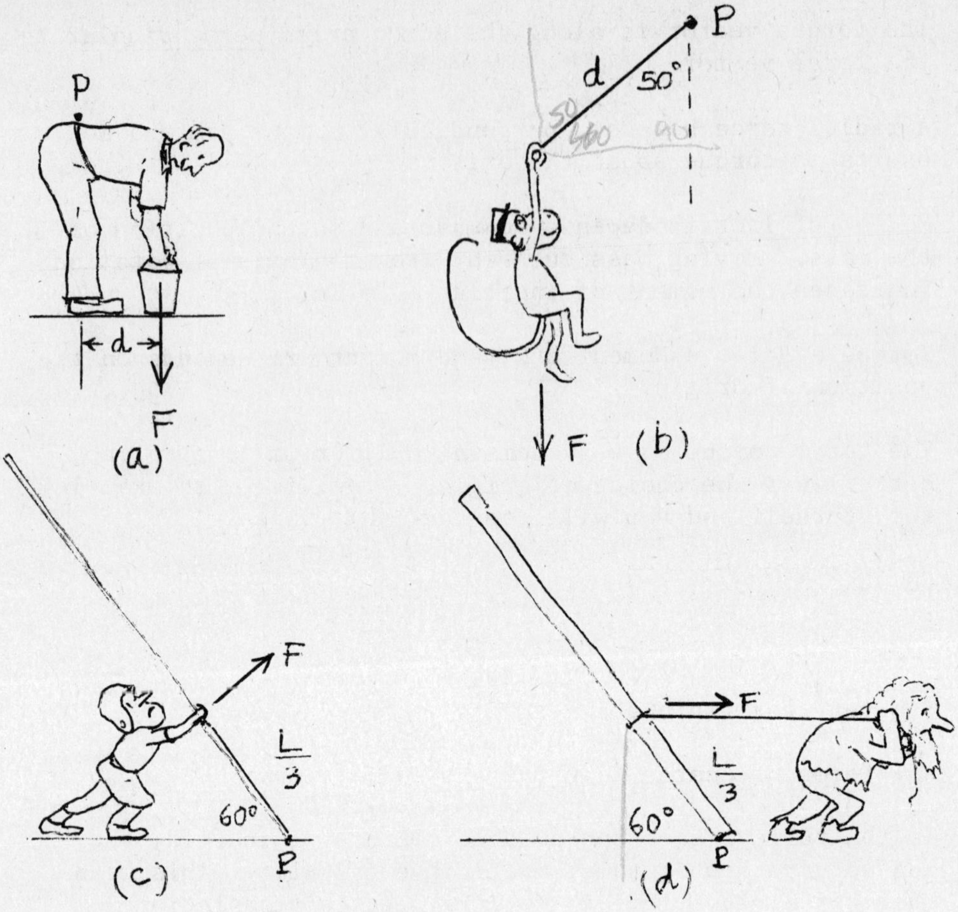

Figure 9.1

9.3 In Figure 8.1 (c) and (d), the pole is stuck in the ground so that it doesn't slide. What force must be exerted in each case to hold the pole upright if it has a mass of 40 kg in each case? Each pole is uniform and has length L.

9.4 A workman wants to tip a large packing crate so that he can put a hand truck under it. The base of the crate is 1 meter x 1 meter and it is 2 meters tall. It weighs 800 n and its center of mass is at its geometrical center. What horizontal force applied at shoulder height (1.4 m above the floor) must be applied, assuming the crate doesn't slip?

9.5 A ladder of length 8 meters leans against a wall 4 meters high. Friction between the ladder and the wall is negligible, but there is a friction force between the ladder and the concrete surface on which it rests. It is found that the ladder will start to slip if the foot of the ladder is more than 3 meters from the base of the wall. What is the coefficient of friction between the ladder and the horizontal concrete surface?

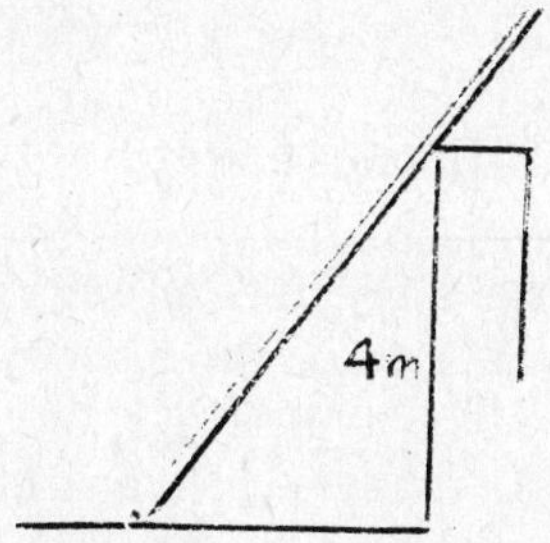

9.6 An architect wants to design a curved archway by laying up bricks one on top of another in such a way that the bricks are in equilibrium without the aid of mortar holding them together. To see how this might work I have sketched the first four bricks down from the top. Notice that the center of mass of the top brick is just above the edge of the brick below it. Similarly, the CM of the combination of the two bricks is just above the edge of the third brick from the top, and so on. Show that for the four bricks shown, the top brick can project a distance of (11/12)L beyond the bottom brick without the aid of mortar. L is the length of one brick.

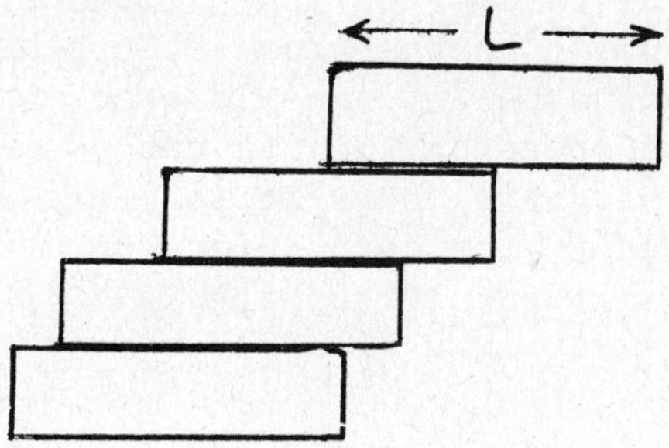

9.7 In a triatomic molecule three atoms each of mass M are located at the vertices of an equilateral triangle of side a. Determine the moment of inertia about (a) an axis perpendicular to the plane of the molecule which passes through the CM and (b) about an axis which passes through the CM and one of the atoms.

9.8 Skyrocketing oil prices have spurred some novel research on new kinds of car engines. One interesting idea works on the same principle as a child's friction car. Energy is stored in a fast spinning flywheel, and by a gear system the flywheel drives the car. One gets the flywheel spinning with an electric motor at home. Sort of like having a squirrel on a treadmill under your hood. Let us see what kind of numbers are involved in such a device. Suppose the flywheel is a solid disk of 1 meter diameter and 100 kg mass. If such a wheel is spun too fast, it will fly apart. Consider the case where the wheel is initially spun with a frequency of 10,000 RPM.

(a) What is the kinetic energy of the wheel. By comparison, a gallon of gasoline has an energy content of about 10^8 J.

(b) It is vital that such a flywheel experience very little friction, otherwise it will quickly run down while you leave your car parked. Some experimental models have been built which will still have most of their energy left after three months. Such wheels are mounted on air bearings. If the frictional force on the shaft of the wheel acted at an effective radius of 1 cm, what is the maximum allowable force if our flywheel is to lose half of its KE in 90 days? At what rate would it then be rotating?

9.9 Once my sons and I tried to rig up a system to haul gold ore out of an old mine. It is sketched here. We diverted water from a stream to fill an oil drum which, when heavy enough, pulled the ore car up the track. One then let out the water and sent the car back down for another load. Suppose that friction on the track and in the pulley (an old truck wheel) is negligible. Determine the speed of the ore car when it reaches the top and the tension in the cable where it is attached to the bucket and where it is attached to the car. Since neglecting friction is only a rough approximation, let us take $g = 10\ m/s^2$ to simplify the arithmetic.

M_c = mass of ore and car = 90 kg

M_w = mass of water and barrel = 50 kg

M_p = mass of pulley = 10 kg

R = radius of pulley = 0.5 m

h = height barrel drops while pulling car up = 10 m

θ = angle of elevation of track = 30°

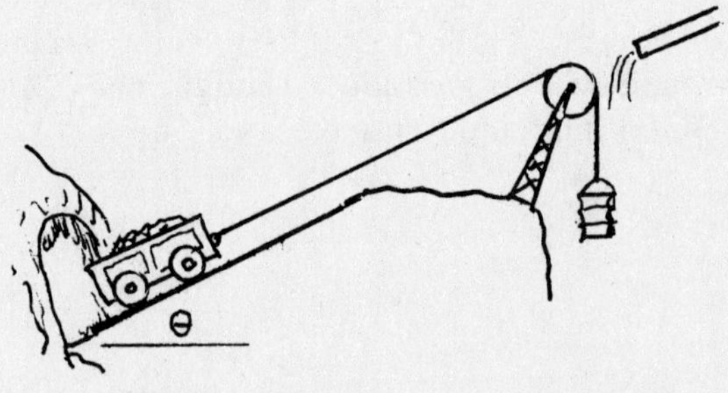

9.10 The ultimate fate of a star depends on its size. If it is small enough it may just fizzle out and end as a cold, hard rock. If it is bigger it may die as a red giant or a supernova. If it is large enough (i.e. a medium sized star) it may collapse to become a neutron star. Even larger stars have such a strong gravity field that they collapse to become black holes. It is not certain just what will become of our star, the sun, but it may someday collapse under its own gravity to become a neutron star. If this happens, its period of rotation will decrease from its present value of 24.7 days to something much less. Rotating neutron stars give off pulses of electromagnetic radiation, much like a rotating beacon at an airport. One such star, in the Crab nebula, rotates about 30 times per second. When these signals were first discovered the LGM (Little Green Men) theory was advanced to explain them, but now more mundane ideas prevail. Suppose the sun were to collapse into a spherical neutron star of 30 km radius. At present it rotates as if it were a sphere of radius of about 2×10^8 m. What would be its new period of rotation and its frequency of rotation?

9.11 A small cylinder of radius r and mass M is released from rest and allowed to roll down a loop-the-loop track of radius R. From what minimum height should it be released if it is never to leave the track? From what height should it be released if it were to slide without rolling? What accounts for the difference in these two results?

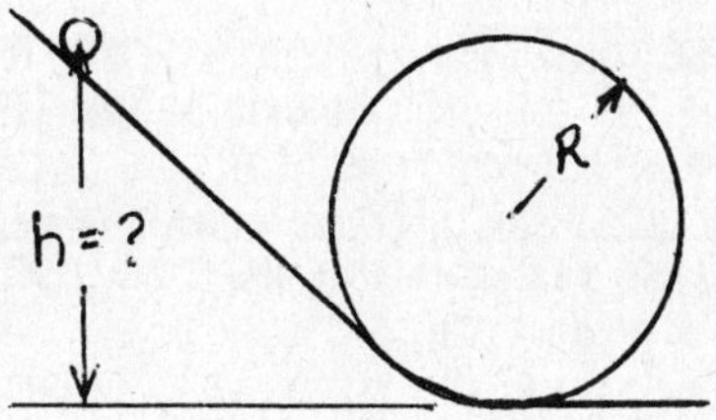

• Problem Solutions

9.1 IN EQUILIBRIUM SO $\tau_{CLOCK} = \tau_{COUNTERCLOCK}$

$$F_L r = F_C R$$
$$F_C = \frac{r}{R} F_L = \left(\frac{1.5\text{ cms}}{6\text{ cms}}\right)(50\text{ n}) = \underline{\underline{12.5\text{ N}}}$$

9.2 (a) $l = d$ (b) $l = d\sin 50° = 0.77d$ (c) $l = \frac{L}{3}$ (d) $l = \frac{L}{3}\sin 60° = 0.29L$

9.3 (c) IN EQUILIBRIUM SO $\tau_c = \tau_{cc}$ ABOUT POINT P

$$(F)\left(\frac{L}{3}\right) = (W)(l)$$
$$= (W)\left(\frac{L}{2}\cos 60°\right)$$
$$W = mg = (40\text{ kg})(9.8\text{ m/s}^2) = 392\text{ n}$$
$$\therefore F = \left(\tfrac{3}{2}\cos 60°\right)(392\text{ n}) = \underline{\underline{294\text{ n}}}$$

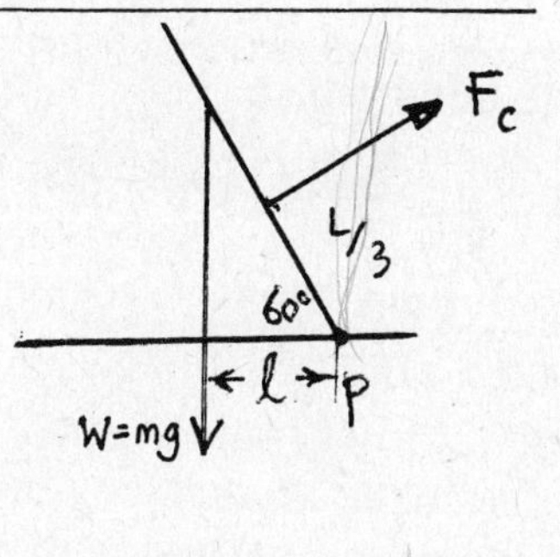

(d) $\tau_c = \tau_{cc}$

$(F_d)(X) = (W)(l)$

$(F_d)(\frac{L}{3}\sin 60°) = (W)(\frac{L}{2}\cos 60°)$

$F_d = \left(\frac{3}{2}\,\frac{\cos 60°}{\sin 60°}\right)(392\,n) = \underline{\underline{339\,n}}$

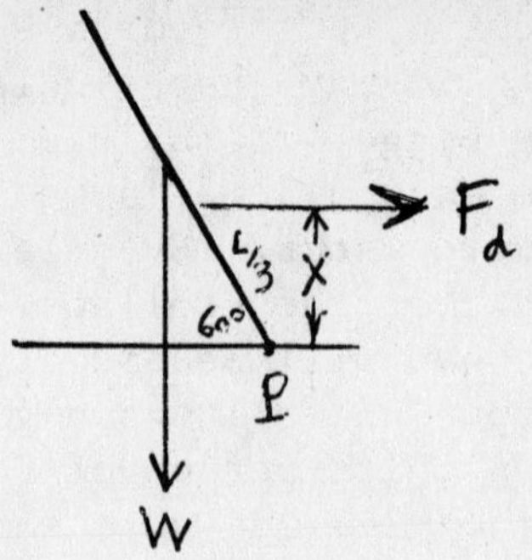

9.4 CRATE WILL TIP ABOUT POINT P.
JUST BEFORE IT TIPS IT IS IN EQUILIBRIUM, SO

$\tau_{CLOCKWISE} = \tau_{COUNTERCLOCKWISE}$

$(1.4\,m)(F) = (0.5\,m)(W)$

$F = \left(\frac{0.5}{1.4}\right)(800\,n) = \underline{\underline{286\,n}}$

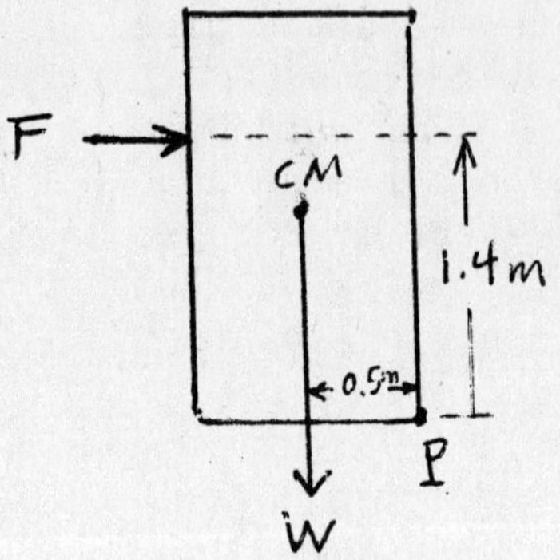

$W = 800\,n$

9.5 JUST BEFORE THE LADDER STARTS TO SLIP IT IS IN EQUILIBRIUM. FIRST IDENTIFY ALL OF THE FORCES ACTING ON THE LADDER. NOTE THAT FORCE OF WALL IS PERPENDICULAR TO LADDER (NO FRICTION THERE).

FIRST RESOLVE F INTO ITS COMPONENTS.

$F_x = F\sin\theta$

$F_y = F\cos\theta$

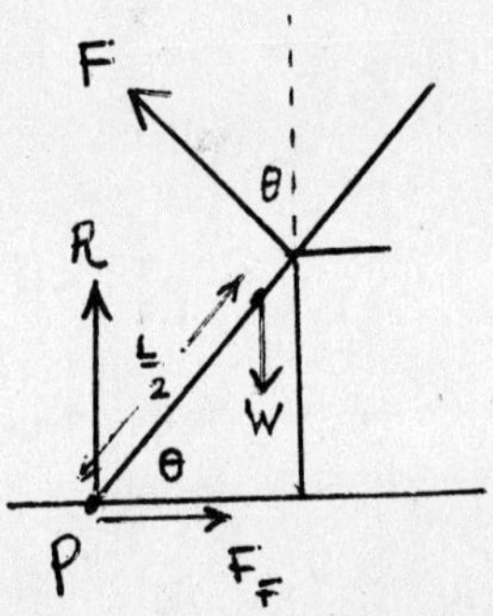

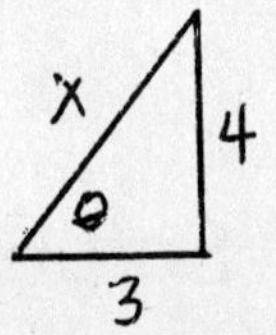

TORQUE ABOUT POINT P IS ZERO:

$(W)(\frac{L}{2}\cos\theta) = (F)(5)$ (1)

FORCE IN Y-DIRECTION IS ZERO:

$F\cos\theta + R - W = 0$ (2)

FORCE IN X-DIRECTION IS ZERO:

$F_F - F\sin\theta = 0$ (3)

$X = \sqrt{3^2 + 4^2} = 5$

$\sin\theta = \frac{4}{5}$

$\cos\theta = \frac{3}{5}$

$F_F = \mu R$. SUBSTITUTE THIS IN (3) AND SOLVE.

$\mu R - F\sin\theta = 0$ (3)

$WL\cos\theta = 10F$ (1) (SUBSTITUT $x = 5\,m$)

$F\cos\theta + R - W = 0$ (2)

$\mu R - \frac{4}{5}F = 0$ (3) $\longrightarrow F = \frac{5}{4}R\mu$

$(W)(8)(\frac{3}{5}) = 10F$ (1) $W = \frac{50}{24}F = \left(\frac{50}{24}\right)\left(\frac{5}{4}R\mu\right)$

$\frac{3}{5}F + R - W = 0$ (2)

$\left(\frac{3}{5}\right)\left(\frac{5}{4}R\mu\right) + R - \left(\frac{50}{24}\right)\left(\frac{5}{4}R\mu\right) = 0$

$\frac{15}{20}\mu + 1 - \frac{250}{96}\mu = 0 \quad , \mu = \dfrac{1}{\frac{250}{96} - \frac{15}{20}} = 0.54 \quad , \mu = \underline{\underline{0.54}}$

9.6 CONSIDER FIRST THE TOP BRICK. IF ITS CM IS POSITIONED TO THE RIGHT OF THE EDGE OF THE BRICK BELOW IT, THE TORQUE ON IT DUE TO GRAVITY WILL CAUSE IT TO FALL OFF. THUS THE MAXIMUM OVERHANG IT CAN HAVE IS $L/2$.

NOW CONSIDER THE TOP TWO BRICKS AS A UNIT. THE CM OF THIS PAIR IS HALFWAY BETWEEN THE CM'S OF EACH SEPARATELY, OR $\frac{1}{4}L$ FROM RIGHT END OF SECOND BRICK DOWN.

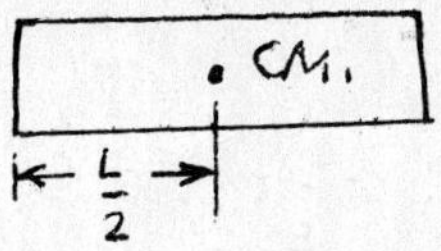

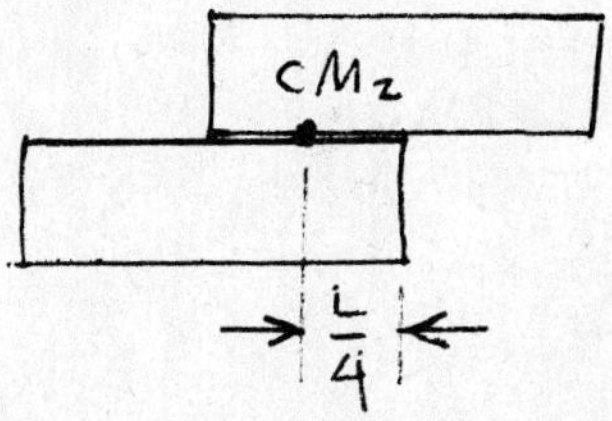

THE CM OF THIS PAIR MUST BE PLACED OVER THE RIGHT EDGE OF THIRD BRICK.
NOW FIND CM_3, CM OF THREE BRICKS TOGETHER.
ITS NOT SO OBVIOUS BY INSPECTION WHERE CM_3 IS, SO I WILL USE THE FORMULA FOR FINDING IT. CM_3 IS A DISTANCE x FROM LEFT END, WHERE

$(3M)(x) = (M)(L/2) + M(3L/4) + M(5L/4)$

USING $x_1 = L/2$, $x_2 = 3L/4$, $x_3 = 5L/4$

SOLVE FOR x: $x = \left(\frac{1}{6} + \frac{1}{4} + \frac{5}{12}\right)L = \left(\frac{2}{12} + \frac{3}{12} + \frac{5}{12}\right)L$
$= \frac{5}{6}L$

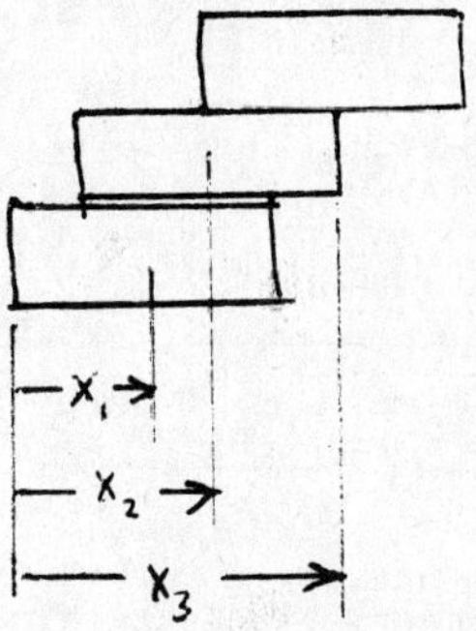

THUS FOURTH BRICK HAS ITS EDGE $\frac{1}{6}L$ FROM RIGHT END OF THIRD BRICK.
TOTAL OVERHANG IS THUS

$O.H. = \frac{L}{6} + \frac{L}{4} + \frac{L}{2} = \left(\frac{2}{12} + \frac{3}{12} + \frac{6}{12}\right)L = \underline{\underline{\frac{11}{12}L}}$

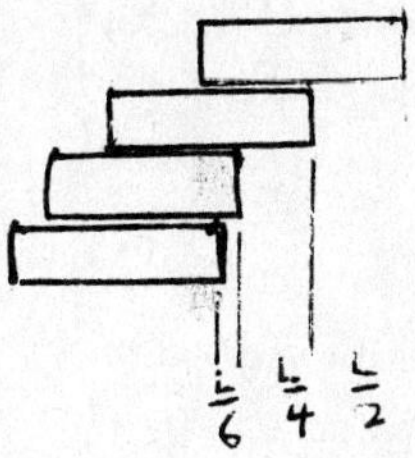

9.7 BY SYMMETRY WE SEE THAT CM MUST LIE AT INTERSECTION OF BISECTORS OF EACH ANGLE. (i.e. THIS IS THE "BALANCE POINT" IF A THIN WEIGHTLESS MEMBRANE WERE STRETCHED OVER THE MOLECULE.)

(a) FOR AN AXIS THROUGH THE CM AND PERPENDICULAR TO THE PAPER,

$I_1 = \sum m_i R_i^2$
$= MR^2 + MR^2 + MR^2$
$= 3MR^2$

$\frac{a/2}{R_1} = \cos 30° = \frac{\sqrt{3}}{2}$, SO $R_1 = \frac{a}{\sqrt{3}}$

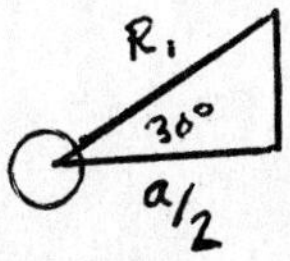

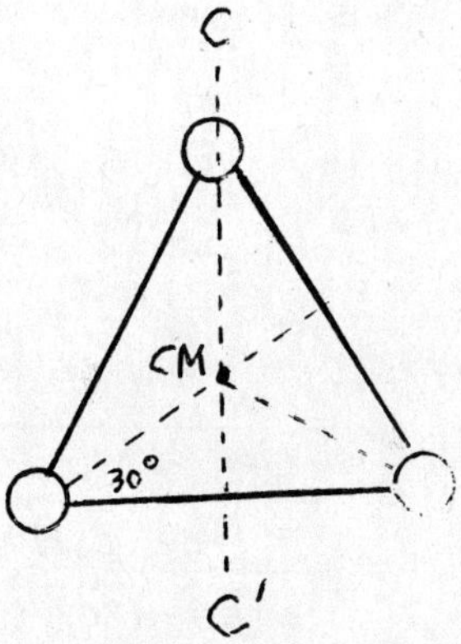

$I_1 = 3MR^2 = 3M(a/\sqrt{3})^2 = Ma^2$, $\underline{\underline{I_1 = Ma^2}}$

(b) FOR AXIS CC′, $I_2 = \sum m_i R_i^2$

$= (M)(0) + (M)(a/2)^2 + (M)(a/2)^2$

$\underline{\underline{I_2 = \frac{1}{2} Ma^2}}$

9.8 (a) $K.E. = \frac{1}{2} I\omega^2$, $I = \frac{1}{2} MR^2$ FOR A DISK , $\omega = 2\pi f$

$K.E. = \frac{1}{2}(\frac{1}{2}MR^2)(2\pi f)^2 = (\frac{1}{4})(100\text{kg})(0.5\text{m})^2(2\pi)^2(10{,}000 \times \frac{1}{60\text{s}})^2$

$\underline{\underline{K.E. = 6.8 \times 10^6\ J}}$

THIS IS ONLY ABOUT AS MUCH ENERGY AS YOU COULD GET FROM A CUP OF GASOLINE. NOT VERY PRACTICAL.

(b) LET ω_1 = INITIAL ANGULAR VELOCITY

AND ω_2 = ANGULAR VELOCITY WHEN $KE_2 = \frac{1}{2} KE_1$

THUS $KE_2 = \frac{1}{2} I\omega_2^2$, $KE_1 = \frac{1}{2} I\omega_1^2$

$KE_2 = \frac{1}{2} KE_1 \longrightarrow \frac{1}{2} I\omega_2^2 = \frac{1}{2}(\frac{1}{2} I\omega_1^2)$

$\omega_2 = \frac{1}{\sqrt{2}}\omega_1$

$\tau = I\alpha \simeq I\frac{\Delta\omega}{\Delta t} \simeq I\left(\frac{\omega_1 - \omega_2}{\Delta t}\right) \simeq I\left(\frac{\omega_1 - \frac{1}{\sqrt{2}}\omega_1}{\Delta t}\right) = I\left(1 - \frac{1}{\sqrt{2}}\right)\frac{\omega_1}{\Delta t}$

$\tau = Fr$. SO $F = \frac{\tau}{r} = (1 - \frac{1}{\sqrt{2}})\frac{I}{r}\frac{\omega_1}{\Delta t}$

$F = (1 - \frac{1}{\sqrt{2}})(\frac{1}{2}MR^2)(\frac{1}{r})\frac{\omega_1}{\Delta t}$

$= (1 - \frac{1}{\sqrt{2}})(\frac{1}{2})(100)(.5)^2(\frac{1}{0.01})\left(\frac{1}{(90)(24)(3600)}\right)\left(2\pi \times \frac{10000}{60}\right) = \underline{\underline{0.05\ n}}$

9.9 $\underbrace{KE_1 + PE_1}_{\text{INITIAL}} = \underbrace{KE_2 + PE_2}_{\text{FINAL}}$ (1)

FOR CONVENIENCE LET US CHOOSE TOTAL PE OF SYSTEM TO BE ZERO INITIALLY. WATER BUCKET THEN DROPS A DISTANCE h (ITS PE BECOMES $-mgh$) AND CAR RISES A DISTANCE h_c, AND THEN ITS PE IS $+mgh_c$. h AND h_c ARE RELATED, BECAUSE WHEN BUCKET DROPS DISTANCE h, CAR MOVES THIS SAME DISTANCE h ALONG TRACK, AND

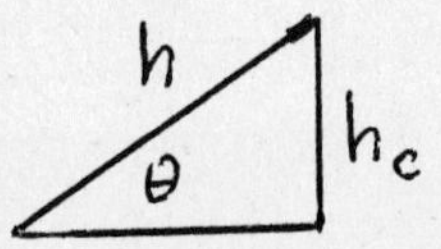

$\frac{h_c}{h} = \sin\theta$

THUS $h_c = h\sin\theta$

EQN. (1) THUS BECOMES

$0 + 0 = \frac{1}{2}M_c v^2 + \frac{1}{2}M_w v^2 + \frac{1}{2}I\omega^2 + M_c g h_c - M_w g h$ (2)

$\frac{1}{2}I\omega^2$ IS THE KE OF THE PULLEY.

$v = R\omega$ AND $I = \frac{1}{2}M_p R^2$

THUS $0 = \frac{1}{2}M_c v^2 + \frac{1}{2}M_w v^2 + \frac{1}{2}(\frac{1}{2}M_p R^2)(\frac{v}{R})^2 + M_c g h_c - M_w g h$

$0 = (2M_c + 2M_w + M_p)v^2 + M_c g h \sin\theta - M_w g h$

$v = \sqrt{\frac{4(M_w - M_c\sin\theta)gh}{2(M_c + M_w) + M_p}}$

$v = \sqrt{\frac{4[50 - (90)(.5)][10m][10m/s^2]}{2(90 + 50) + 10}} = \underline{\underline{2.6\ m/s}}$

TO DETERMINE THE TENSION IN THE CABLE FIRST OBSERVE THAT TENSION T_W IN CABLE ATTACHED TO WATER BUCKET WILL BE GREATER THAN TENSION IN PIECE OF CABLE ATTACHED TO CAR. THIS IS POSSIBLE BECAUSE THERE IS FRICTION BETWEEN THE CABLE AND THE PULLEY, OTHERWISE THE CABLE WOULD MERELY SLIP OVER THE PULLEY AND IT WOULD NOT TURN. AS THE BUCKET FALLS IT ACCELERATES DOWNWARD, WHICH MEANS THAT THE PULLEY IS GOING TO HAVE AN ANGULAR ACCELERATION. THE DIFFERENCE IN TENSION IN THE TWO PARTS OF THE CABLE GIVES RISE TO THE TORQUE NECESSARY TO GIVE THE PULLEY ANGULAR ACCELERATION. OBSERVE THAT THE BUCKET AND THE CAR EACH HAVE THE SAME LINEAR ACCELERATION, AND THEIR ACCELERATION IS RELATED TO THE ANGULAR ACCELERATION α OF THE PULLEY BY $a = R\alpha$.

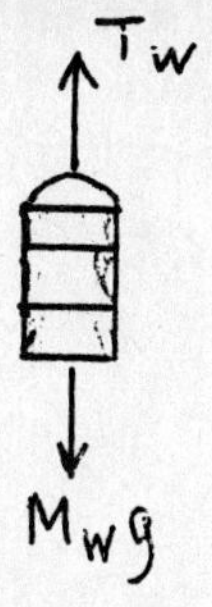

LOOKING AT THE WATER BUCKET AND APPLYING $F = M_W a$ YIELDS

$$M_W g - T_W = M_W a \qquad (3)$$

NOW LOOK AT THE ORE CAR:

$$T_c - M_c g \sin\theta = M_c a \qquad (4)$$

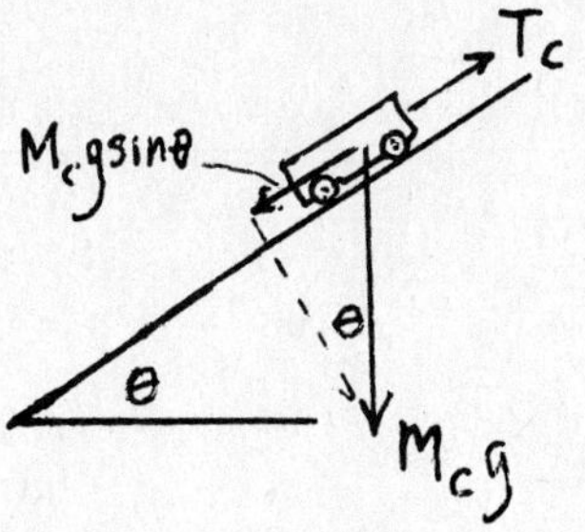

NOW LOOK AT THE PULLEY AND APPLY $\tau = I\alpha$

$$RT_W - RT_c = I\alpha$$

$$I = \tfrac{1}{2} M_p R^2, \quad \alpha = \frac{a}{R}$$

$$RT_W - RT_c = \tfrac{1}{2} M_p R^2 \left(\frac{a}{R}\right)$$

OR $$T_W - T_c = \tfrac{1}{2} M_p a \qquad (5)$$

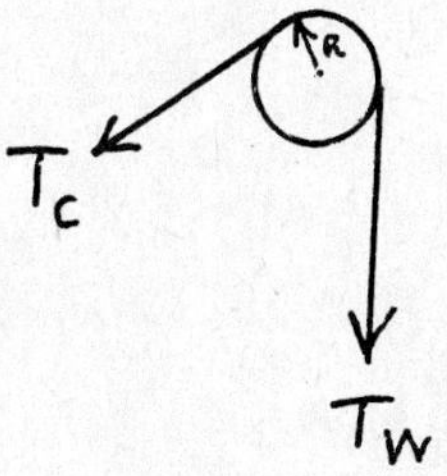

EQNS. (3), (4) AND (5) ARE THREE EQUATIONS FOR THE THREE UNKNOWNS T_W, T_c AND a. WE CAN SOLVE THEM SIMULTANEOUSLY. IT IS EASIEST TO DO SO IF WE PUT IN NUMERICAL VALUES AT THIS TIME, REMEMBERING THAT WE ARE TAKING $g \simeq 10\ m/s^2$.

EQN. (3) THUS BECOMES $(50)(10) - T_W = 50a$

EQN. (4) BECOMES $T_c - (90)(10)(0.5) = 90a$

EQN (5) BECOMES $T_W - T_c = \frac{1}{2}(10)a$

SIMPLIFY THESE:

$$500 - T_W = 50a \longrightarrow T_W = 500 - 50a$$

$$T_c - 450 = 90a \longrightarrow T_c = 450 + 90a$$

$$T_W - T_c = 5a$$

SOLVE. $(500 - 50a) - (450 + 90a) = 5a$

$$145a = 50$$

$$a = 0.34\ m/s^2$$

SUBSTITUTE BACK TO FIND T_W AND T_c.

$T_W = 500 - 50a = 500 - 50(0.34) = \underline{\underline{483\ n}}$

$T_c = 450 + 90a = 450 + 90(0.34) = \underline{\underline{481\ n}}$

THE DIFFERENCE IN TENSION HERE IS NOT VERY LARGE BECAUSE THE MOMENT OF INERTIA TO THE PULLEY HERE IS RELATIVELY SMALL. FOR A BIG FLYWHEEL IN A MACHINE THIS EFFECT COULD BE IMPORTANT.

9.10 NO EXTERNAL TORQUE ACTS ON THE SUN, SO ITS ANGULAR MOMENTUM REMAINS CONSTANT AS IT COLLAPSES.

$$L = I_1\omega_1 = I_2\omega_2$$

AS THE SUN COLLAPSES ITS MOMENT OF INERTIA I DECREASES, SO THE ANGULAR VELOCITY ω INCREASES.

$\omega_2 = 2\pi f_2 = 2\pi\left(\frac{1}{T_2}\right)$, T_2 = FINAL PERIOD OF REVOLUTION.

$\omega_1 = 2\pi\left(\frac{1}{T_1}\right)$

$I = \frac{2}{5}MR^2$ FOR A SPHERE.

THUS $\left(\frac{2}{5}MR_1^2\right)\left(\frac{2\pi}{T_1}\right) = \left(\frac{2}{5}MR_2^2\right)\left(\frac{2\pi}{T_2}\right)$

$$T_2 = \left(\frac{R_2}{R_1}\right)^2 T_1 = \left(\frac{30\times10^3\,m}{2\times10^8\,m}\right)^2 (24.7)(24)(3600\,s)$$

$$T_2 = \underline{0.048\ \text{sec.}}$$

$$f_2 = \frac{1}{T_2} = \underline{\underline{21\ \text{REVOLUTIONS PER SECONDS}}}$$

9.11 ENERGY IS CONSERVED.

$$\underbrace{KE_1 + PE_1}_{\text{INITIAL POSITION}} = \underbrace{KE_2 + PE_2}_{\text{AT TOP OF TRACK}}$$

$$0 + Mgh = \frac{1}{2}Mv^2 + \frac{1}{2}I\omega^2 + Mg(2R) \qquad (1)$$

$\omega = \frac{v}{r}$, $I = \frac{1}{2}Mr^2$

IF CYLINDER IS JUST ABOUT TO LEAVE TRACK REACTION FORCE OF TRACK IS ZERO, SO

$$Mg = \frac{Mv^2}{R} = \text{CENTRIPETAL FORCE}$$

OR $v^2 = Rg$

EQN. (1) BECOMES

$$Mgh = \frac{1}{2}Mv^2 + \frac{1}{2}\left(\frac{1}{2}Mr^2\right)\left(\frac{v}{r}\right)^2 + 2MgR$$

$$gh = \frac{1}{2}Rg + \frac{1}{4}Rg + 2Rg$$

$$\underline{\underline{h = \frac{11}{4}R = 2.75R}}$$

IF NO ROLLING OCCURS EQN. (1) IS

$$0 + Mgh = \frac{1}{2}Mv^2 + Mg(2R)$$

$$Mg = Mv^2/R, \quad v^2 = Rg$$

$$Mgh = \frac{1}{2}MRg + 2MRg, \quad \underline{\underline{h = \frac{5}{2}R = 2.5R}}$$

IN ORDER TO STAY ON THE TRACK THE CYLINDER MUST HAVE A TRANSLATIONAL SPEED $V = \sqrt{Rg}$. WHEN IT ROLLS SOME OF THE PE LOST GOES INTO ROTATIONAL KE, SO FOR GIVEN h IT IS NOT TRANSLATING AS FAST. HENCE IT MUST BE STARTED AT A GREATER HEIGHT h WHEN ROLLING IN ORDER TO AQUIRE THE NEEDED SPEED V.

Mechanical Properties of Matter

10

• Summary of Important Ideas, Principles, and Equations

1. A molecule of a substance is the smallest unit which displays the chemical properties of the substance itself. A molecule may consist of a single atom or of a finite number of atoms arranged in a unique pattern. Not all of the atoms in a molecule need be the same kind (e.g. H_2O).

   Matter can exist in solid, liquid and gaseous phases. Phase transitions between these states occur at well defined conditions of temperature and pressure.

2. Solids are materials which deform elastically. A deformation is elastic if it is proportional to the force causing it and if it vanishes when the force is removed.

3. Definition: When a force F is exerted over an area A it creates a stress S.

$$\boxed{\text{Stress } S = \frac{F}{A}\ \left(\frac{n}{m^2}\right)} \qquad (10.1)$$

   If the force F is applied parallel to the area A, as when you slide a piece of sandpaper along a surface, the stress is called a shear stress.

   If the force is applied perpendicular to A and is squashing the material, the stress is called a compressional stress. If the force is perpendicular to A and is stretching the material, the stress is a tensile stress.

4. Fluids are materials which flow when subjected to a shear stress, whereas solids deform elastically when subjected to a shear stress. Liquids and gases are fluids. A gas expands to fill the entire container in which it is placed. A liquid does not. The

molecules in a fluid are randomly distributed, whereas in a solid they have a regular arrangement, at least for very small pieces of solid.

5. Definition: Density ρ is

$$\rho = \frac{\text{mass}}{\text{volume}} = \frac{M}{V} \qquad (\text{kg/m}^3) \qquad (10.2)$$

$$\text{Specific gravity} = \frac{\text{Density}}{\text{Density of water at } 4^{\circ}\text{C}}$$

The density of water at 4°C is 1,000 kg/m^3 = 1 gram/cm^3.

The density of a material does not depend on the size of the piece of material being considered, but is rather a characteristic of the particular kind of substance.

6. Strain is the response of a material to an applied stress. It is defined as the fractional change in length or volume.

$$\text{Tensile strain} = \frac{\Delta L}{L_o}$$

$$\text{Compressional strain} = \frac{\Delta V}{V_o}$$

$$\text{Shear strain} = \frac{\Delta x}{L} \qquad (10.3)$$

For an elastic deformation strain is proportional to the stress causing it (Hooke's Law).

Stress = (constant) x Stress

or

$$\frac{\text{Stress}}{\text{Strain}} = \text{constant}$$

The ratio of stress to strain is called a modulus. Young's modulus, the bulk modulus and the shear modulus characterize the response of a material to tensile, compressional and shear stress, respectively.

Remember, a stress is a force per unit area. A tensile strain is a stretch per unit length.

7. Cohesive forces between molecules cause surface tension in liquids. A liquid thus acts like it is covered with a thin rubber skin. It

takes energy to stretch this skin and increase the surface area of the liquid, and the energy needed for unit increase in area is the surface tension γ.

$$\boxed{\gamma = \frac{\Delta E}{\Delta A}} \qquad (10.4)$$

Imagine you draw a line of length L on the rubbery skin and try to pull the skin apart along this line with a force F. γ also is the force needed per unit length to pull the skin apart.

$$\boxed{\gamma = \frac{F}{L} \qquad (n/m)} \qquad (10.5)$$

8. Adhesion is the tendency of liquid molecules to be attracted to the molecules of a solid in contact with the liquid. If the adhesive force is greater than the cohesive force, the liquid wets the solid.

Adhesion and surface tension give rise to capillary action, the tendency for liquids to be pulled up into narrow openings. A liquid of density ρ and surface tension γ will rise to a height h in a tube of radius R.

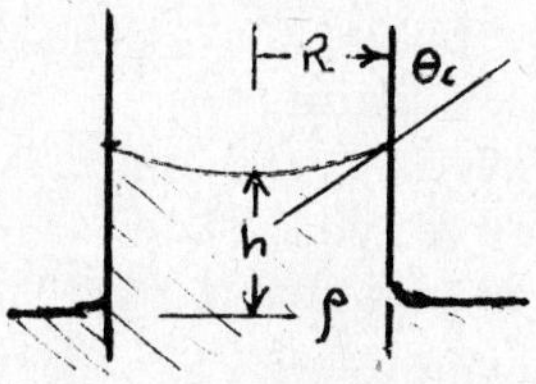

$$\boxed{h = \frac{2\gamma \cos \theta_c}{\rho R g}} \qquad (10.6)$$

• Qualitative Questions

10M.1 When a weight W is suspended from a section of muscle L millimeters long it is observed to stretch elastically by an amount δ. Thus if the weight had been suspended from a section of muscle of length 2L, the stretch we would expect is

A. $\frac{1}{4}\,\delta$ C. δ E. $4\,\delta$
B. $\frac{1}{2}\,\delta$ D. $2\,\delta$

10M.2 If a piece of nylon monofilament fishing line 1 meter long is found to break when subjected to a force of 10 pounds, we would expect that a piece of the line 2 meters long would break when subjected to a force of

A. 5 lbs. C. 10 lbs. E. 20 lbs.
B. 7.1 lbs. D. 14 lbs. F. 40 lbs.

10M.3 A characteristic of a fluid is that it

A. deforms elastically when subjected to a shear stress.
B. flows when subjected to a shear stress.
C. has a very regular arrangement of molecules.
D. is very readily compressed.
E. cannot exert a compressive stress on anything else.

10M.4 In the sense that we are using it, "stress" has a meaning most closely related to

A. one of the fundamental properties of a material, which are often tabulated in books as some kind of "modulus".
B. fractional strength.
C. the force per unit area on a material.
D. the tendency of a material to break after it has been bent or stretched many times.
E. the change in size of something.

10M.5 The density of an object

A. depends on the mass of the object.
B. depends on the volume of the object.
C. has no meaning if the object is in the gaseous phase.
D. depends on the material from which the object is made.
E. depends only on the number of atoms in the material and on their spacing, not on what kind of atoms are involved (i.e. one atom is as good as another; or, an atom is an atom is an atom, as in "up 'n atom").

10M.6 When an object is compressed, its

A. mass increases.
B. volume increases.
C. density increases.
D. more than one of the above are true.
E. none of the above is true.

10M.7 A rope of length L and diameter D will support a maximum weight W before breaking. We would expect that a piece of the same kind of rope of length 2L and diameter $\frac{1}{2}$D would thus be able to support a maximum weight of

A. $\frac{1}{4}$W
B. $\frac{1}{2}$W
C. W
D. 2W
E. 4W
F. 8W

10M.8 The molecules in a liquid all attract each other, and this gives rise to

A. adhesion.
B. fluidity.
C. capillary action.
D. surface tension.
E. a Young's modulus for the liquid.

10M.9 If you place a carnation with its stem in a jar of water colored with green food coloring, you will find that in a day or so the entire flower has turned green. We can best understand why this happens in terms of

A. the distinction between liquids and fluids.
B. the distinction between liquids and solids.
C. elastic deformation.
D. surface tension.
E. capillary action.

10M.10 Have you ever tried to float a razor blade or a needle in a cup of water? Try it. Amazing. You have to set the object in carefully to keep from breaking the "skin" on the water. You can also float flat little metal boats this way. Sketched here are some objects cut from a roll of aluminum foil bought at the market. Which one has the best chance of floating?

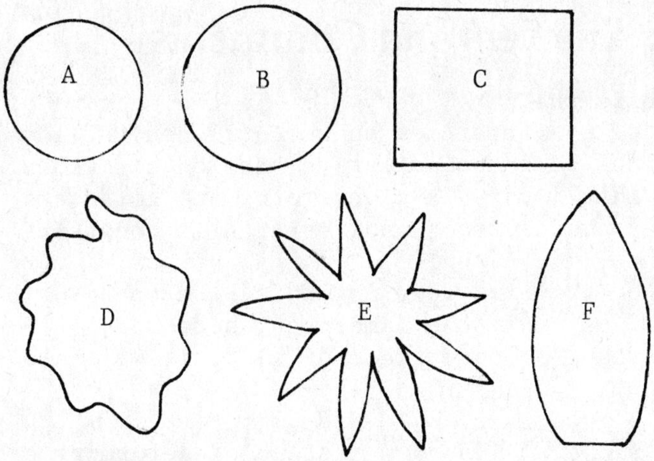

• True–False Questions

10TF.1 There is actually no difference between a gas and a liquid, since both are fluids.

10TF.2 The density of 20 kilograms of aluminum is greater than the density of 10 kilograms of iron.

10.TF.3 Liquids have densities comparable to those of solids.

10TF.4 When an object is stretched to increase its length, it is experiencing a compressional strain.

10TF.5 The ratio of stress to strain is constant for an elastic deformation.

10TF.6 When a substance undergoes a phase change from liquid to gas under everyday circumstances, the spacing of the molecules does not change appreciably.

10TF.7 If water rises 3 cms in a capillary of 1 mm diameter, it will rise 1.5 cms in a capillary of 0.5 mm diameter.

10TF.8 A liquid will wet a solid surface providing adhesive forces are greater than cohesive forces.

10TF.9 We all know that the surface area of a sphere, say, is proportional to the square of its radius, whereas its volume is proportional to the cube of its radius. This fact helps us to understand why small bugs can walk on water, but big ones can't.

10TF.10 Suppose water will rise 10 cms in a particular capillary tube which is 20 cms long. If a piece of this same diameter tube and only 5 cms long is dipped in water, a small fountain of water squirting out of the top will be observed.

• Multiple Choice: Answers and Comments

10M.1 D The tension is the same everywhere in the section of muscle. Thus the top 1 cm stretches an amount δ and the lower 1 cm also stretches δ, thus the entire section stretches a distance 2δ. That is, for a given force and given cross-sectional area, the fractional stretch is constant.

10M.2 C The length of the line does not influence the tension in the line. The tension is the same everywhere in the line, and if 10 lbs breaks a short piece of line, it will also break a long piece.

10M.3 B A is false since fluids <u>flow</u> instead of deforming elastically under shear stresses.
C is false. The molecules are randomly arranged in fluids.
D is false. Some fluids (e.g. gases) are readily compressed whereas others (e.g. liquids) are not.
E is false. A fluid can push on something and exert a compressive stress.

10M.4 C

10M.5 D A and B are false, but the reasoning is a little tricky, because from the equation (Density) = (Mass)/(Volume) it appears that density depends on both mass and volume. This is not true, however, since density depends only on the ratio of mass to volume, not on the magnitude of either. The density of copper, say, is a fixed value, independent of the mass or volume of the piece of copper under consideration. E is false, since for given spacing massive atoms like uranium will give rise to a denser material than will lighter atoms like carbon.

10M.6 C

10M.7 A The weight a rope will support depends only on its cross-sectional area. Since $A_1 = \frac{1}{4}\pi D^2$, if D is made half as large area will become $\frac{1}{4}$ as large, i.e.

$$A_2 = \frac{1}{4}\pi\left(\frac{D}{2}\right)^2 = \frac{1}{4}A_o$$

10M.8 D

10M.9 E

10M.10 E The surface tension force supporting the object is proportional to the length of the perimeter of the object. The weight is proportional to the surface area of the object (for foil of given thickness). The object most likely to float is thus the one with the greatest ratio of perimeter length to surface area, which in this case is the starfish-like object.

• True–False: Answers and Comments

10TF.1 F A gas completely fills the chamber in which it is placed. A liquid does not.

10TF.2 F The density of a pure material does not depend on the size of the sample, since it is the ratio of mass to volume.

10TF.3 T

10TF.4 F A stretched object is experiencing a tensile strain.

10TF.5 T

10TF.6 F When a substance is changed from a liquid to a gas (i.e. boiled) at constant pressure, the spacing between molecules increases greatly. Gases are much less dense than liquids at the same pressure.

10TF.7 F The rise in a capillary is inversely proportional to the diameter, so if the diameter is made half as big, the height will double.

10TF.8 T

10TF.9 T A bug's weight is proportional to his volume. The force due to surface tension is proportional to the length of the interface between him and the water, and this is proportional to his area. Thus if bug size is doubled, the ratio of area (like force up) to volume (like force down) decreases by a factor of ¼. If this doubling process is continued one sees that eventually the bug becomes too heavy to be supported by surface tension.

10TF.10 F Water rises only when it is pulled up along the sides of the tube. If the tube ends, water is no longer pulled up and will not squirt out.

• Problems

10.1 Lead has a density of 11.3 gms/cm^3. A fisherman wants to make a lead sinker whose mass is 50 grams. What is the diameter of a lead sphere of this mass?

10.2 A commercial steel alloy is advertised to have a Young's modulus of 20×10^{10} n/m^2. By how much will a rod of this material of length 2 meters and diameter 1 mm stretch when subjected to a load of 200 n?

10.3 Methyl alcohol has a surface tension of 23×10^{-3} n/m and a specific gravity of 0.8. How high will it rise in a glass capillary of diameter 0.05 mm diameter? Assume the contact angle is zero.

10.4 A six legged insect standing on water makes a depression of radius 2 mm with each leg. The depression makes an angle of 60° with the vertical. What is the mass of the insect?

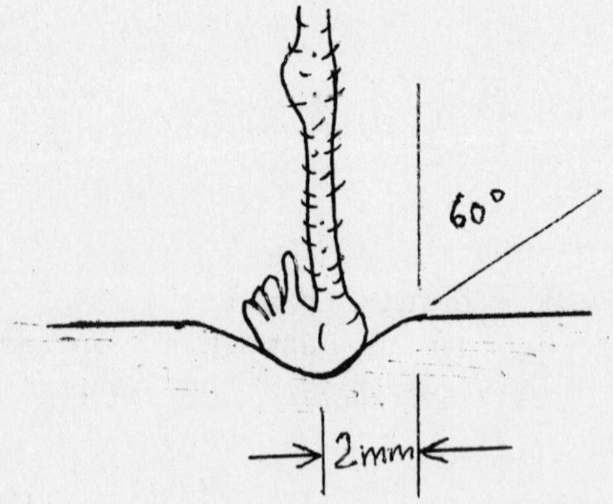

• Problem Solutions

10.1 $\rho = \frac{M}{V}$, $V = \frac{4}{3}\pi R^3 = \frac{M}{\rho}$, $R = \left[\frac{3M}{4\pi\rho}\right]^{1/3} = \left[\frac{(3)(50\,gm)}{(4\pi)(11.39\,gm/cm^3)}\right]^{1/3}$

$D = 2R = 2\,cms$

10.2 $y = \frac{F/A}{\Delta L/L_o} = \frac{FL_o}{A\Delta L}$

$\Delta L = \frac{FL_o}{Ay} = \frac{FL_o}{\frac{\pi}{4}D^2 y} = \frac{(200\,n)(2\,m)}{(\frac{\pi}{4})(10^{-3}\,m)^2(20\times10^{10}\,n/m)}$

$\Delta L = 0.0025\,m = 0.25\,cms. = 2.5\,mm.$

10.3 $h = \frac{2\gamma\cos\theta_c}{\rho R g} = \frac{(2)(23\times10^{-3}\,n/m)(1)}{(0.8\times10^3\,kg/m^3)(0.025\times10^{-3}\,m)(9.8\,m/s^2)}$

$h = 0.23\,m = 23\,cms.$

10.4 ONLY THE VERTICAL COMPONENT, $T\cos 60°$, SUPPORTS THE INSECT. THUS

$F = mg = (6)(2\pi R)(\gamma\cos 60°) = (12\pi)(2\times10^{-3}\,m)(70\times10^{-3}\,n/m)(\cos 60°)$

$F = 2.6\times10^{-3}\,kg = 2.6\,gms$

---

11 Hydrostatics and Hydrodynamics

• Summary of Important Ideas, Principles, and Equations

1. Atmospheric pressure at a distance h above the earth's surface is

$$P_A(h) = P_A(0)e^{-\frac{h}{h_o}} \qquad (11.1)$$

Here $P_A(0)$ is the pressure at the earth's surface, h is measured in kilometers, and h_0 is a constant with the value 8.6 km. A useful approximation is $P_A(h + \Delta h) \simeq P_A(h)[1 - \frac{\Delta h}{h_o}]$ if $\Delta h << h_o$.

2. Pressure in a liquid of density ρ at a depth h below the surface is

$$P = P_A + \rho gh \qquad (11.2)$$

Here P_A is the atmospheric pressure at the surface of the liquid. Pressure is sometimes measured in "atmospheres", where 1 atm. is approximately the atmospheric pressure at sea level.

$$1 \text{ atm} = 1.01 \times 10^5 \text{ N/m}^2 = 1.01 \times 10^5 \text{ Pa}$$

A column of mercury (Hg) 760 mm high will give rise to a pressure of 1 atm, and pressure is thus sometimes measured in units of "mm of Hg". The pressure in Pascals, the SI unit, can be found using a ratio.

$$\frac{P \text{ (in Pa)}}{P \text{ (in mm of Hg)}} = \frac{1.01 \times 10^5}{760}$$

3. Important principles of hydrostatics are

(a) At equilibrium the pressure at a point in a fluid is the same in all directions.

(b) Pascal's principle. At equilibrium a change in the pressure at one point in an incompressible fluid is transmitted undiminished to all points of the fluid.

4. Archimedes' principle states that in a gravitational field an object immersed in a fluid experiences a buoyant force equal to the weight of the displaced fluid. The buoyant force acts through the center of buoyancy, which is at the center of mass of the displaced fluid.

5. Laminar flow (in layers, or "smooth") occurs at small flow velocities. Turbulent flow (such as with whirlpools) occurs at high velocities.

6. Bernoulli's equation results from applying the conservation of energy principle to a fluid experiencing laminar flow.

$$P + gh + \tfrac{1}{2}v^2 = \text{constant} \qquad (11.3)$$

Thus where velocity is large, pressure is small, and vice versa.

7. Viscosity opposes fluid flow and is described by the viscosity coefficient.

$$\eta = \frac{F/A}{v/\ell} \qquad (11.4)$$

F/A is the shear stress and v is the difference in velocity of two layers of fluid separated by a distance ℓ. η is measured in units of Pa · s or poises (1 poise = 0.1 Pa · s).

8. Laminar flow rate Q in m^3/s is given by Poisseuille's equation,

$$Q = \frac{\pi R^4 \Delta P}{8L\eta} \qquad (11.5)$$

Here R = radius of pipe, P = pressure difference over length L of pipe, and η = viscosity coefficient.

9. Turbulent flow will occur in a smooth pipe when the Reynolds number R exceeds a value of about 2000.

$$R = \frac{\rho v \ell}{\eta} \qquad (11.6)$$

η is the viscosity, ρ is fluid density, v the velocity, and ℓ a characteristic length of the vessel carrying the flow. For a circular pipe ℓ is the radius.

10. An object falling through a viscous fluid reaches a final constant terminal velocity because it experiences a velocity dependent retarding force. The retarding force is proportional to the velocity v at low velocities and proportional to v^2 at high velocity.

    For small particles the terminal velocity is called the sedimentation velocity. The sedimentation velocity can be greatly increased by placing the fluid in a centrifuge.

• Qualitative Questions

11M.1 At a given depth a scuba diver finds that the pressure on his ear drums is

A. greatest when his head is vertical.
B. greatest when the side of his head is horizontal.
C. independent of how his head is oriented.
D. no greater than at the surface provided he clears his Eustachian tubes.
E. None of the above.

11M.2 The water pressure on an immersed object is

A. greatest on the top of the object.
B. greatest on the sides of the object.
C. greatest on the bottom of the object.
D. the same on all surfaces of the object.
E. greatest on the surface with the greatest area.

11M.3 Sketched here is a dam used for generating hydroelectric power. The dam is temporarily shut down, and no water is flowing through it. At which of the labelled points is the pressure greatest?

A. Point A
B. Point B
C. Point C
D. Point D
E. Point E
F. The maximum pressure, among the points labelled, occurs at more than one of the points.

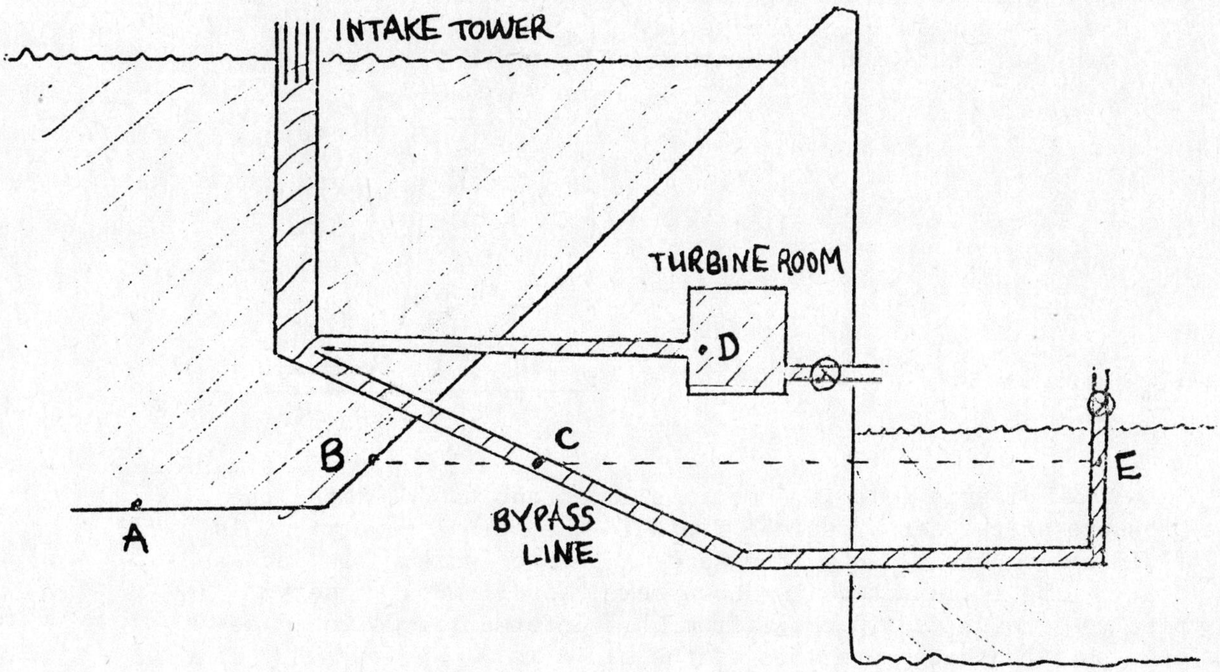

11M.4 For which of the water-filled flasks shown here is the pressure at point P greatest? Each flask is open at the top.

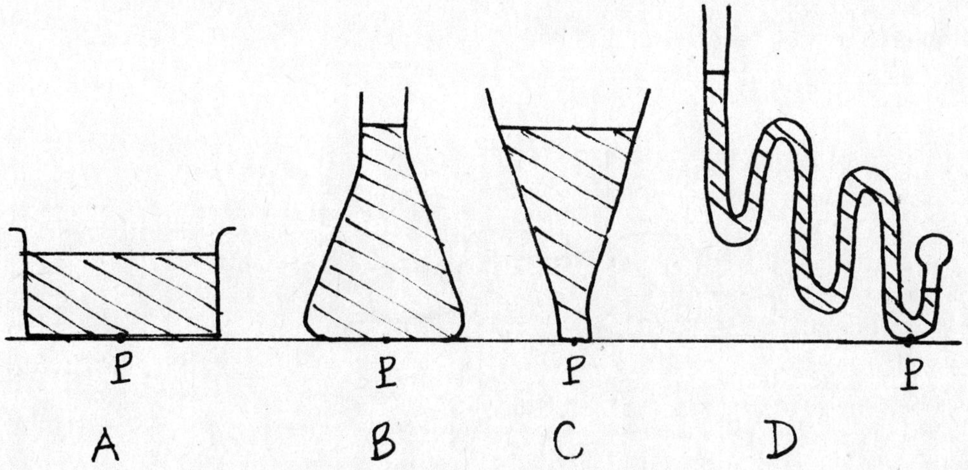

11M.5 Suppose that you decide to demonstrate Pascal's Principle to some friends at a party. You collect a hot water bottle, a length of tubing, and a board. These are to be assembled as shown here, and a volunteer is chosen to stand on the board. The idea is to lift him up into the air by pouring water into the little tube. In choosing your apparatus, which of the following "design considerations" would be most important to you?

A. Whether the volunteer would stand on one foot or on two feet.
B. Whether you will use cold water or hot water.
C. The length of the tubing used.
D. The diameter of the tubing used.
E. Both the length <u>and</u> the diameter of the tubing used.

11M.6 In medical diagnosis it is sometimes important to determine the pressure of the cerebrospinal fluid in the spinal cavity. This fluid has a density about the same as that of water. In such a test a tube is inserted, as shown here, and fluid is observed to rise to a height of 10 cms. From this information we would deduce that the gauge pressure of the fluid is

A. 7.4 mm Hg
B. 10 cm Hg
C. 136 cm Hg
D. 136 cm H_2O
E. not possible to determine without knowing the tube diameter.

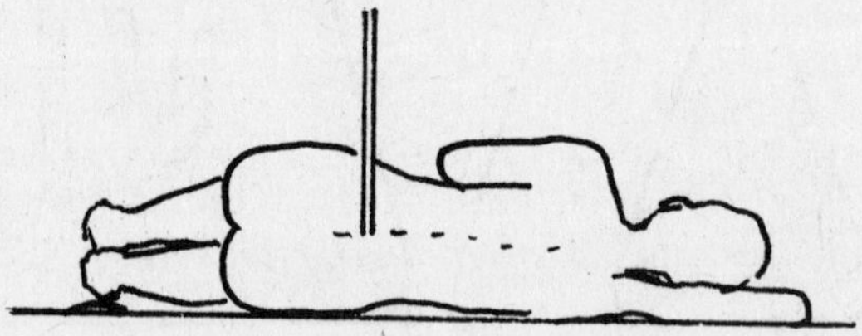

11M.7 A barge loaded with iron ore floats in a lock. If some ore is thrown overboard, the water level in the lock will

A. stay the same.
B. drop.
C. rise.
D. sometimes drop, sometimes rise, depending on the size of the boat.

11M.8 If you have ever had to wade across a rocky creek while hiking in the mountains, you have probably noticed that by the time you get to the deep water in the center of the creek the rocks don't seem to hurt your bare feet so much. What is the reason for this?

A. The greater pressure on one's feet in deep water means the rocks cannot dig in so much.
B. The velocity of the water is less in deep regions than in shallow regions.
C. The velocity of the water is greater in deep regions than in shallow regions.
D. One tends to stand on tiptoe in deep water, thereby reducing the area of the foot in contact with the rocks.
E. Deeper water is colder, and hence more dense, than shallow water.
F. One experiences a greater buoyant force in deeper water.

11M.9 Canals were vital arteries of communication in England in the 19th century. In places they crossed rivers by means of overhead viaducts not unlike those built much earlier by the Romans. I recently saw a film of one at Pont-Cysylltou which goes high over the River Dee. It was built in 1795, and I wonder that it has stood so long. When a heavily laden barge crosses the viaduct,

A. the force on the viaduct would increase by the weight of the boat.
B. the force on the viaduct would increase by a force (not zero) somewhat less than the weight of the boat.
C. the force on the viaduct would not increase.
D. the force on the viaduct would depend on whether the boat stopped on the viaduct or kept moving.

11M.10 A real-life hydraulic jack is enclosed in a housing which prevents us from seeing the cylinders and pistons. However, we observe that a force of 100 N on the "input" piston of diameter d will lift a load of 3,600 N on the output piston of diameter d_i. From this information alone we can reason that

A. $d_i \geq 6\ d_o$
B. $d_i \leq 6\ d_o$
C. $d_i \geq 36\ d_o$
D. $d_i \leq 36\ d_o$
E. $d_o \geq 6\ d_i$
F. $d_o \leq 6\ d_i$
G. $d_o \geq 36\ d_i$
H. $d_o \leq 36\ d_i$
I. None of the above can be deduced on the basis of the information given.

11M.11 Bernoulli's Principle describes the relationship between pressure and velocity in a fluid.

A. It applies only to viscous fluids.
B. It applies to gases but not to liquids.
C. It explains the circulation of air by convection.
D. It is simply a statement of the conservation of energy as applied to fluids.

11M.12 Which of the following statements best explains what happens when the roof blows off a house in a violent wind storm?

A. The wind creates reduced pressure inside the house.
B. The wind creates a low pressure region just above the peak of the roof.
C. The wind exerts a force perpendicular to the surface of the roof, thereby blowing it off.
D. The house implodes.
E. The wind exerts a viscous drag force parallel to the surface of the roof, and this "drags" the roof off.

11M.13 A non-viscous liquid is flowing in a tube which has a single constriction, as sketched here. Open tubes are connected at points X and Y and the liquid rises to the levels shown. What will happen when the stopcock is closed?

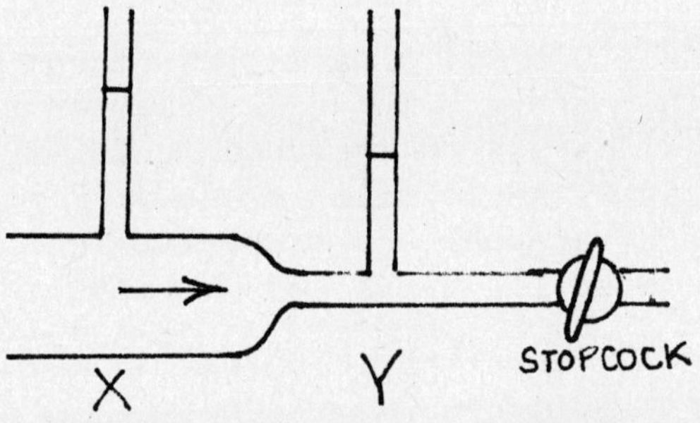

A. The levels will not change.
B. The level in each tube will rise, but the difference in the levels will not change.
C. The level in each tube will drop, but the difference in the levels will not change.
D. Both levels will rise, reaching a common level.
E. The levels will become equal at the level originally present in tube Y.
F. The levels will become equal at a level intermediate between that originally at X and at Y.

11M.14 The heart problems associated with arteriosclerosis (due to deposits of cholesterol) are illustrated by the fact that if the diameter of a vessel were reduced by a factor of 2, the flow of fluid through it could only be maintained by increasing the pressure difference across the vessel by a factor of

A. 2
B. 4
C. 8
D. 16

11M.15 My sons and I enjoy a game of diving for objects thrown into a pool. You have probably done this and have noticed that large objects sink more quickly than do small ones. This happens because the water exerts a viscous drag proportional to the speed of the sinking object. For a five gram marble on which the drag force is .025 v (where v is the velocity of the marble), what is the maximum velocity with which the marble will sink if buoyancy effects are neglected?

A. 1 m/s
B. 1.4 m/s
C. 2 m/s
D. 2.8 m/s
E. 5 m/s

• Multiple Choice: Answers and Comments

11M.1 C is correct, since pressure is the same in all directions.

11M.2 C is correct, since pressure increases with depth.

11M.3 Pressure is greatest at A, since depth below surface is greatest here.

11M.4 Pressure is greatest at D, since depth below surface is greatest here.

11M.5 A is incorrect, since the pressure from the hot water bottle is transmitted to the board of fixed area.
B is not the best answer, since cold water is only slightly more dense than hot water.
C is correct, since the pressure achieved depends primarily on the height of the tube.
D is incorrect, since the pressure doesn't depend on the diameter of the tube.

11M.6 A

$$\rho_{H_2O}\, g\, h_{H_2O} = \rho_{Hg}\, g\, h_{Hg}$$

$$\text{so } h_{Hg} = \left(\frac{\rho_{H_2O}}{\rho_{Hg}}\right)(h_{H_2O}) = \left(\frac{1}{13 \cdot 6}\right)(10 \text{ cms})$$

$$h_{Hg} = 0 \cdot 74 \text{ cms.} = 7 \cdot 4 \text{ mm.}$$

11M.7 B Suppose, for example, one had 1000 kg of ore. This would push the boat down and displace 1000 kg of water. Now take the ore out of the boat and place it on the dock. The water level will drop just as if 1000 kg of water had been removed. Now throw the ore into the water. If the density of the ore is, say, twice that of water, the volume of the 1000 kg of ore is just half that of 1000 kg of water, so throwing the ore in the water causes the level to rise as if 500 kg of water had been added. Thus taking ore out of the boat causes the level to drop as if 1000 kg of water were removed, and throwing ore in the water causes the level to rise as if 500 kg of water were thrown in, so the net effect is to lower the water level.

11M.8 F is correct, since buoyant force is equal to the weight of water displaced. When you are in deep water you displace more water, so buoyant force lifting you is greater and hence the force you exert downward on the rocks is reduced.

11M.9 C is correct. The barge displaces water which has the same weight as the barge, so the net force down on the viaduct does not change. It is simply as if one interchanged the barge (when it is 5 km away) with a piece of water (of the same weight) above the viaduct.

11M.10 E Apply Pascal's principle. $P_i = P_o$, so $\frac{f_i}{A_i} = \frac{f_o}{A_o}$

$$A_i = \frac{\pi d_i^2}{4}, \quad A_o = \frac{\pi d_o^2}{4}$$

$$\frac{f_i}{f_o} = \frac{A_i}{A_o} = \frac{d_i^2}{d_o^2} = \frac{100\text{N}}{3600\text{N}} = \frac{1}{36}$$

$$\frac{d_i}{d_o} = \frac{1}{6}, \quad d_o = 6\,d_i \text{ in ideal case}$$

If friction is present we reason that $d_o \geqslant 6d_i$

11M.11 D

11M.12 B is correct. The wind velocity is high at the peak of the roof, and from Bernoulli's principle we see that this means that the pressure must be low there. The higher pressure in the still air inside the house thus causes the house to "explode", blowing the roof off.

11M.13 D is correct. Imagine the source of fluid is a large reservoir where the fluid velocity is approximately zero. When the fluid is flowing the pressure in the large tube is slightly less than in the reservoir because of the Bernoulli effect, and is less yet in the small tube where the velocity is greatest. When flow stops the pressure everywhere will rise to the reservoir value, and thus the fluid in each standpipe will rise to the same level.

11M.14 D According to Poiseuille's equation, eq. 10.5, flow rate Q is

$$Q = \frac{\pi R^4 \Delta P}{8L\eta}$$

IF Q = CONSTANT, $R_1^4 \Delta P_1 = R_2^4 \Delta P_2$

OR $\Delta P_2 = \left(\frac{R_1}{R_2}\right)^4 \Delta P_1$

$$= \left(\frac{R_1}{\frac{1}{2}R_1}\right)^4 \Delta P_1$$

$$\Delta P_2 = 16 \Delta P_1$$

11M.15 C Neglecting buoyancy the net downward force on the marble is

$$F = mg - 0.025\ v$$

At terminal velocity v has increased so that F = 0,

$$0 = mg - 0.025\ v \quad \text{or} \quad v = \frac{mg}{0.025} = \frac{(0.005)(9.8)}{0.025}$$

$$v = 2\ \text{m/s}$$

• Problems

11.1 If the pressure at sea level is 10^5 Pa, what atmospheric pressure would a balloonist experience at an elevation of 2000 meters?

11.2 The gauge pressure in a bicycle tire is 70 lbs/in^2. What is this pressure in Pascals and in atmospheres?

11.3 A viewing port in an aquarium is circular with a radius of 50 cms. Its center if 4 meters below the surface of a large freshwater pool. The surface of the port is inclined at 45° to vertical. What is the force acting on the port?

11.4 Here is a tricky way to weigh your car. Observe the area of the portion of the tire in contact with the ground. You can do this by sliding a piece of newspaper in against the tire front and back, and by measuring the width of the tire track left after driving through a puddle of water on your driveway. With a tire gauge measure the pressure in the tire. Now it is easy to calculate the weight of the car. For example, suppose that the pressure in each tire is 30 lbs/in^2 and that each tire presses down on an area 10 cms x 14.2 cms. What is the mass of the car?

11.5 A polar bear stands on an ice floe 10 cms thick. What is the minimum area of the floe if it is not to sink below the surface of the sea? The specific density of ice is 0.92 and that of sea water is 1.03. The bear's mass is 250 kg.

11.6 A block of wood floats in water with 80% of its volume submerged. If the volume of the block is 3000 cm^3, what is its mass?

11.7 A typical average blood flow through an aorta of cross-section 0.8 cm^2 is 3.5 liters per minute. It eventually spreads out to flow through a system of about 5 million fine capillary vessels each of area about 5×10^{-7} cm^2 (just big enough to allow passage of a red blood cell).

(a) What is the velocity of the blood in the aorta?

(b) What is the velocity of the blood in a capillary?

11.8 Air flows past the upper surface of an airplane wing at 300 m/s and past the lower surface at 250 m/s. The surface area of the wing is 16 m^2 and the plane is flying at an elevation where the density of air is 1.0 kg/m^3. What is the upward lift force on the wing?

• Problem Solutions

11.1 $P_A = P_A(0)\exp\left(-\frac{h}{h_0}\right)$ eqn (10.1)

$P_A = (10^5\ \text{Pa})\exp\left(-\frac{2\ \text{km}}{8.6\ \text{km}}\right) = \underline{\underline{0.79 \times 10^5\ \text{Pa}}}$

11.2 $P = 70\ \text{lbs/in}^2 = 70\frac{(4.45\ \text{N})}{\left(\frac{1}{39.4}\text{m}\right)^2} = \underline{\underline{4.8 \times 10^5\ \text{Pa}}} = \frac{4.8\times10^5}{1.01\times10^5}\text{Atm} = 4.78\ \text{Atm}$

11.3 THE AVERAGE PRESSURE ON THE PORT IS JUST THAT AT ITS CENTRE. THE FACT THAT THE PORT IS INCLINED AT 45° HAS NO EFFECT.

$F = PA$, $P = \rho gh$, SO $F = \rho ghA$

$= (1000\ \text{kg/m}^3)(9.8\ \text{m/s}^2)(4\ \text{m})(\pi)(.5\ \text{m})^2$

$F = \underline{\underline{3.1 \times 10^4\ \text{N}}}$

11.4 $F = PA_{TOT} = P(4A_1)$ A_1 = AREA UNDER ONE TIRE

$1\ \text{lb/in}^2 = 6900\ \text{Pa}$, $30\ \text{lbs/in}^2 = (30)(6900\ \text{Pa}) = 2.1 \times 10^5\ \text{Pa}$

$F = 4PA_1 = (4)(2.1\times10^5\ \text{Pa})(.1\ \text{m})(0.142\ \text{m})$

$F = 11750\ \text{N}$

$F = mg$, $m = \frac{F}{g} = \frac{1175\text{N}}{9.8\ \text{m/s}^2} = \underline{\underline{1200\ \text{kg}}}$

11.5 MASS OF BEAR PLUS MASS OF ICE MUST EQUAL MASS OF WATER DISPLACED WHEN ICE IS JUST BARELY COMPLETELY SUBMERGED.

IF V = VOLUME OF ICE, $m_{ICE} = \rho_{ICE}V$, $m_{H_2O} = \rho_{H_2O}V$

IF m = MASS OF BEAR, $m + m_{ICE} = m_{H_2O}$, $m + \rho_{ICE}V = \rho_{H_2O}V$

$V = tA = 0.1A$

$250\text{kg} + (0.92)(1000\text{kg/m}^3)(0.1\text{m})(A) = (1.03)(1000\text{kg/m}^3)(.1\text{m})A$

$A = \frac{250}{103 - 92} = \underline{\underline{22.7\ \text{m}^2}}$

11.6 VOLUME SUBMERGED IS $V_S = 0.8V = (.8)(3000\ \text{cm}^3)$

$V_S = 2400\ \text{cm}^3$

MASS OF DISPLACED WATER IS THUS $m_{H_2O} = \rho_{H_2O}V_S = \left(1\frac{\text{gm}}{\text{cm}^3}\right)(2400\text{cm}^3)$

$m_{H_2O} = 2400\text{gm} = 2.4\text{kg}$

FOR A FLOATING BLOCK $m_{BLOCK} = m_{H_2O} = \underline{\underline{2.4\ \text{kg}}}$

11.7 (a) $Q_A = A_AV_A$ $V_A = \frac{Q_A}{A_A} = \left(\frac{3.5\times10^3\ \text{cm}^3}{60\text{s}}\right)\left(\frac{1}{0.8\text{cm}^2}\right)$

$V_A = 73\ \text{cms/s}$

(b) $Q_c = A_cV_c$, $V_c = \frac{Q_c}{A_c} = \left(\frac{3.5\times10^3\ \text{cm}^3}{60\text{s}}\right)\left(\frac{1}{5\times10^6\times5\times10^{-7}\text{cm}^2}\right)$

$V_c = 23\ \text{cm/s}$

THUS VELOCITY IN CAPILLARIES IS ABOUT $\frac{1}{3}$ THAT IN AORTA.

11.8 BY BERNOULLI'S PRINCIPLE, $\frac{1}{2}\rho v_1^2 + P_1 = \frac{1}{2}\rho v_2^2 + P_2$

THUS PRESSURE DIFFERENCE BETWEEN TOP AND BOTTOM OF WING IS $\Delta P = P_2 - P_1 = \frac{1}{2}\rho v_1^2 - \frac{1}{2}\rho v_2^2 = \frac{1}{2}\rho(v_1^2 - v_2^2)$

$$= \left(\frac{1}{2}\right)(1\ \text{kg/m}^3)\left[(300\ \text{m/s})^2 - (250\ \text{m/s})^2\right]$$

$$\Delta P = 1.38 \times 10^4\ \text{N/m}^2$$

LIFT FORCE IS $F = (\Delta P)(A) = (1.38 \times 10^4\ \text{N/m}^2)(16\ \text{m}^2)$

$$F = \underline{\underline{2.2 \times 10^5\ \text{N}}}$$

Thermal Properties, Calorimetry, and the Mechanical Equivalent of Heat

12

• Summary of Important Ideas, Principles, and Equations

1. Temperature is a measure of how hot something is and of how rapidly atoms are moving. Temperatures measured on the Celsius, Kelvin and Fahrenheit scales are related by

$$T_F = 32 + \frac{9}{5}T_C \qquad T_C = \frac{5}{9}(T_F - 32) \qquad T_K = T_C + 273.2 \tag{12.1}$$

2. Thermal expansion occurs when most materials are heated. If a rod of length L_0 at temperature T_0 is heated to a temperature T, its length will change by an amount ΔL,

$$\Delta L = \alpha L_0 \Delta T \tag{12.2}$$

where $\Delta T = T - T_0$

α = linear thermal expansion coefficient $(^oC^{-1})$

Similarly, surface area and volume change on heating, such that

$$\Delta A \simeq 2\alpha A_0 \Delta T \text{ and } \Delta V \simeq 3\alpha V_0 \Delta T \text{ if } \alpha << 1 \tag{12.3}$$

3. The energy associated with the random motion of the atoms in a substance is called thermal energy. Energy which is transferred from one object to another by virtue of their difference in temperature is called heat. Thus work, energy and heat are all measured in the same units, the joule or kilocalorie.

When an amount of heat ΔQ is added to m grams of material the temperature will rise by an amount ΔT, where

$$\Delta Q = mc\Delta T \tag{12.4}$$

C is the specific heat of the material (kcal/^{o}C·kg).

1 kilocalorie is the heat needed to raise the temperature of 1 kilogram of water by 1 ^{o}C, and 1 kcal = 1000 calories.

4. A system of several objects is in thermal equilibrium when all parts are at the same temperature. For an isolated system, the heat lost by the hotter parts is equal to the heat gained by the colder parts, i.e. thermal energy is conserved.

5. Matter can exist in solid, liquid or gaseous phases. A definite amount of energy, called the latent heat, must be added (as in melting or boiling) or removed (as in freezing or condensation) in order for 1 kg of matter to undergo a change of phase. These phase changes occur at definite temperatures.

6. Heat transport can occur via convection, conduction or radiation. In convection heat is transported by the physical transport of hot material, as in a hot air furnace.

The thermal energy transported per second by conduction is

$$\boxed{\frac{\Delta Q}{\Delta t} = -KA\frac{\Delta T}{\Delta x}} \qquad (12.5)$$

ΔT = temperature difference present over distance Δx.

A = cross-sectional area through which heat is flowing.

K = thermal conductivity of material $\left(\frac{\text{kcal}}{\text{m}\cdot\text{s}\cdot{}^{o}\text{C}}\right)$

Radiative heat transfer occurs because every object is giving off electromagnetic waves (such as microwaves, infrared, visible light, etc.). The amount of heat radiated per second from a surface of area A whose absolute temperature is T_K is

$$\boxed{P_r = \sigma\varepsilon A T_K^4} \qquad (12.6)$$

Here $\sigma = 5.67 \times 10^{-8}$ W/m^{2}· ^{o}K^{4} and the emissivity ε is a factor between 0 and 1 which depends on the nature of the surface.

A good radiator of radiation is also a good absorber. Black surfaces absorb all visible light striking them and thus also are good visible radiators. A substance with ε = 1 is called a blackbody.

• Qualitative Questions

12M.1 "Heat" is the same as

A. temperature.
B. kinetic energy.
C. thermal energy in transit.
D. potential energy
E. radiation.

12M.2 "Heat" is measured in the same units as

A. force.
B. temperature
C. work.
D. entropy.
E. power.

12M.3 Heat is

A. a measure of how hot an object is.
B. qualitatively the same as temperature.
C. the energy which flows from a hot body to a cold one when they are placed in contact.
D. More than one of the above is true.
E. None of the above is true.

12M.4 Which of the following is an accurate statement?

A. In principle there is no lower limit to the temperature one could reach, provided a sufficiently good refrigerator could be built.
B. If the average kinetic energy of the molecules in a solid increases, this means the temperature increases.
C. If heat flows into an object, its temperature will increase.
D. "Absolute zero" refers to the ice point on the Celsius temperature scale.
E. None of the above is true.

12M.5 A temperature of 72 ^{o}F is equal to

A. 8 ^{o}C
B. 22.2 ^{o}C
C. 40 ^{o}C
D. 56.6 ^{o}C

12M.6 A temperature of -20 ^{o}C is the same as

A. -20 ^{o}F
B. -4 ^{o}F
C. 20.9 ^{o}F
D. 36 ^{o}F
E. 68 ^{o}F

12M.7 If heat is added to a substance

A. it will always expand.
B. its temperature will rise.
C. it will eventually melt.
D. it will change phase.
E. None of the above is true.

12M.8 Consider a metal plate in which are drilled some bolt holes. If the plate is heated, the diameter of the holes will

A. increase.
B. decrease.
C. remain the same.
D. either increase or be decreased, depending upon whether the plate was initially at a temperature above or below room temperature.

12M.9 If you wanted to know how high the mercury in a thermometer would rise for a given temperature increase, which of the following would be most useful to know (in addition to the dimensions of the thermometer)?

A. Coefficient of linear expansion of mercury.
B. Specific heat of mercury.
C. Thermal conductivity of mercury.
D. Melting point of mercury.
E. Density of mercury.

12M.10 When 100 gms of copper at 100 ^{o}C is added to 100 gms of water at 20 ^{o}C, the resulting temperature of the mixture will be

A. between 20 ^{o}C and 60 ^{o}C.
B. 60 ^{o}C.
C. between 60 ^{o}C and 100 ^{o}C.
D. dependent on the heat of vaporization of water.
E. None of the above is true.

12M.11 Consider two objects made of different materials and different masses. The amount of heat required to raise the temperature of each by 10 ^{o}C will be

A. greater for the larger object.
B. the same for both objects.
C. greater for the smaller object.
D. possibly greater for either the larger or smaller object, depending on what they are made of.
E. None of the above is true.

12M.12 Which of the following best accounts for the fact that water acts as a thermostat in controlling the earth's temperature?

A. Water is a poor thermal conductor.
B. Water is a good thermal conductor.
C. Water has an extremely low freezing point.
D. Water has an extremely high boiling point.
E. Water has a very high specific heat.
F. Water has a very low specific heat.

12M.13 If you wanted to know how much the temperature of a particular piece of material would rise when a known amount of heat was added to it, which of the following would be most helpful to know?

A. thermal conductivity
B. density
C. initial temperature
D. coefficient of linear expansion
E. specific heat

12M.14 A sample of a pure compound is contained in a closed, well-insulated container. Heat is added at a constant rate and the sample temperature is recorded. The resulting data is sketched below. Which of the following conclusions is justified from the data given?

A. The sample was initially liquid.
B. After 5 minutes the sample was a mixture of solid and liquid.
C. The sample never boiled.
D. The heat capacity of the solid phase was greater than that of the liquid phase.
E. The heat of fusion is greater than the heat of vaporization.
F. After 20 minutes the solid was all liquid.

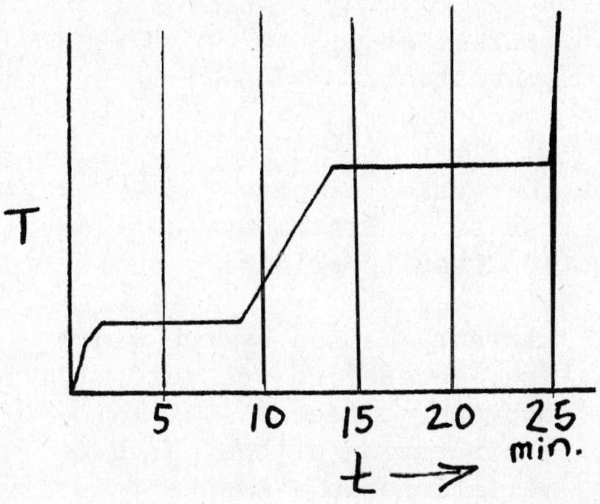

12M.15 You may have noticed when out skiing that sometimes the temperature drops down and down, perhaps well below 0 °C, and then it starts to snow. When the snow starts to fall it often seems to me that it gets warmer.

A. This is true, mainly because the snow reduces the effect of the wind.
B. This is a purely psychological effect; the formation of snowflakes have no influence on the air temperature.
C. This is a real effect. Heat is given off when snowflakes form, and this warms the air.
D. The original assertion is in error. When it starts to snow, the air always becomes colder, not warmer.
E. This is true because the snow absorbs some of the cold from the air. Were is not for the snow, the cold would have flowed into your body.

12M.16 Which of the following is not considered to be an important mechanism of heat transfer?

A. radiation.
B. convection
C. conduction
D. diffusion

12M.17 If the absolute temperature of the filament of a light bulb were doubled, the energy radiated per second by the filament would

A. remain the same.
B. double.
C. increase by a factor of four.
D. increase by a factor of eight.
E. increase by a factor of sixteen.

12M.18 A forced air heating system in a house transfers heat throughout the building primarily by

A. radiation.
B. convection.
C. conduction.
D. thermal reactance.
E. None of the above is true.

12M.19 The convection of water in lakes is vital to marine life, since it results in distribution of nutrients and oxygen. This convection results primarily because of

A. currents caused by the wind.
B. the increase in pressure with increasing depth.
C. buoyancy effects.
D. the rotation of the earth.
E. variations in atmospheric pressure.

• Multiple Choice: Answers and Comments

12M.1 C

12M.2 C

12M.3 C

12M.4 B A is false, since it is not possible to reach a temperature lower than absolute zero.
C is false, since the heat could be causing a change of phase while the temperature remains constant.

12M.5 B $T_C = \frac{5}{9}(T_F - 32) = \frac{5}{9}(72 - 32) = 22.2\ °C$

12M.6 B $T_F = \frac{9}{5}T_C + 32 = \frac{9}{5}(-20) + 32 = -4°C$

12M.7 E A is false. Some substances, such as water near 4 °C, contract when heated.
B is false, since a change of phase may be taking place.
C is false, since the substance may already be a liquid or gas.
D is false, since it is not necessary that a phase change occur.

12M.8 A Imagine that instead of a hole in the plate one merely drew a circle on a solid plate of metal. The metal inside the circle is like a metal coin. When the plate is heated, this "coin" inside the circle will expand along with the rest of the plate. After heating the plate imagine the inside of the circle removed. You can now see that the hole has become larger than it was.

12M.9 A

12M.10 A Water has a larger specific heat than does copper. The copper gives up a certain amount of heat, say Q, and the water absorbs this same amount of heat. If the specific heats were equal the final temperature would be 60 °C, halfway between 20 °C and 100 °C. However, since water has a larger specific heat, it does not rise in temperature as much as does the copper.

$$Q = m_W c_W \Delta T_W = m_c c_c \Delta T_c \,, \quad m_W = m_c$$

$$\text{so } \Delta T_W = \left(\frac{c_c}{c_W}\right) \Delta T_c \quad \text{where } \frac{c_c}{c_W} < 1$$

12M.11 D If the smaller object has a very large heat capacity (like water) and the larger object has a very small heat capacity, it may take more heat to raise the temperature of the small object.

12M.12 E

12M.13 E

12M.14 B The first plateau must indicate melting and the second boiling. Thus the sample was initially solid. The second plateau is longer than the first, so more heat was needed to boil than to melt and heat of vaporization is greater than heat of fusion. Cont'd. . . .

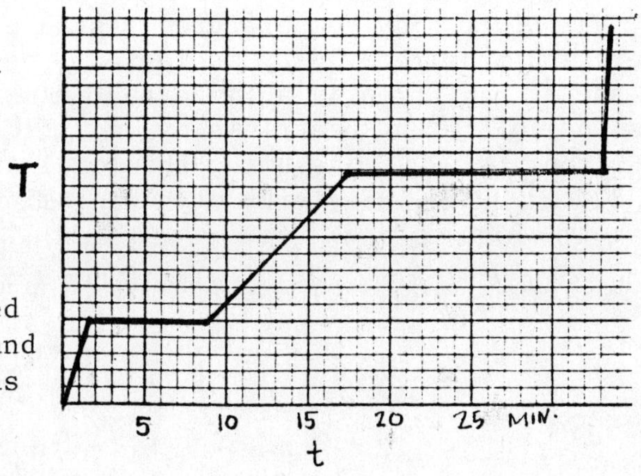

12M.14 Continued . . . After 20 minutes sample was a mixture of liquid and gas.

12M.15 C

12M.16 D

12M.17 E The power radiated from a hot body is proportional to T^4, and $2^4 = 16$.

12M.18 B

12M.19 C

• Problems

12.1 A steel surveyor's tape has been calibrated so that it is accurate when used at 20 °C. On a day when the temperature is 35 °C it is used to measure a building lot. If the depth of the lot is found to be 202.05 meters, what is the true dimension of the lot?

12.2 In a family like ours with seven sons it was not unusual to run out of hot water just when someone was taking a bath, so sometimes I would heat some water on the stove to add to my tepid bath water. What temperature can I reach by adding 4 liters of boiling water at 100 °C to 40 liters of water at 32 °C?

12.3 How much will the temperature of a 200 g cup of coffee at 90 °C drop when you add 10 grams of ice cubes to it?

12.4 The thermal conductivity of brick is approximately 0.04 W/m·°C. At what rate is heat lost through a brick wall in a house on a day when the outside temperature is 0 °C and the inside temperature is 20 °C? The wall is 10 cm thick and measures 2.5 m x 5 m in area.

12.5 When resting an average adult uses energy at a rate of about 120 W. Since body temperature is stable at about 28 °C, this energy must be removed by radiation, convection or conduction. To see if radiation is an important mechanism for heat dissipation calculate the heat loss by radiation for an unclothed person in a room at 22 °C, assuming the total surface area of the body to be 2.0 m^2. The emissivity of human skin for infrared radiation is very close to one.

• Problem Solutions

12.1 CONSIDER A PIECE OF THE TAPE WHICH IS 202·05 m LONG AT 20°C. WHEN HEATED 15°C (TO 35°C) THIS PIECE OF METAL HAS A LENGTH EQUAL TO THAT OF THE BUILDING LOT, L.

THUS $L = L_{20}(1 + \alpha \Delta T) = (202{\cdot}05\,m)\left[1 + (12 \times 10^{-6}\,{}^{\circ}C^{-1})(15\,{}^{\circ}C)\right]$

$L = \underline{\underline{202{\cdot}09\,m}}$

12.2 ONE LITER OF WATER HAS 1 kg. MASS.

HEAT GAINED BY TEPID WATER = HEAT LOST BY HOT WATER

$m_1 c(T - 32) = m_2 c(100 - T)$, T = FINAL TEMPERATURE

$40(T - 32) = 4(100 - T)$

$44T = 1680$, $T = \underline{\underline{38{\cdot}2^{\circ}}}$

12.3 FIRST CALCULATE HOW MUCH HEAT IS ABSORBED WHEN ICE MELTS.

$Q_1 = m_1 L = (0{\cdot}01\,kg)(80\,kcal/kg) = 0{\cdot}8\,kcal$

HEAT LOST BY COFFEE = HEAT GAINED BY ICE PLUS ICE WATER

$m_2 c(90 - T) = m_1 c(T - 0) + 0{\cdot}8\,kcal.$

$(0{\cdot}2\,kg)(1\,kcal/kg\,{}^{\circ}C)(90 - T) = (0{\cdot}01\,kg)(1\,kcal/kg\,{}^{\circ}C)(T) + {\cdot}8\,kcal$

$18 - {\cdot}2T = {\cdot}01T + {\cdot}8$

$T = 81{\cdot}9\,{}^{\circ}C$. SO $\Delta T = 90 - 81{\cdot}9 = \underline{\underline{8{\cdot}1\,{}^{\circ}C}}$

12.4 FROM EQN. (11·5) $\frac{\Delta Q}{\Delta t} = -kA\frac{\Delta T}{\Delta x} = -(0{\cdot}04\,\frac{W}{m\,{}^{\circ}C})(2{\cdot}5\,m)(5\,m)\frac{(0\,{}^{\circ}C - 20\,{}^{\circ}C)}{(0{\cdot}1\,m)}$

$\frac{\Delta Q}{\Delta t} = \underline{\underline{100\,W}}$

12.5 FROM EQN. (11·6), $P_R = \sigma \epsilon A T_B^4$ = POWER RADIATED FROM BODY AT TEMPERATURE T_B

$P_A = \sigma \epsilon A T_R^4$ = POWER ABSORBED IF SURROUNDING ROOM IS AT TEMPERATURE T_R

THUS NET RATE OF ENERGY LOSS IS $P = P_R - P_A = \sigma \epsilon A (T_B^4 - T_R^4)$

BE SURE TO EXPRESS TEMPERATURES IN °K.

$P = (5{\cdot}67 \times 10^{-8}\,W/m^2\,{}^{\circ}K^4)(1)(2\,m^2)\left[(301\,{}^{\circ}K)^4 - (295\,{}^{\circ}K)^4\right]$

$P = (931 - 859)\,W = \underline{\underline{72\,W}}$

THUS WE SEE THAT RADIATED ENERGY ACCOUNTS FOR SLIGHTLY MORE THAN HALF OF THE 120 W LOSS FROM THE BODY, AND THUS RADIATION IS AN IMPORTANT MECHANISM HERE.

---

13 The Ideal Gas Law and Kinetic Theory

• Summary of Important Ideas, Principles, and Equations

1. One mole of a substance is 6.02×10^{23} molecules (Avogadro's number, N_A) and has a mass in grams equal to the molecular weight of the substance.

2. An ideal gas consists of non-interacting point particles and is described by the equation of state

$$\boxed{PV = nRT} \qquad (13.1)$$

Here R = gas constant = 2 cal/mol·°K = 8.31 J/mol·°K
n = number of moles of gas
T = absolute temperature
P = pressure (in Pascals) and V = volume (in m^3)

R may be written $R = N_A k$, where $k = 1.38 \times 10^{-123}$ J/°K = Boltzmann's constant.

3. Pressure in a gas is due to motion of the molecules and may be expressed in terms of $<\varepsilon_t>$, the average translational KE of a single molecule.

$$\boxed{PV = \frac{2}{3} N_A n <\varepsilon_t>} \qquad (13.2)$$

The average kinetic energy per molecule per degree of freedom is $\frac{1}{2}kT$.

Thus we see that absolute temperature is a measure of the kinetic energy of the molecules, and this gives us good intuitive insight into what is meant by "temperature".

A monatomic molecule (consisting of a single atom) has 3 degrees of freedom. More complex molecules have more degrees of freedom. For a molecule with ν degrees of freedom the average KE per molecule is

$$\langle \varepsilon \rangle = \tfrac{1}{2}\nu kT \qquad (13.3)$$

4. The heat capacity per mole of an ideal gas is different when pressure or volume is held constant, with the relation

$$c_p - c_v = R \qquad (13.4)$$

c_v is $\frac{3}{2}R$ for a monatomic gas and $\frac{5}{2}R$ for a diatomic gas.

5. The pressure exerted by a mixture of gases is the sum of the partial pressures exerted by each component gas, where we treat each component gas as an independent ideal gas.

• Qualitative Questions

13M.1 The atomic weight of oxygen is 16. Thus in ½ mole of O_2 gas has a mass of

A. 4 gms
B. 8 gms
C. 16 gms
D. 32 gms
E. 64 gms

13M.2 When the temperature of a steel tank of gas is doubled, the pressure will

A. be reduced by a factor of ½.
B. be unchanged.
C. double.
D. increase, but we cannot determine by how much without knowing the volume of the tank.

13M.3 A container holds a mixture of two gases at pressure P. The partial pressures of each gas will be

A. equal to ½P in all cases.
B. equal to P in all cases.
C. equal to ½P if equal masses of the two gases are present.
D. equal to ½P if the same number of moles of each gas is present.
E. equal to P if the same number of moles of each gas is present.

13M.4 We associate the absolute temperature of an ideal gas most closely with

A. the pressure in the gas.
B. the speed of the molecules.
C. the average kinetic energy of the molecules.
D. the number of moles of gas present.

13M.5 Which of the following is a true statement?

A. When a gas is heated it will expand.
B. One mole of every ideal gas at a pressure of 2 atmospheres with a temperature of 500 ^{o}C occupies the same volume.
C. If a gas is heated from 100 ^{o}C to 200 ^{o}C its pressure will double.
D. At a given temperature all molecules in a gas move at the same speed.

• True–False Questions

13TF.1 In using the ideal gas law it is necessary that temperature be expressed in degrees Celsius.

13TF.2 One mole of gas contains 6 x 10^{23} molecules <u>only</u> if the gas is a monatomic gas.

13TF.3 When making calculations with the ideal gas law pressure must be expressed in atmospheres.

13TF.4 If in a mixture of two gases the partial pressures of the gases differ, their temperatures must also differ.

13TF.5 Although air contains a larger fraction of nitrogen than oxygen, the average speeds of nitrogen and oxygen molecules in a particular sample of air are equal.

13TF.6 The ideal gas law assumes that the gas will never condense to a liquid, no matter how much the temperature is lowered.

13TF.7 When heat is added to a gas its temperature will rise by an amount independent of whether or not the pressure or volume is held constant.

13TF.8 A real gas behaves more like an ideal gas at high temperatures than at low temperatures.

• Multiple Choice: Answers and Comments

13M.1 C The molecular weight of O_2 is 32, so ½ mole of O_2 has a mass of 16 grams.

13M.2 C In a steel tank volume is constant, so

$$PV = nRT$$

$$\left(\frac{P}{T}\right) = \frac{nR}{V} = \text{constant}$$

13M.3 D

13M.4 C

13M.5 B A is false, since a gas in a container of fixed volume cannot expand.
C is false. Pressure will double if absolute temperature (not ^{o}C) is doubled.
D is false. The molecules in a gas always have a distribution of speeds.

• True–False: Answers and Comments

13TF.1 F Temperature must be expressed in degrees absolute. Do not forget this. Failure to do so is a common error.

13TF.2 F Whether the gas is monatomic or not makes no difference.

13TF.3 F The pressure must be expressed in the SI unit, the Pascal.

13TF.4 F All of the components of a mixture of gases are at the same temperature.

13TF.5 F At a given temperature the average KE of the N_2 is equal to the average KE of the O_2. Since KE = $\frac{1}{2}mv^2$, and the mass of an N_2 molecule is less than that of an O_2 molecule, the nitrogen molecules have higher average speed.

13TF.6 T In order for a gas to condense there must be an attractive force between the molecules, and the ideal gas model assumes no interactions.

13TF.7 F The temperature rise will be greater if volume is held constant than if pressure is held constant, since $c_v < c_p$.

13TF.8 T At high temperatures the KE of the molecules is large compared to any energy of interaction, so the assumption of non-interacting molecules is more nearly fulfilled and the ideal gas law will hold better.

• Problems

13.1 How many hydrogen atoms and how many oxygen atoms are in an ice cube of mass 8 gms?

13.2 A sealed gas laser tube is heated from 20 ^{o}C to 60 ^{o}C. By what factor will the pressure of the gas change?

13.3 Outer space is not a perfect vacuum. Instead, it is a very rarefied gas of about one hydrogen atom per cm^3 at a temperature of about 3 ^{o}K. What is the pressure, in atmospheres, of this gas?

13.4 Helium gas is used in heliarc welding to reduce oxidation of the metal surfaces. A typical tank of gas has a volume of 0.04 m^3 and contains gas at a pressure of 4000 psi at 20 ^{o}C. What mass of helium is contained in the tank?

13.5 In an experiment on plant respiration you wish to use a mixture of inert argon gas and carbon dioxide (CO_2) such that the partial pressure of the CO_2 is 0.1 atm and the partial pressure of the argon is 0.9 atm. What is the mass composition (in %) of Ar, C and O in such a gas?

• Problem Solutions

13.1 THE MOLECULAR WEIGHT OF H_2O IS $(2\times 1+16)=18$. THUS 8 GMS OF H_2O IS $\frac{8}{18}=0.44$ MOLE.

ONE MOLE IS 6.02×10^{23} MOLECULES, SO 0.44 MOLE IS $(0.44)(6.02\times 10^{23})=2.68\times 10^{23}$ MOLECULES. IN EACH H_2O THERE IS ONE OXYGEN ATOM AND TWO HYDROGEN ATOMS, SO IN 8 GMS OF ICE THERE ARE $\underline{2.68\times 10^{23}}$ OXYGENS AND $\underline{5.36\times 10^{23}}$ HYDROGENS.

13.2 $P_1V_1=nRT_1$, $P_2V_2=nRT_2$ V IS CONSTANT

$\frac{P_1V_1}{P_2V_2}=\frac{nRT_1}{nRT_2}$, SO $\frac{P_1}{P_2}=\frac{T_1}{T_2}$ OR $P_2=\left(\frac{T_2}{T_1}\right)P_1$

T MUST BE EXPRESSED IN °K, SO $P_2=\left(\frac{273+60}{273+20}\right)P_1=\underline{1.14P_1}$

13.3 $PV=nRT$

$n=\frac{1}{6.02\times 10^{23}}$ MOLE

$T=3°K$

$V=1\ cm^3=1\times(10^{-2}m)^3=10^{-6}m^3$

$P=\frac{nRT}{V}=\left(\frac{1}{6.02\times 10^{23}}\text{MOL}\right)\left(8.31\ J/\text{MOL}\ °K\right)\left(3°K\right)\left(\frac{1}{10^{-6}m^3}\right)$

$=4.1\times 10^{-17}\ J/m^3=4.1\times 10^{-17}\ N/m^2$ $(1J\equiv 1N\cdot m)$

1 ATM $=1.01\times 10^5\ N/m^2$. SO $P=\frac{4.1\times 10^{-17}}{1.01\times 10^5}$ ATM $=\underline{4.1\times 10^{-22}\ \text{ATM}}$.

THUS OUTER SPACE IS A PRETTY GOOD VACUUM EVEN THOUGH THERE ARE A FEW HYDROGEN ATOMS FLOATING AROUND THERE.

13.4 $PV=nRT$, $n=\frac{PV}{RT}$; 1 ATM. $=14.7$ psi $=1\times 10^5\ N/m^2$

$$n = \frac{PV}{RT} = \frac{[(4000/14.7)\times 10^5\ N/m^2](0.04\ m^3)}{(8.31\ J/MOL\ ^\circ K)(293\ ^\circ K)} = 447\ \text{MOLES}$$

THE MOLECULAR WEIGHT OF HELIUM IS 4 . SO 447 MOLES HAS A MASS $m = (4)(447)\,g = 1788\,g = \underline{\underline{1.8\,kg}}$

13.5 FOR EACH MOLE OF CO_2 THERE MUST BE PRESENT 9 MOLES OF ARGON. ATOMIC WT. OF CARBON IS 12, OF OXYGEN 16, AND OF ARGON 40. THUS IN 1 MOLE OF CO_2 THERE ARE 32g OF OXYGEN AND 12g OF CARBON.

IN 9 MOLES OF ARGON ARE 360g OF ARGON. THUS FRACTIONAL MASS DISTRIBUTION IS

$$f_L = \frac{12}{12+32+360} = 0.03 = \underline{\underline{3\%}}$$

$$f_O = \frac{32}{12+32+360} = 0.08 = \underline{\underline{8\%}}$$

$$f_A = \frac{360}{12+32+360} = 0.89 = \underline{\underline{89\%}}$$

---

14 Thermodynamics

• Summary of Important Ideas, Principles, and Equations

1. Two systems are in thermal equilibrium when they are at the same temperature. In this case no heat will flow between the systems.

   The Zeroth Law of Thermodynamics states that if A is in equilibrium with C and B is also in equilibrium with C, then A is also in equilibrium with B.

   In an isothermal process temperature doesn't change.

   In an isochoric process volume doesn't change.

   In an isobaric process pressure doesn't change.

   In an adiabatic process there is no flow of heat.

2. The First Law of Thermodynamics is a statement of the principle of conservation of energy, written as follows:

$$\boxed{\Delta Q = \Delta U + \Delta W} \qquad (14.1)$$

   where ΔQ = heat supplied to the system
   ΔU = change in internal energy of the system
   ΔW = work done by the system

3. A heat engine works in a cycle, extracting heat from a hot reservoir at temperature T_h, doing work, and rejecting remaining heat into a cold reservoir at temperature T_c. The maximum possible efficiency of such an engine (called the Carnot efficiency) is

$$\boxed{\eta = 1 - \frac{Q_{exh}}{Q_{in}} = 1 - \frac{T_c}{T_h}} \qquad (14.2)$$

where Q_{in} = heat put into the engine

Q_{exh} = heat exhausted into the cold reservoir

4. The Second Law of Thermodynamics may be stated in any of the following ways:

   a. Heat does not, of itself, flow from a cooler to a hotter body.

   b. One cannot convert heat completely into work if no other changes take place in the system or the environment.

   c. In any process in an isolated system the entropy will either stay the same (reversible process) or increase. The entropy of an isolated system and of the entire universe tends to increase.

5. The entropy of a system is a state function of the system. That is, a system has a definite entropy, S, just as it has a definite internal energy, U.

   The entropy of a system increases when heat flows into the system and decreases when heat flows out. The change, ΔS, in entropy is related to the heat which flows in, ΔQ, by

$$\boxed{\Delta S = \frac{\Delta Q}{T}} \qquad (14.3)$$

   It can be shown that entropy is a measure of disorder in a system. Increasing disorder corresponds to increasing entropy.

• Qualitative Questions

14M.1 Consider the following relation: $\Delta W = F\Delta X = (\frac{F}{A})(A\Delta X)$

This leads us to recognize that

A. ΔQ is the heat that flows into a system in an isobaric process.
B. $P\Delta V$ is the work done by a gas when it expands a small amount ΔV.
C. work and heat are measured in the same units.
D. $P\Delta V$ represents the heat flow into a system at constant pressure.

14M.2 The first law of thermodynamics tells us that

A. the decrease in internal energy of a system is equal to the work done by the system in an adiabatic process.
B. the decrease in internal energy of a system is equal to the work done by the system in all cases.
C. the decrease in internal energy of a system is equal to the work done by the system in an isochoric process.
D. the decrease in internal energy of a system is equal to the heat flow out of the system in all cases.

14M.3 A gas initially at P_1, V_1 is caused to change its volume and pressure reversibly such that it moves along the path sketched here. In one cycle the net work done by the gas is thus

A. $P_2(V_2 - V_1)$

B. $V_2(P_2 - P_1)$

C. P_1V_1

D. $(P_2 - P_1)(V_2 - V_1)$

E. $P_2V_2 - P_1V_1$

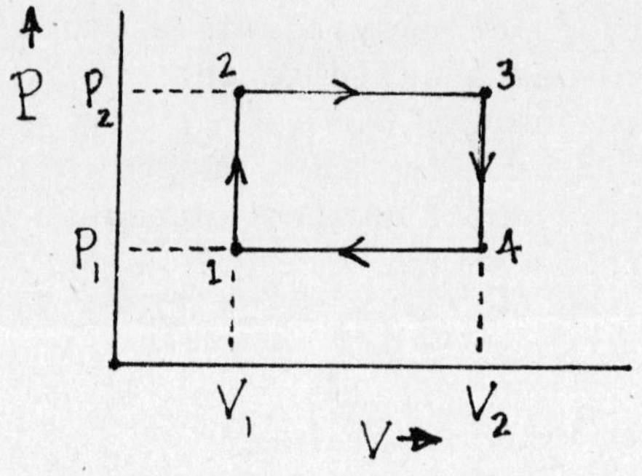

14M.4 During an adiabatic expansion of a gas

A. the temperature remains constant.
B. the pressure remains constant.
C. no work is done by the gas.
D. no heat flows into or out of the gas.

14M.5 Suppose that an ideal gas is compressed to half its original volume. Would the work done be greater for an isothermal process, for an adiabatic process, or would the work done be independent of the kind of process?

A. The work done would be the same for both isothermal and adiabatic processes.
B. The work done would be greater for an isothermal process.
C. The work done would be greater for an adiabatic process.
D. It is not possible to give a definitive answer without more information.

14M.6 For an ideal gas the internal energy U depends only on

A. pressure.
B. volume.
C. temperature.
D. heat content.
E. entropy.

14M.7 What is the maximum efficiency of an engine which extracts heat from a reservoir at 600 ^{o}C and rejects heat at a reservoir at 100 ^{o}C?

A. 17%
B. 43%
C. 57%
D. 83%
E. 87%

14M.8 Suppose that in a room you had an electric refrigerator and an old-fashioned ice box containing a cake of ice. You consider cooling the room by leaving the door of one or both of these appliances open. Can you effectively cool a room in this way?

A. Yes, it would be best to leave both open.
B. No, it won't work to leave either open.
C. It will work if you leave the ice box open, but not the refrigerator.
D. It will work if you leave the refrigerator open, but not the ice box.

14M.9 If you place some cold water in a well-insulated thermos bottle and close the lid tightly, you will find that by vigorous shaking you can warm the water appreciably. This is because in this process

A. heat flowed into the water.
B. work was done on the water.
C. the first law of thermodynamics does not apply, since the process was not a cyclic one.
D. some of the water evaporates.

14M.10 When one freezes water to make ice cubes, the entropy of the water used

A. increases.
B. decreases.
C. remains constant.
D. may either increase or decrease, depending upon how rapidly the freezing occurs.

• Multiple Choice: Answers and Comments

14M.1 B We recognize F/A as pressure and AΔx as ΔV, the change in volume. Remember, a gas must change volume in order to do work.

14M.2 A In an adiabatic process $\Delta Q = \Delta U + \Delta W = 0$, so $\Delta W = -\Delta U$.

14M.3 D The gas does no work in going from 1 to 2, since there is no volume change along this path. The work done <u>by</u> the gas in going from 2 to 3 is $P_2(V_2 - V_1)$. No work is done in going from 3 to 4, since again there is no volume change. In going from 4 back to 1 work is done <u>on</u> the gas in the amount $P_1(V_2 - V_1)$. Thus the net work done by the gas in completing the cycle is

$$\Delta W = P_2(V_2 - V_1) - P_1(V_2 - V_1)$$

$$\Delta W = (P_2 - P_1)(V_2 - V_1)$$

We see that this is just the area enclosed by the cyclic path in the P-V diagram.

14M.4 D This is by definition what is meant by an adiabatic process.

14M.5 C When you compress an ideal gas it tends to get hotter. For a given volume, the hotter the gas the higher the pressure, as we can see from the ideal gas law. In an adiabatic compression no heat is removed and the gas gets very hot and one must work against high pressures. On the other hand, in an isothermal compression one must remove heat during the compression to keep the temperature from rising. This in turn means the pressure will not be so large and less work will be done in the compression. One can also deduce this answer by observing that the adiabats are steeper curves than the isotherms in a P-V diagram.

14M.6 C This is an important result to remember. Recall that a good intuitive meaning of temperature is that it is a measure of the average KE per molecule in an ideal gas.

14M.7 C $\text{Efficiency} = 1 - \frac{T_c}{T_h} = 1 - \frac{373\ ^oK}{873\ ^oK} = 57\%.$

Be careful to use temperature in ^oK, not ^oC.

14M.8 C Heat will be absorbed from the room in melting the ice, so leaving the door of the ice box open will cool the room. It will not work, however, to leave the refrigerator door open, since the motor of the refrigerator will dissipate heat in room and further the heat extracted from the contents of the refrigerator will be dumped into the room. Thus, leaving the refrigerator door open will cause the refrigerator to run all the time and actually heat up the room. This, incidentally, is effectively the way a heat pump works. A heat pump (as opposed to a furnace) is like a refrigerator which is cooling off the outdoors. The waste heat in the motor <u>plus</u> the heat removed from outdoors is dumped into your house, thereby heating it very effectively. The only disadvantage is that such devices are somewhat more expensive.

14M.9 B No heat flows in, since the bottle is well insulated. The first law of thermodynamics always applies.

14M.10 B The molecules in ice are arranged in more orderly fashion than in liquid water, so this is a state of lower entropy. Note that heat must be removed to freeze water, and when heat is removed entropy decreases, since $\Delta S = \Delta Q/T$.

• Problems

14.1 Consider two four-sided dice. Each is a regular tetrahedron with each side an equilateral triangle. The faces are numbered 1, 2, 3 and 4. When the dice are thrown the side which ends up face down on the table is the one which is read. Make a table showing the possible microstates and macrostates of this system where a macrostate is characterized by a particular sum for the two dice. What is the most probable value for the sum of the two faces thrown? What is the probability that the sum of the two sides is 4?

14.2 Consider flipping a coin four times in a row. What is the probability of flipping only one head, or of flipping heads only twice, or of flipping heads four times? What are the macrostates and microstates of this system? (Characterize a macrostate by the number of heads thrown.)

14.3 A cylinder of ideal gas is slowly compressed to ¼ of its original volume. Fifty J of work are required to accomplish this compression.

a) By how much did the internal energy of the gas change in this process?

b) How much heat flowed into or out of the gas in this process?

14.4 A cylinder of ideal gas is slowly compressed adiabatically to ¼ of its original volume. Fifty J of work are required to accomplish this compression.

a) By how much did the internal energy of the gas change in this process?

b) How much heat flowed into or out of the gas in this process?

14.5 An ideal monatomic gas is contained at 27 °C in a cylinder of volume 4 liters. Heat is slowly added to the gas and it is allowed to expand against a constant pressure of one atmosphere until its volume has doubled. Determine the number of one atmosphere until its volume has doubled. Determine the number of moles of gas in the container, the final temperature, the amount of work done by the gas, and the amount of heat supplied to the gas.

14.6 In Idaho they are trying to utilize geothermal energy to generate electricity. A hole about 2000 m deep is being drilled in an area where the rock temperature at this depth is about 130 °C. The plan is to run a heat engine using the hot rock as the hot reservoir and surface air at 20 °C as the cold reservoir. This engine would then drive an electric generator. What is the maximum efficiency of such a generator?

14.7 When 100 g of water are boiled, what is the change in entropy?

14.8 Half a mole of neon, a monatomic gas, at 300 °K is contained in a cylinder of volume 8 L. The gas is heated slowly and allowed to expand isobarically until its volume

• Problem Solutions

14.1

MACROSTATE (TOTAL SUM) →	2	3	4	5	6	7	8
MICROSTATES	1,1	2,1	1,3	1,4	4,2	4,3	4,4
		1,2	3,1	4,1	2,4	3,4	
			2,2	2,3	3,3		
				3,2			

BY COUNTING I SEE THAT THERE ARE 16 POSSIBLE MICROSTATES. THE MACROSTATE WHICH IS MOST PROBABLE IS THAT FOR WHICH THE SUM OF THE TWO DICE IS 5 SINCE THERE ARE 4 MICROSTATES BELONGING TO THIS MACROSTATE.

THE PROBABILITY THAT THE SUM WILL BE 5 IS

$$P_5 = \frac{\text{NUMBER OF MICROSTATES IN MACROSTATE 5}}{\text{TOTAL NUMBER OF MICROSTATES}} = \frac{4}{16} = \underline{\underline{0.25}}$$

THE PROBABILITY THAT THE SUM IS 4 IS

$$P_4 = \frac{3}{16} = \underline{\underline{0.19}}$$

MACROSTATE	0	1	2	3	4
MICROSTATE	TTTT	HTTT THTT TTHT TTTH	HHTT HTHT HTTH THHT THTH TTHH	HHHT HHTH HTHH THHH	HHHH

THE SYSTEM HAS 16 MICROSTATES.
ONLY ONE MICROSTATE HAS 4 HEADS, SO PROBABILITY OF THIS IS

$$P_4 = \frac{1}{16} = 0.063$$

4 MICROSTATES BELONG TO THE MACROSTATE WITH ONE HEAD, SO

$$P_1 = \frac{4}{16} = 0.25$$ = PROBABILITY OF FLIPPING ONE HEAD.

SIMILARLY, PROBABILITY FOR 2 HEADS IS $P_2 = \frac{6}{16} = 0.375$

14.3 (a) SINCE TEMPERATURE REMAINED CONSTANT INTERNAL ENERGY U DOESN'T CHANGE.

(b) $\Delta Q = \Delta U + \Delta W = 0 + \Delta W$, $\Delta Q = \Delta W = 50\,J$

14.4 (a) IN AN ADIABATIC PROCESS $\Delta Q = 0$

THUS $\Delta Q = \Delta U + \Delta W = 0$. $\Delta U = -\Delta W = 50\,J$

($\Delta W = -50\,J$ = WORK DONE BY GAS)

(b) $\Delta Q = 0$ (ADIABATIC)

14.5 $P_1 V_1 = nRT_1$. $n = \frac{P_1 V_1}{RT_1} = \frac{(10^5\,Pa)(4\times10^{-3}\,m^3)}{(8.31\,J/MOL°K)(300°K)} = 0.16$ MOLE

P IS CONSTANT, SO $\frac{P}{nR} = \frac{T_1}{V_1} = \frac{T_2}{V_2}$, $T_2 = \frac{V_2}{V_1} T_1 = (2)(300°K)$

$T_2 = 600°K$

$\Delta W = P\Delta V = (10^5\,N/m^2)(4\times10^{-3}\,m^3) = 400\,J$

$\Delta U = nC_p\Delta T$, n = NO. MOLES OF GAS

$C_p = \frac{5}{2}R$ PER MOLE (MONATOMIC GAS)

$\Delta U = (0.16\,MOLE)(\frac{5}{2})(8.31\,J/MOL\,°K)(300°K) = 997\,J$

$\Delta Q = \Delta U + \Delta W = 997\,J + 400\,J = 1397\,J \simeq 1400\,J$

14.6 EFFICIENCY $\eta = 1 - \frac{T_c}{T_h} = 1 - \frac{293°K}{403°K} = 0.27 = 27\%$

14.7 TO BOIL 1 gm OF WATER REQUIRES 539 CAL, SO 53.900 CAL, OR 53.9 KCAL, REQUIRED TO BOIL 100g OF WATER.

$$\Delta S = \frac{\Delta Q}{T} = \frac{53.9\ \text{KCAL}}{373°\text{K}} = \underline{\underline{0.145\ \text{KCAL}/°\text{K}}}$$

14.8 $PV = nRT$, $P_1 = nR\frac{T_1}{V_1} = \frac{(0.5\ \text{MOL})(8.31\ \text{J/MOL}°\text{K})(300°\text{K})}{(8\times10^{-3}\ \text{m}^3)} = \underline{\underline{1.56\times10^5\ \text{Pa}}}$

PROCESS IS ISOBARIC, SO $P_2 = P_1 = \underline{\underline{1.56 \times 10^5\ \text{Pa}}}$

$$\frac{P}{nR} = \frac{T_1}{V_1} = \frac{T_2}{V_2}\ ;\ T_2 = \frac{V_2}{V_1}T_1 = \left(\frac{9}{8}\right)(300°\text{K})$$

$$\underline{\underline{T_2 = 337.5°\text{K}}}$$

$\Delta Q = \Delta U + P\Delta V$

$\Delta U = mC\,\Delta T$, $C_p = C_V + R = \frac{3}{2}R + R = \frac{5}{2}R$ FOR A MONATOMIC GAS.

C_p = MOLAR SPECIFIC HEAT

$c_p = C_p/(MW)$ = SPECIFIC HEAT PER GRAM, WHERE MW = MOLEC. WT.

$m = \frac{1}{2}(MW)$ FOR 1/2 MOLE OF GAS

$$\Delta Q = \frac{mC_p}{(MW)}\Delta T + P\Delta V$$

$$= \frac{(\frac{1}{2}MW)(\frac{5}{2}R)(\Delta T)}{(MW)} + P\Delta V$$

$$= \left(\frac{5}{9}\right)(8.31\ \text{J/MOL}°\text{K})(37.5°\text{K}) + (1.56\times10^5\ \text{N/m}^2)(10^{-3}\ \text{m}^3)$$

$\Delta Q = 545.5\ \text{J}$

$\Delta S = \frac{\Delta Q}{T}$ SINCE T IS CHANGING IT IS NECESSARY TO USE CALCULUS TO CALCULATE ΔS EXACTLY. HOWEVER, SINCE $\Delta T \ll T$, WE CAN OBTAIN AN APPROXIMATE RESULT BY USING $T \approx T_{AVE}$.

$$T_{AVE} \approx T_1 + \tfrac{1}{2}\Delta T = 300 + \tfrac{1}{2}(37.5) \approx 319°\text{K}$$

$$\Delta S \approx \frac{545.5\ \text{J}}{319°\text{K}} = \underline{\underline{1.7\ \text{J}/°\text{K}}}$$

Oscillatory Motion

15

• Summary of Important Ideas, Principles, and Equations

1. An object is in equilibrium when it is subject to no net force or torque.

   If an object in equilibrium is displaced a small distance and released, three things can happen.

   It can experience no force and remain where it is. In this case the object was in neutral equilibrium.

   It can experience a force which causes it to return to its original equilibrium position. In this case the object was in stable equilibrium.

   It can experience a force which moves it even farther away from its original equilibrium position. In this case the object was in unstable equilibrium.

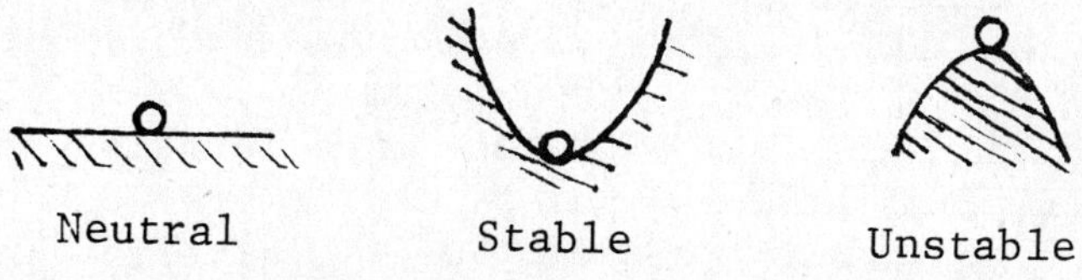

2. Whenever a system in stable equilibrium is displaced a very small amount x from equilibrium it will experience a restoring force F,

$$F = -kx \quad \text{(Hooke's Law)} \tag{15.1}$$

   Once displaced and released the system will undergo simple harmonic motion (SHM).

The number of oscillations per second is called the frequency, f. Frequency is measured in Hertz, where one Hertz is one cycle per second.

The time required for one oscillation is the period, T, in seconds.

$$f = \frac{1}{T} \qquad (15.2)$$

The displacement x varies between maximum values +A and -A, where A is called the amplitude of the motion.

The variation of x with time is given by

$$x = A \sin(2\pi ft + \phi) = A \sin(\omega t + \phi) \qquad (15.3)$$

Here $\omega = 2\pi f$ = angular frequency, measured in radians/second.

ϕ = phase angle.

The value of the phase angle depends on the instant one chooses for t = 0. We will usually choose this time origin so that $\phi = 0$. In some books the displacement is written $x = A\cos(\omega t + \phi)$. This is equivalent to (14.3) and merely reflects a different choice of time origin, since $\sin(\omega t + \pi/2) = \cos\omega t$.

The velocity and acceleration of an object undergoing SHM (taking $\phi = 0$) are

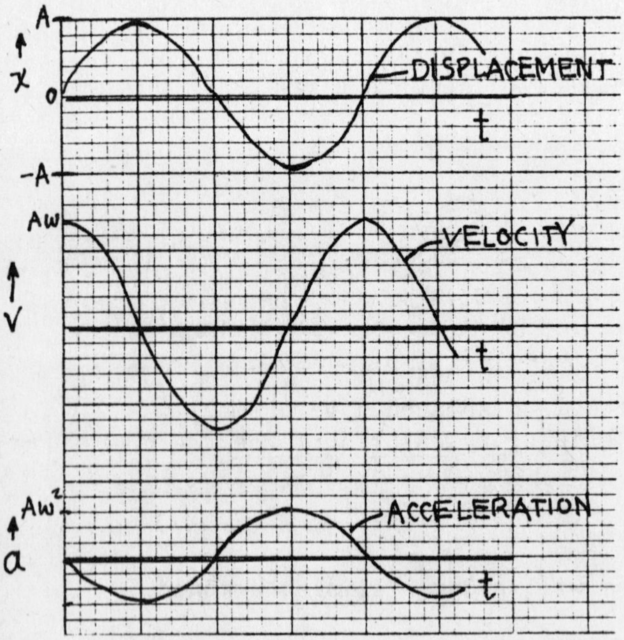

$$v = A\omega\cos\omega t$$

$$a = -A\omega^2\sin\omega t \qquad (15.4)$$

We say that velocity leads displacement by 90° in phase. Acceleration and displacement are 180° out of phase.

3. A mass M attached to a spring oscillates in SHM. The restoring force acting on the mass is $F = -kx$, where k is the spring constant of the spring and x is the displacement from equilibrium. The stiffer the spring, the larger is k.

The frequency of oscillation f is called the natural frequency or the resonance frequency.

$$f = \frac{1}{2\pi}\sqrt{\frac{k}{M}}, \quad \omega = \sqrt{\frac{k}{M}}, \quad T = \frac{1}{f} = 2\pi\sqrt{\frac{M}{k}} \tag{15.5}$$

4. A mass M attached to a string of length L is a simple pendulum which oscillates in SHM.

$$f = \frac{1}{2\pi}\sqrt{\frac{g}{L}}, \quad \omega = \sqrt{\frac{g}{L}}, \quad T = 2\pi\sqrt{\frac{L}{g}} \tag{15.6}$$

5. The energy of an oscillating system changes back and forth between KE and PE. For a mass on a spring

$$E = \tfrac{1}{2}mv^2 + \tfrac{1}{2}kx^2 = \text{constant} \tag{15.7}$$

When x is a maximum, $x = A$, the velocity is zero, so $E = PE_{max} = \tfrac{1}{2}kA^2$.

When $x = 0$ velocity is a maximum and $E = KE_{max} = \tfrac{1}{2}Mv_{max}^2$

Thus

$$KE_{max} = PE_{max} \tag{15.8}$$

6. If an oscillatory system is driven by an oscillating force, the displacement will be largest when the driving force is applied at the natural frequency of the system. In this case the system is being driven at resonance.

Friction limits the maximum amplitude which can be achieved.

The amplitude of a freely oscillating system will gradually decay due to friction according to the relation $A = A_{max}\exp(-t/\tau)$, where τ is a parameter which measures the amount of friction present.

• Qualitative Questions

15M.1 A pencil balanced on its point is an example of

A. neutral equilibrium.
B. stable equilibrium.
C. unstable equilibrium.

15M.2 An equation of the form $F = -kx$ can describe the force acting on any system displaced a small distance from stable equilibrium (such as a mass on a spring). What is the significance of the minus sign in this equation?

A. The minus sign is needed because k is negative.
B. The minus sign is needed because x is negative.
C. The minus sign is needed because F is a restoring force.
D. The minus sign is of no importance, and the equation could just as well be written without it.

15M.3 If a system is oscillating with a frequency of 5 Hertz, the period of the oscillation is

A. not possible to determine unless we know if the system is undergoing SHM.
B. 10 seconds
C. 5 seconds
D. 0.2 second

15M.4 When a system is undergoing SHM, its amplitude

A. is constant.
B. varies as $\sin \omega t$.
C. varies as $\cos \omega t$.
D. is dependent on the frequency of the motion.
E. More than one of the above is true.

15M.5 Consider a mass M suspended on a spring of spring constant k.

A. If the mass were increased the frequency would not change.
B. If k were increased the frequency would decrease.
C. If the spring were cut in half the frequency of oscillation would change.
D. The amplitude of the motion depends on the period.
E. The PE of the system is constant.

15M.6 The speed of a particle undergoing SHM is

A. a minimum when the particle is passing through the equilibrium position.
B. zero when the particle is subject to the greatest force.
C. a maximum when the force is a maximum.
D. constant.
E. 180° out of phase with the displacement.

15M.7 The frequency of a simple pendulum

A. is independent of the mass.
B. is independent of the length of the pendulum.
C. varies with time according to cos ωt.
D. is porportional to its period.
E. is also called the "angular frequency" because it is the angle of inclination of the pendulum which is oscillating.

15M.8 If the length of a simple pendulum is doubled, the frequency will change by a factor of

A. ¼
B. ½
C. $1/\sqrt{2}$
D. $\sqrt{2}$
E. 2
F. 4
G. 1, i.e. it won't change.

15M.9 Sometimes when I am in the woods cutting firewood I try to fell a tall snag which is too big to be cut through all the way with my saw. To try to finish it off I have on occasion pushed it sideways with a periodic driving force. If the sideways displacement can be made large enough she will snap off and down she will come. (My forestry students say this is an idiotic way to fell a snag, and it isn't too safe, but it does work.) Anyway, what is the best way to do this?

A. It would be best simply to push with a steady force in one direction, as opposed to applying an oscillating force.
B. Apply an oscillating force with the highest frequency you can generate.
C. Apply an oscillating force at the natural frequency of the tree when it is waving back and forth.
D. Apply an oscillating force at the natural frequency of your body as you push back and forth.
E. Apply an oscillating force, but the frequency is not important.

• Multiple Choice: Answers and Comments

15M.1 C

12M.2 C

15M.3 D $T = 1/f = 1/5s^{-1} = 0.2s$

15M.4 A Amplitude is the <u>maximum</u> displacement, and this remains constant while the displacement at any instant varies sinusoidally.

15M.5 C A and B are false, since $f = \frac{1}{2\pi}\sqrt{\frac{k}{m}}$.

D is false, since amplitude is constant.
E is false, PE = $\frac{1}{2}kx^2$ and x is varying.

15M.6 B A is false. The speed is greatest when the particle passes through the equilibrium position.
B is true, since the force is greatest when the spring is experiencing maximum stretch, and at this point the particle stops and turns around.
E is false because velocity is 90° out of phase with displacement.

15M.7 A $f = \frac{1}{2\pi}\sqrt{\frac{g}{L}}$

15M.8 C

15M.9 C If the driving force is applied at the natural frequency the system will be driven at resonance and the displacement will become very large, thereby increasing the chance of breaking off the snag.

• Problems

15.1 A system oscillates with a frequency of 300 Hz with an amplitude of 4 cms. Write an expression for the displacement x as a function of time, and sketch x vs. t, for the following cases:

(a) x(0) = 0
(b) x(0) = 4 cms
(c) x(0) = -4 cms

15.2 The displacement of an object undergoing SHM is x(t) = 2 sin 20t. What are the amplitude, frequency, angular frequency, period, maximum velocity, and maximum acceleration?

15.3 When a 200 g mass is attached to a spring it causes the spring to stretch 4 cms. Suppose that now the mass is displaced 2 cms from its equilibrium position (in the positive direction) and released. Write an expression for the subsequent displacement as a function of time. What is the maximum velocity and maximum acceleration?

15.4 A mass M is attached to two identical springs, each with spring constant k, in the three ways sketched here. What is the effective spring constant and the frequency of oscillation in each case?

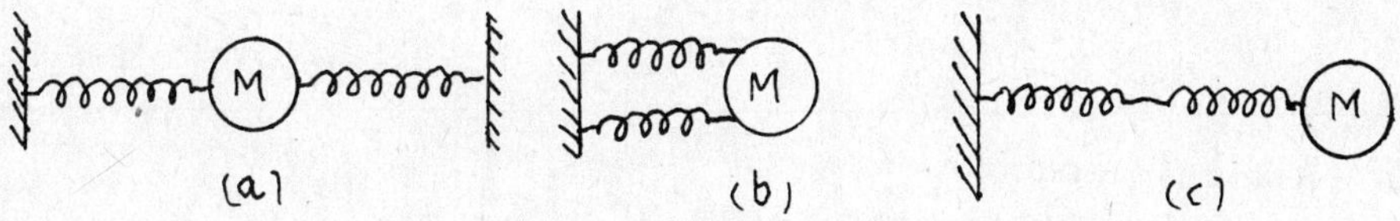

15.5 What is the natural frequency of a child of mass 20 kg who swings from a rope 5 meters long?

15.6 Suppose that in an old fashioned airplane one of the landing gear, of mass 8 kg, is attached to the wing with bolts which can exert a maximum force of 4000 N. If the wing vibrates at a frequency of 10 Hz what is the maximum amplitude allowable at the point of attachment if the bolts are not to snap off?

15.7 A mass M attached to a spring is oscillating with SHM. At what displacement, in terms of the amplitude A, is KE equal to twice the PE?

15.8 A child in a swing of length 6 m is displaced through an angle of 25° and released. After swinging down and back twice her maximum displacement is only 23° because of friction. What would you estimate her maximum displacement to be after 10 swings?

15.9 A mass of liquid M fills a U-tube to a height h. Show that if the liquid is displaced slightly from equilibrium and released it will undergo SHM. Determine the frequency of the motion. Assume friction is negligible.

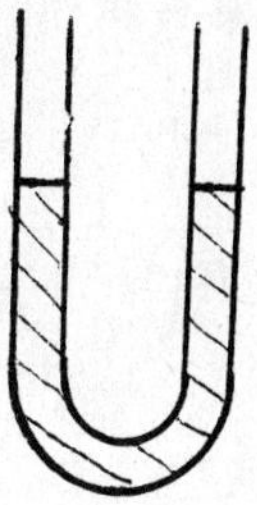

• Problem Solutions

15.1 $x = A\sin(2\pi f t + \phi)$

(a) $x(0) = 0$, $0 = .04\sin(0+\phi)$, $\sin\phi = 0$, $\therefore \phi = 0$

$x(t) = .04\sin(600\pi t)$ METERS

(b) $x(0) = 4$ cms. $0.04 = 0.04\sin(0+\phi)$

$\sin\phi = 1$. $\therefore \phi = 90^{\circ}$ OR $\frac{\pi}{2}$ RADIANS

$x(t) = 0.04\sin(600\pi t + \pi/2)$ METERS

(c) $x(0) = -4$ cms,

$-0.04 = 0.04\sin(0+\phi)$

$\sin\phi = -1$, $\phi = -90^{\circ}$ OR $-\pi/2$ RADIANS

$x(t) = 0.04\sin(600\pi t - \pi/2)$

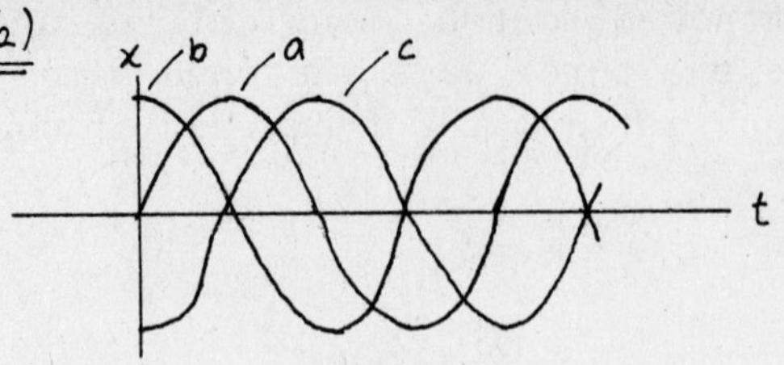

15.2 $x = 2\sin 20t = A\sin 2\pi f t$

$\omega = 20$ RAD/S , $f = \omega/2\pi = 20/2\pi$ Hz $= 3.2$ Hz , $T = \frac{1}{f} = 0.31$ s , $A = 2$ m

$V_{max} = \omega A = (20)(2)$ m/s $= 40$ m/s , $a_{max} = A\omega^2 = 2(20)^2$ m/s^2 $= 800$ m/s^2

15.3 $F = -kx$. SO $k = -\frac{Mg}{x} = -\frac{(.2\,kg)(9.8\,m/s^2)}{(-0.04\,m)} = 49$ N/m

$\omega = \sqrt{\frac{k}{M}} = \sqrt{\frac{49\,N/m}{0.2\,kg}} = 15.7$ RAD/S

$x = A\sin(\omega t + \phi) = \underline{0.02\sin(15.7t + \pi/2)}$

$V_{max} = \omega A = \underline{.313\ m/s}$, $a_{max} = \omega^2 A = \underline{4.9\ m/s^2}$

15.4 IMAGINE MASS IS MOVED SMALL DISTANCE X TO THE RIGHT.

(a) $F = -kx - kx = -2kx$

IF $F = -k_a x$, $k_a = 2k$ = EFFECTIVE SPRING CONSTANT.

$\omega_a = \sqrt{\frac{k_a}{M}} = \underline{\sqrt{\frac{2k}{M}}}$

(b) $F = -kx - kx = -2kx$

IF $F = -k_b x$, $k_b = 2k$, $\omega_b = \sqrt{\frac{k_b}{M}} = \underline{\sqrt{\frac{2k}{M}}}$

(c) $F = -k\left(\frac{1}{2}x\right)$ EACH SPRING IS STRETCHED ONLY $\frac{1}{2}x$, SO FORCE ON M DUE TO SPRING ATTACHED TO IT IS $-\frac{1}{2}kx$.

$= k_c x$

SO $k_c = \frac{1}{2}k$

$\omega_c = \sqrt{\frac{k}{M}} = \underline{\sqrt{\frac{k}{2M}}}$

15.5 $f = \frac{1}{2\pi}\sqrt{\frac{g}{L}} = \frac{1}{2\pi}\sqrt{\frac{9.8\,m/s^2}{5\,m}} = \underline{0.22\ Hz}$

15.6 $F_{max} = m a_{max} = m\omega^2 A$

* $A = \frac{F_{max}}{m\omega^2} = \frac{4000\,N}{(8\,kg)(2\pi)^2(10\,Hz)^2} = 0.127\,m = \underline{12.7\ cm}$

15.7 $PE = \frac{1}{2}kx^2 = \frac{1}{2}k(A\sin\omega t)^2$

$KE = \frac{1}{2}mv^2 = \frac{1}{2}m(A\omega\cos\omega t)^2$ FROM EQ. (14.4)

$= \frac{1}{2}m\omega^2 A^2\cos^2\omega t$

GIVEN $KE = 2PE$,

$$\frac{1}{2} m\omega^2 A^2 \cos^2 \omega t = 2 \cdot \frac{1}{2} k A^2 \sin^2 \omega t$$

SUBSTITUTE $\omega^2 = \frac{k}{m}$, $\cos^2 \omega t = 2 \sin^2 \omega t$

$$\left(\frac{\sin \omega t}{\cos \omega t}\right)^2 = \frac{1}{2}, \quad \frac{\sin \omega t}{\cos \omega t} \equiv \tan \omega t = \frac{1}{\sqrt{2}}$$

THUS $\sin \omega t = \frac{1}{\sqrt{3}}$

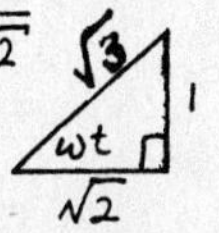

$\therefore x = A \sin \omega t = \frac{1}{\sqrt{3}} A$

SEE THIS BY DRAWING A RIGHT TRIANGLE WITH ωt ONE ANGLE. USE PYTHAGOREAN TH^{m} TO FIND HYPOTENUSE.

15.8 PERIOD IS $T = 2\pi\sqrt{\frac{L}{g}} = 2\pi\sqrt{\frac{6m}{9.8 m/s^2}} = 4.9 s$

INITIALLY AMPLITUDE IS θ_0, AND AFTER TWO PERIODS $(2T)$ AMPLITUDE IS θ_2, SO

$$\theta = \theta_0 e^{-t/\tau}, \quad \theta_2 = \theta_0 e^{-2T/\tau}$$

AFTER 10 PERIODS AMPLITUDE IS θ_{10}, $\theta_{10} = \theta_0 e^{-10T/\tau}$

?

THUS $\theta_0^5 = (\theta_0 e^{-2T/\tau})^5 = \theta_0^5 e^{-10T/\tau}$

OR $e^{-10T/\tau} = \left(\frac{\theta_2}{\theta_0}\right)^5$

SUBSTITUTE THIS IN θ_{10}: $\theta_{10} = \theta_0\left(\frac{\theta_2}{\theta_0}\right)^5 = 25°\left(\frac{23°}{25°}\right)^5$

$$\theta_{10} = 16.5°$$

16 Mechanical Waves

• Summary of Important Ideas, Principles, and Equations

1. A wave is a disturbance which propagates through a medium. A complicated wave may be viewed as a superposition of sinusoidal waves.

   If the disturbance is measured by a quantity y (which might be the displacement of a string, the pressure, or the electric field) a sinusoidal travelling wave moving in the +x direction is written

   $$y = A\sin(2\pi\frac{x}{\lambda} - 2\pi ft) = A\sin(kx - \omega t) \qquad (16.1)$$

   where A = amplitude of the wave
   f = frequency of the wave
   $T = 1/f$ = period
   ω = angular frequency
   λ = wavelength
   $k = 2\pi/\lambda$ = wave number

   The wavelength λ is the separation in space of equivalent displacements (i.e. the distance from one crest of the wave to the next).

   The period T is the separation in time between equivalent displacement at a particular point in space.

   The quantity in parenthesis above, $(kx - \omega t)$, is called the phase. A point of constant phase moves with the wave velocity v, where

   $$v = f\lambda \qquad (16.2)$$

   A wave crest, for example, is a point where the phase is always $\pi/2$.

   The polarization of the wave tells the direction of the vibrations. For example, in a longitudinally polarized wave the vibrations are parallel to the direction of travel of the wave, whereas in a transversely polarized wave the vibrations are perpendicular to this direction.

2. The principle of superposition states that the displacement caused by the two waves at a given point in space is just the vector sum of the displacement due to each wave separately.

3. An overtone is any resonance frequency that is higher than the fundamental (lowest) resonance frequency of a system. A harmonic is an overtone whole frequency is a multiple of the fundamental frequency, f_o. The frequency of the nth harmonic is thus nf_o.

4. Standing waves result from the superposition of two waves of the same frequency and wavelength propagating in opposite directions. Points where the displacement is always zero are nodes, whereas points where the displacement takes its largest values are antinodes. Standing waves will be set up in a string fixed at both ends whenever the length of the string is a multiple of one half wavelength of the wave.

$$L = n\left(\frac{\lambda_n}{2}\right), \qquad n = 1,2,3, \ldots \qquad (16.3)$$

$$f_n = \frac{v}{\lambda_n} = n\left(\frac{v}{2L}\right)$$

Here $f_1 = v/2L$ = fundamental frequency of the string.

The wave velocity is

$$v = \sqrt{\frac{F}{\mu}} \qquad (16.4)$$

where F = tension in the string
μ = mass/length of string

5. Two waves of slightly different frequencies f_1 and f_2 when superimposed give rise to a wave whose frequency is approximately $\frac{1}{2}(f_1 + f_2)$ and whose amplitude varies slowly at the beat frequency $(f_1 - f_2)$.

6. The power carried by a mechanical wave is proportional to vf^2A^2, where f = wave frequency, v = wave velocity, A = amplitude.

• Qualitative Questions

16M.1 Consider the transverse vibrations of a string. The wave velocity of such a string wave refers to

A. the velocity of the particles of the string as they move back and forth perpendicular to the direction of propagation of the wave.
B. the velocity of the particles of the string as they move back and forth along the direction of propagation of the wave.
C. the velocity with which a disturbance of given phase (e.g. a crest of the wave) moves perpendicular to the direction of propagation.
D. the velocity with which a disturbance of given phase (e.g. a crest) moves parallel to the direction of propagation of the wave.
E. the velocity of sound produced by the vibrating string.

16M.2 When a pulse disturbance in a string is incident on the end of the string that is tied to a post,

A. the pulse is flipped upside down on reflection.
B. the pulse is reflected right side up.
C. whether the pulse is flipped over or not depends on its phase at the moment it reaches the end of the string.
D. there is no reflected pulse because the end of the string is tied and cannot move.
E. the nature of the reflected pulse will be the same whether or not the end of the string is tied to the post or allowed to slide up and down (i.e. as if tied to a ring on the post).

16M.3 The wavelength of a wave is

A. the distance between a crest of the wave and an adjacent trough.
B. the distance between two successive crests or two successive troughs.
C. the same as the period of the wave.
D. dependent on the amplitude of the wave.
E. independent of wave velocity for a given frequency.

16M.4 The waves which travel down a stretched string which is plucked

A. have longitudinal polarization.
B. have transverse polarization.
C. are not polarized.
D. do not carry energy.
E. travel at a velocity which is independent of the velocity of the particles which make up the string.

16M.5 Suppose that two strings of different thicknesses are joined together smoothly at the ends to make a composite string. This composite string is now stretched taut and a transverse wave is propagated along it. When this wave goes from one string to the other,

A. the wavelength will not change.
B. the velocity will not change.
C. the frequency will not change.
D. the energy carried per second by the wave will not change.
E. the amplitude of the wave will not change.

16M.6 The little cross-ridges under the strings on a guitar are called frets. When you press a string down against a fret you decrease its effective length. In so doing, you change

A. the wave velocity in the string.
B. the fundamental frequency, but not the fundamental wavelength.
C. the fundamental wavelength, but not the fundamental frequency.
D. both the fundamental frequency and the fundamental wavelength.
E. the overtone frequencies but not the fundamental frequency.

16M.7 If you were to double the tension of a violin string the fundamental frequency of the string would change by a factor of

A. 1, i.e. it would not change.
B. $\frac{1}{2}$
C. $1/\sqrt{2}$
D. $\sqrt{2}$
E. 2

16M.8 In order for standing waves to be set up in a string, two waves must be superimposed. It is necessary that these two waves

A. have the same frequency.
B. have the same wavelength.
C. travel at the same speed.
D. travel in opposite directions.
E. All of the above are true.
F. None of the above is true.

16M.9 If you wished to excite the second harmonic ($f_2 = 2f_0$) of a violin string, where should you bow the string?

A. At the middle.
B. One fourth of the wave from one end.
C. One third of the way from one end.
D. Close to, but not right at, one end.
E. It makes no difference where you bow the string, since exciting the fundamental is what excites all of the harmonics.

• True–False Questions

16TF.1 A wave transports both energy and mass.

16TF.2 The expression $y = A \sin(kx + \omega t)$ represents a wave travelling in the negative x direction.

16TF.3 The frequency of a wave is independent of its amplitude.

16TF.4 The crest of a standing wave remains at one position and does not move.

16TF.5 Wavelength is to period as space is to time.

16TF.6 The resonant frequencies of a vibrating string are all integral multiples of the fundamental frequency.

16TF.7 If you bow a violin string at its center you will excite all odd harmonics.

16TF.8 When a wave travels from one medium to another its frequency does not change.

16TF.9 If two musical instruments in an orchestra are not tuned so that they play the same fundamental frequency for a given note, a faint "beating" can be heard. This is the result of superposition of waves.

16TF.10 If one doubles the amplitude of a wave the energy carried by the wave will also be doubled.

• Multiple Choice: Answers and Comments

16M.1 D

16M.2 A We say that the reflected wave has undergone a 180° phase shift in this case. If reflected back from a "loose" end (e.g. one tied to a sliding ring) the pulse comes back right side up, i.e. it suffers no phase change. This kind of effect occurs when sound waves and light waves are reflected.

16M.3 B

16M.4 B

16M.5 C A and B are false, since $v = \sqrt{F/\mu}$ and the linear density μ is larger for the thicker string. Thus v decreases when the wave goes from the thin string to the thick one. Frequency does not change. Since $v = f\lambda$, λ must decrease when v does. D and E are false, since some of the wave will be reflected back by the discontinuity.

16M.6 D The length of the string determines the fundamental wavelength, $\lambda = L/2$. Shortening the string decreases λ_0. The wave velocity is determined by the tension, so this stays constant. Since $f_0\lambda_0 = v$, decreasing λ_0 must cause f_0 to increase if the product of the two is to stay constant.

16M.7 D

$$f_0\lambda_0 = \sqrt{F/\mu}$$

λ is fixed by string length, so if F is doubled frequency will increase by $\sqrt{2}$.

16M.8 E

16M.9 B The displacement of the second harmonic is greatest one fourth of the way from one end, so this is the best place to excite it. This harmonic can be excited to some extent by bowing the string elsewhere, but not at the middle, where the displacement is always zero.

Displacement is large here

• True–False: Answers and Comments

16TF.1 F

16TF.2 T A wave is characterized by the fact that a point of fixed phase moves along at velocity v. The phase is $(kx + \omega t)$, and as time passes t becomes a large positive value. If the phase is to remain constant x must become a large negative value, i.e. the point of constant phase is moving toward negative x.

16TF.3 T

16TF.4 T This is why the standing wave is "standing" instead of "travelling".

16TF.5 T Wavelength is the repeat unit in space and period is the repeat unit in time.

16TF.6 T $f_n = nf_0$

16TF.7 T All odd harmonics have maximum displacement at the center.

16TF.8 T

16TF.9 T This is the phenomenon of "beats".

16TF.10 F Energy carried by a wave is proportional to the square of the wave amplitude, so doubling the amplitude increases the energy by a factor of 4.

• Problems

16.1 Waves in the deep ocean can have very long wavelengths, and a distance between crests of 400 m is not unusual. If one such crest passes a buoy each 16 seconds, what is the velocity of the waves?

16.2 The tension in a violin string is adjusted so that the velocity of a wave travelling along the string is 300 m/s. What is the fundamental frequency of vibration for a string 50 cms long? What is the frequency of the second harmonic?

16.3 A power cable of mass 0.2 kg is stretched tightly between two poles 10 meters apart. When one of the poles is struck a blow it is observed that the transverse pulse is reflected back from the second pole in 0.5 s. What is the tension in the cable?

16.4 Two wires have the same lengths but their diameters differ by a factor of two. If they are adjusted to the same tension, what will be the ratio of their fundamental frequencies of vibration?

16.5 A 250 Hz tuning fork produces 3 beats per second when sounded in conjunction with the fundamental vibration of a cello string. The beat frequency decreases when the tension in the string is decreased slightly. What is the original frequency of the string? By what fraction should the tension be reduced to tune the string exactly to 250 Hz?

16.1 $v = \lambda f = \lambda\left(\frac{1}{T}\right) = (400\,\text{m})\frac{1}{16\,\text{s}} = 25\ \text{m/s}$

16.2 $\lambda = 2L$, $f_0 = \frac{v}{\lambda_0} = \frac{v}{2L} = \frac{300\,\text{m/s}}{(2)(0.5\,\text{m})} = 300\ \text{Hz}$

$f_2 = 2f_0 = 600\ \text{Hz}$

16.3 $v = \sqrt{\frac{F}{\mu}} = \frac{2L}{t} = \frac{\text{DISTANCE PULSE TRAVELS}}{\text{TIME}}$, $\mu = \frac{M}{L}$

$$\frac{F}{M/L} = \left(\frac{2L}{t}\right)^2, \quad F = \frac{4LM}{t^2} = \frac{(4)(10\,m)(0.2\,kg)}{(0.5\,s)^2} = \underline{\underline{32\,N}}$$

16.4 $f_0 = \frac{v}{\lambda_0} = \frac{v}{2L}, \quad v = \sqrt{\frac{F}{\mu}}$

$$\frac{f_0}{f_0'} = \frac{v/2L}{v'/2L} = \frac{v}{v'} = \frac{\sqrt{F/\mu}}{\sqrt{F/\mu'}} = \sqrt{\frac{\mu'}{\mu}}$$

$$\mu = \frac{M}{L} = \frac{(\rho)(\pi \frac{d^2}{4})(L)}{L} = \left(\frac{\rho\pi}{4}\right)d^2 \qquad \rho = \frac{\text{MASS}}{\text{VOLUME}} = \text{DENSITY}$$

$$\text{SO } \frac{f_0}{f_0'} = \sqrt{\frac{(\frac{\rho\pi}{4})d'^2}{(\frac{\rho\pi}{4})d^2}} = \frac{d'}{d} = \underline{\underline{2}} \qquad \text{IF } d' = 2d$$

THICKER WIRE WILL HAVE FUNDAMENTAL FREQUENCY HALF AS GREAT AS THAT FOR THINNER WIRE.

16.5 $f - 250)\,Hz = 3\,Hz$, SO $\underline{\underline{f = 253\,Hz}}$

$$f = \frac{v}{\lambda} = \frac{v}{2L} = \frac{1}{2L}\sqrt{\frac{F}{\mu}}$$

LET f_0 = DESIRED FUNDAMENTAL FREQUENCY = 250 Hz

$f = f_0 + \Delta f$

F = TENSION WHICH PRODUCES FREQUENCY f

$F_0 = F - \Delta F$ = TENSION WHICH PRODUCES f_0

$$f = \frac{1}{2L}\sqrt{\frac{F}{\mu}}, \quad f_0 = \frac{1}{2L}\sqrt{\frac{F_0}{\mu}}$$

$$\frac{f_0}{f} = \frac{\frac{1}{2L}\sqrt{F_0/\mu}}{\frac{1}{2L}\sqrt{F/\mu}} = \sqrt{\frac{F_0}{F}}$$

$$F_0 = \left(\frac{f_0}{f}\right)^2 F = \left(\frac{250}{253}\right)^2 F = \underline{\underline{0.976\,F}}$$

NOTE THAT FREQUENCY CHANGES BY 3 PARTS IN 250 OR 1.2% WHEN TENSION CHANGES BY TWICE AS MUCH, 2.4%

17 Sound

• Summary of Important Ideas, Principles, and Equations

1. Sound waves are longitudinal compressional waves. They are longitudinal because the pressure variations occur along the direction of propagation. They can travel through solids, liquids or gases, but not vacuum. Audible sound has frequency from about 20 Hz to 20,000 Hz. Infrasonics are frequencies lower than 20 Hz, and ultrasonics are frequencies higher than 20,000 Hz.

2. The velocity of sound in a gas is independent of pressure and depends on temperature according to

$$\boxed{v = \sqrt{\frac{\gamma kT}{m}}} \qquad (17.1)$$

Here γ is a constant between 1.0 and 1.67 that depends on the molecular structure of the gas, $k = 1.38 \times 10^{-23}$ J/^{o}K, T is absolute temperature and m is the mass of a gas molecule.

In air at 20 ^{o}C the velocity of sound is 343 m/s.

3. Standing waves can be set up by sound waves bouncing back and forth in a pipe. The standing wave frequencies (also called resonant frequencies) for a pipe open at both ends or closed at both ends are given by

$$\boxed{f_n = \frac{nv}{2L}} \qquad n = 1,2,3, \ldots \qquad (17.2)$$

where L is the length of the pipe and v is the velocity of sound in air.

For a pipe open at one end and closed at the other

$$f_m = \frac{mv}{4L} \qquad m = 1,3,5,... \qquad (17.3)$$

4. The intensity of a wave is the power (energy per second) that passes through unit area perpendicular to the direction of travel of the wave. If sound radiates uniformly in all directions from a source its intensity decreases with the distance R from the source at a rate proportional to $1/R^2$.

Sound intensity is measured in decibels (dB). The minimum sound intensity detectable by the human ear is approximately $I_0 = 10^{-12}$ W/m^2, and this level is used to define the intensity in dB.

$$\beta(\text{in dB}) = 10 \log_{10} \frac{I}{I_0}$$

5. The Doppler effect is the change in frequency of sound as perceived by an observer moving with respect to the source of sound. The source emits a frequency f and the observer hears a frequency f' which is larger than f if he is approaching the source and less than f if receding from the source. If v_0 is the velocity of the observer, v_s the velocity of the source and v the velocity of sound,

$$f' = f\left(\frac{1 + v_0/v}{1 - v_s/v}\right) \quad \text{Moving together} \qquad (17.4)$$

$$f' = f\left(\frac{1 - v_0/v}{1 + v_s/v}\right) \quad \text{Moving apart} \qquad (17.5)$$

If the relative speed of the observer with respect to the source is v_r and $v_r \ll v$, the speed of sound, the magnitude of the frequency shift $|f' - f| = \Delta f$ may be written approximately

$$\frac{\Delta f}{f} = \frac{v_r}{v} \qquad (17.6)$$

When $v_r << c$, where c = speed of light = 3×10^8 m/s, eq. 16.6 also applies to electromagnetic radiation, including light.

6. The ability of an object to scatter a wave depends on the size of the object in comparison to the wavelength. An object small compared to the wavelength will cause no appreciable scattering. If the object size is comparable to the wavelength the scattering depends strongly on wavelength, with shorter wavelengths being more strongly scattered.

Bats use the reflection of ultrasonic echoes to locate prey.

• Qualitative Questions

17M.1 A sound wave is

A. a transverse wave.
B. not able to transmit energy.
C. a compression wave.
D. able to travel through solids, liquids, gases or vacuum.
E. a type of very low frequency electromagnetic wave.

17M.2 Suppose you are swimming under water and someone in the air above you blows a whistle with a frequency of 600 Hz. The frequency you would hear would be

A. less than 600 Hz.
B. 600 Hz.
C. greater than 600 Hz.
D. a mixture of frequencies, some greater than 600 Hz and some less.
E. You would not hear any sound because sound will not travel from one medium into another.

17M.3 When a sound wave goes from air into steel

A. its frequency increases.
B. its velocity decreases.
C. its wavelength increases.
D. it becomes more intense.
E. it changes from a longitudinal wave into a transverse wave.

17M.4 When a 600 Hz sound wave travels through a certain type of rock it is found to have a wavelength of 10 meters. Thus a 400 Hz sound wave in the same material would have a wavelength of

A. 5 m
B. 6.7 m
C. 10 m
D. 13.3 m
E. 15 m

17M.5 In aerodynamics it is useful to measure the speed of a rapidly moving object such as a jet plane in terms of its Mach number, defined as the ratio of the speed of the plane to the speed of sound. In view of this, we would expect that a plane flying at Mach 2 at an elevation of 35,000 ft (where the air is very cold) would have a ground speed which is

A. about the same as for a plane at Mach 2 near sea level.
B. less than that of a plane at Mach 2 near sea level.
C. greater than that of a plane at Mach 2 near sea level.
D. 680 m/s.

17M.6 Which of the following is the most accurate statement?

A. The frequency of a sound wave is the primary factor determining the sensation of loudness when the sound is heard.
B. The velocity of a sound wave is proportional to the number of times per second that the sound makes your ear drum vibrate.
C. The wavelength of a sound wave travelling through air will increase if the frequency is increased.
D. The velocity of a sound wave refers to the speed of the vibrating molecules in the medium through which the sound is propagating.
E. The velocity of a sound wave refers to the speed at which the pressure disturbance moves through the medium.

17M.7 Shown here are two oscilloscope traces displaying two different sound waves detected by a microphone. The horizontal scales and the vertical scales are the same in both cases. From this information we can deduce that

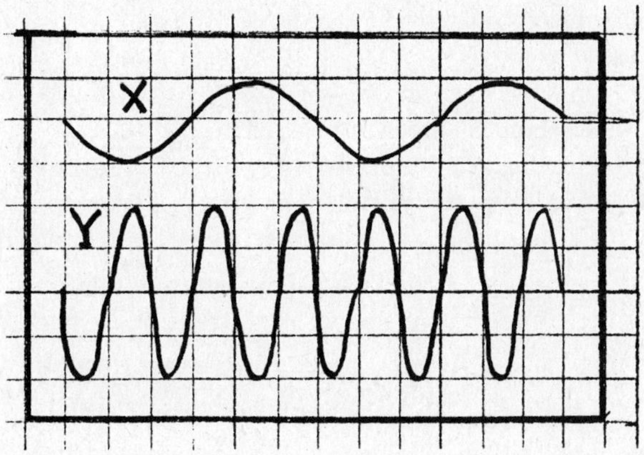

A. X has a higher pitch than does Y.
B. X has greater amplitude than does Y.
C. the sound wave which generated X would travel faster than the sound which generated Y.
D. if one of these was generated by the fundamental resonance of a closed organ pipe, the other could be due to one of the overtones.
E. both of these sound waves would sound the same to the human ear.

17M.8 For the oscilloscope trace in question 16M.7 the horizontal time scale is 1 ms per division (1 ms = 10^{-3}s). What is the frequency of sound wave Y?

A. 400 Hz
B. 500 Hz
C. 667 Hz
D. 1000 Hz
E. 2000 Hz

17M.9 Sound intensity is measured in

A. meters
B. meters/second
C. Hertz
D. watts
E. watts/m^2
F. Joules

17M.10 The loudness of a sound wave is dependent primarily on the wave's

A. velocity.
B. wavelength.
C. frequency.
D. amplitude.
E. polarization.

17M.11 Suppose that a source of sound is emitting waves uniformly in all directions. If an observer at a point a distance R from the source hears a sound of frequency f, an observer at a distance 2R from the source will hear a frequency

A. f
C. $\frac{1}{2}$f
B. $\frac{1}{4}$f
D. 2f
E. 4f

17M.12 One may most readily deduce that sound intensity decreases proportionally to $1/R^2$ as the distance R from a uniformly radiating source of sound increases. One may most readily deduce this result by making use of

A. the relation $v = f\lambda$.
B. Newton's Laws of Motion.
C. the law of conservation of energy.
D. the law of conservation of momentum.
E. the fact that sound is a longitudinal wave.

17M.13 The fundamental frequency of an organ pipe is primarily dependent on

A. the pipe diameter.
B. the pipe length.
C. the kind of material from which the pipe is made.
D. the amplitude of the fundamental vibration.
E. the way in which the vibrations are excited.
F. more than one of the above.

17M.14 If you were to inhale a few breaths of inert helium gas you would probably experience an amusing change in your voice. You would probably sound like a close relative of Donald Duck and Alvin the chipmunk. What is the cause of this curious effect?

A. The helium causes your vocal cords to tighten and vibrate at a higher frequency.
B. For a given frequency of vibration of your vocal cords, the wavelength of sound is less in helium than it is in air.
C. Your voice box is resonating at the second harmonic, rather than at the fundamental frequency.
D. Low frequencies are absorbed in helium gas, leaving the high frequency components which result in the funny squeaky sound.
E. Sound travels faster in helium than in air at a given temperature.

17M.15 Sketched here are some wavefronts emitted by a source of sound S. This picture can help us to understand

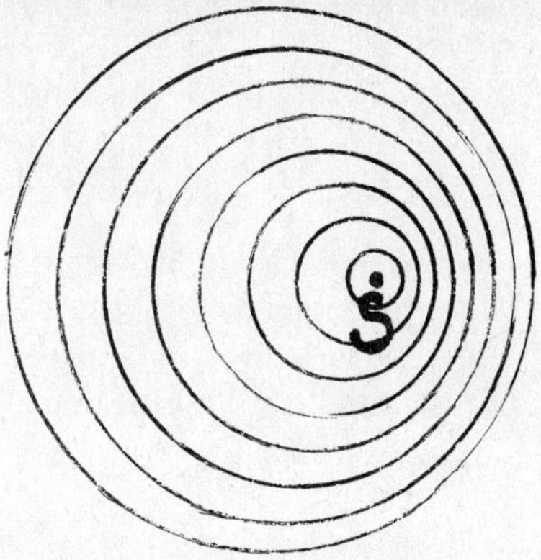

A. why the siren on a police car changes its pitch as it races past us.
B. why a sound grows quieter as we move away from the source.
C. how sonar works.
D. the phenomenon of beats.
E. why it is that our hearing is best near 3000 Hz.

17M.16 Suppose that while out birdwatching with your portable sound detector you record a sound intensity of 48 dB when an eagle cries out while flying overhead. If the bird had been twice as far away, what level would you have detected?

A. 12 dB
B. 24 dB
C. 42 dB
D. 45 dB
E. 48 dB

• Multiple Choice: Answers and Comments

17M.1 C A sound wave can transmit energy. It cannot travel in a vacuum.

17M.2 B The frequency does not change when sound goes from one medium to another. Think of the process by which sound is transmitted from one material to another (like someone standing outside and pounding rhythmically on your door.) The frequency you hear is the same as the frequency with which he pounds.

17M.3 C When sound goes from air into steel the frequency stays constant and the velocity increases, and thus since $v = f\lambda$, the wavelength λ must increase.

17M.4 E $v = f_1\lambda_1 = f_2\lambda_2$. Thus $\lambda_2 = \frac{f_1}{f_2}\lambda_1 = (\frac{600\text{ Hz}}{400\text{ Hz}})(10\text{ m}) = 15\text{ m}$

17M.5 B The speed of sound in a gas is $v = \frac{\sqrt{\gamma kT}}{m}$, thus at high elevation where the air is cold the velocity of sound is less than it is at sea level and so a Mach 2 plane has lower ground speed than does one at sea level.

17M.6 E

17M.7 D The period of X is 6 divisions and that of Y is 2 divisions, thus

$$\frac{f_x}{f_y} = \frac{T_y}{T_x} = \frac{2}{6}, \text{ so } f_Y = 3f_x$$

Since f_Y is an integral multiple of f_X, it could be one of the harmonics of f_X.

17M.8 B From the sketch we see T = 2 ms = 0.002 s, f = 1/T = 500 Hz.

17M.9 E

17M.10 D

17M.11 A Intensity, but not frequency or wavelength, changes with distance from the source.

17M.12 C Imagine two spheres, of radii R_1 and R_2, surrounding a source. All of the energy which passes out through one of them in one second must also pass out through the other if energy is conserved. If I is the intensity (in W/m^2) and $4\pi R^2$ is the surface area of a sphere,

$$I_1 \cdot 4\pi R_1^2 = I_2 \cdot 4\pi R_2^2$$

$$\text{or} \quad \frac{I_1}{I_2} = \frac{1/R_1^{\,2}}{1/R_2^{\,2}}$$

17M.13 B

17M.14 E If you try this, don't take too many breaths. You may faint.

17M.15 A A stationary source emits a series of wave crests which travel out in concentric circles. The source which emitted the crests drawn here was moving to the right. If you were positioned off to the right the separation between crests would be smaller than normal, and hence the frequency you would hear would be higher than normal, since f = v = constant. This is the Doppler effect. If you were positioned off to the left you would hear a lower frequency.

17M.16 C $\beta_1 = 10\log\frac{I_1}{I_0}$ $\quad \frac{I_1}{I_2} = \frac{1/R_1^2}{1/R_2^2} = \left(\frac{R_2}{R_1}\right)^2 = \left(\frac{2R_1}{R_1}\right)^2 = 4$

$\beta_2 = 10\log\frac{I_2}{I_0} = 10\log\frac{\frac{1}{4}I_1}{I_0} = 10\log\frac{I_1}{I_0} + 10\log\frac{1}{4}$

$\beta_2 = 10\log\frac{I_1}{I_0} - 10\log 4 = 48\,dB - 6\,dB = 42\,dB$

USING $\log AB = \log A + \log B$ and $\log\frac{1}{C} = -\log C$

• Problems

Assume the speed of sound in air is 340 m/s at normal temperature unless otherwise indicated.

17.1 What is the speed of sound in H_2 gas at 20 °C? For H_2 we find $\gamma = 1.4$ from tables of heat capacities, $\gamma = cp/cv$.

17.2 A boy makes a whistle from a tube 20 cms long, closed at both ends. He makes a small slit in it through which he can blow to excite standing sound waves. What would be the frequency of the fundamental and of the first overtone?

17.3 Suppose that a stereo amplifier delivers 20 W of power to a loudspeaker which has an efficiency of 10 percent (i.e. 10 percent of the power is converted into sound and the rest is converted into heat). If this sound were radiated out uniformly in all directions (a first approximation) what sound intensity level would a person 4 meters from the speaker experience? Express your result in W/m² and in dB.

17.4 A bat emits an ultrasonic shreik which returns to him in 20 ms. How far away was the object from which the sound pulse was reflected?

17.5 A musician standing on a train platform hears the whistle of an approaching train. Having a well calibrated ear he notes that the frequency of the whistle on the approaching train is 740 Hz, but as the train rushes past and vanishes down the track the frequency drops to 600 Hz. How fast was the train going?

• Problem Solutions

17.1 $v = \sqrt{\frac{\gamma kT}{m}}$, $m = 2m = 2 \times 1.67 \times 10^{-27}\ \text{kg} = 3.34 \times 10^{-27}\ \text{kg}$

$$v = \sqrt{\frac{(1.4)(1.38 \times 10^{-23}\ \text{J/°K})(293\text{°K})}{3.34 \times 10^{-27}\ \text{kg}}} = \underline{\underline{1302\ \text{m/s}}}$$

17.2 FROM EQ. 16.2 $f = \frac{nv}{2L}$

FOR $n = 1$ (FUNDAMENTAL) $f_1 = \frac{340\ \text{m/s}}{(2)(.2\ \text{m})} = \underline{\underline{850\ \text{Hz}}}$

FOR $n = 2$ (FIRST OVERTONE) $f_2 = 2f_1 = \underline{\underline{1700\ \text{Hz}}}$

17.3 10% OF 20W = 2W = SOUND POWER = P_s

IF SOUND IS RADIATED OUT IN ALL DIRECTIONS, INTENSITY ON THE SURFACE OF A SPHERE OF RADIUS 4 m IS I, WHERE

$$4\pi R^2 I = P_s \quad (4\pi R^2 = \text{AREA OF SPHERE})$$

$$I = \frac{P}{4\pi R^2} = \frac{2\text{W}}{(4\pi)(4\text{m})^2} = \underline{\underline{10^{-3}\ \text{W/m}^2}}$$

$$\beta = 10 \log \frac{I}{I_0} = 10 \log \frac{10^{-3} W/m^2}{10^{-12} W/m^2} = 10 \log 10^9 = 90 \log 10 = \underline{\underline{90 \text{ dB}}}$$

NOTE: IN MOST COUNTRIES WORKERS ARE NOT ALLOWED TO WORK IN AN ENVIRONMENT WHERE "NOISE POLLUTION" LEVEL EXCEEDS 85dB. BECAUSE TO DO SO IS HAZARDOUS TO ONE'S HEALTH.

17.4 $2d = vt$, $d = \frac{vt}{2} = \frac{(340 \text{ m/s})(20 \times 10^{-3} \text{J})}{2} = \underline{\underline{3.4 \text{ m}}}$

17.5

$$f + \Delta f = 740 \text{ Hz}$$
$$f - \Delta f = 660 \text{ Hz}$$

SUBTRACT: $0 + \Delta f = 80 \text{ Hz}$, $\Delta f = 40 \text{ Hz}$

$$f = 700 \text{ Hz}$$

$$\frac{\Delta f}{f} = \frac{V_s}{V}$$

$$V_s = \frac{\Delta f}{f} V = \left(\frac{40 \text{ Hz}}{700 \text{Hz}}\right)(340 \text{ m/s})$$

$$V_s = \underline{\underline{19.4 \text{ m/s}}}$$

Electrostatics: Charges, Electric Fields, and Potentials

18

• Summary of Important Ideas, Principles, and Equations

1. Electric charge describes an attribute of certain subatomic particles, of which the most important are electrons and protons. Charge always appears in nature as a multiple of a fundamental unit of charge which in SI units is assigned the value

$$e = 1.6 \times 10^{-19} \text{ Coulomb}$$

The charge on a proton is +e and the charge on an electron is -e. We thus say that electric charge is quantized (i.e. exists only in discrete little chunks).

The nucleus of an atom of atomic number Z contains Z protons and has a net electric charge of +Ze. Surrounding the nucleus is a cloud of Z electrons with a net charge of -Ze. Thus an atom is electrically neutral. Furthermore, it appears that the universe as a whole is electrically neutral.

Experimentally we observe that charge cannot be created or destroyed. We say that charge is thus conserved.

When a particle moves very rapidly (at speeds approaching the velocity of light, 3×10^8 m/s) its mass increases. Thus mass is not relativistically invariant. The charge on a particle does, however, remain invariant when a particle speeds up. This invariance of electric charge with respect to motion has very important consequences in the theory of relativity.

2. Conductors (such as metals) are materials in which some of the electrons (called free electrons) can move relatively freely. In insulators, such as glass or paper, the electrons cannot move about freely. In semiconductors, such as silicon, there are a few free electrons, and their number depends on temperature and specimen purity.

When we say that a material "conducts electricity" we mean that electric charge (usually that on free electrons) is moving through the material.

3. Charged bodies exert forces on each other. Like charges repel, opposite charges attract, with the magnitude of the force given by Coulomb's law:

$$F = k\frac{q_1 q_2}{R^2} \quad (18.1)$$

Here F is the force acting on charge q_1 due to charge q_2 (or vice versa). F is in Newtons and q is in Coulombs. The constant k has the experimental value $k = 9 \times 10^9\ N\cdot m^2/C^2$.

The electrostatic force due to several charges is the vector sum of forces due to each charge individually.

4. Objects may be charged when electric charge flows on to them (conduction). Objects also can be charged by induction. If a positively charged rod, say, is brought near a large metal object, electrons on the object will move to the side closest to the positive rod, leaving a net positive charge on the opposite side of the object. These are called induced charges.

"Ground" is a term used to describe any very large conductor (such as the earth).

5. A charge +q and another charge -q separated by a distance d constitute an electric dipole moment. The magnitude of the dipole moment is defined to be

$$p = qd \quad (18.2)$$

The electric dipole moment is a vector quantity, directed from -q to +q.

Electric dipoles can be permanent (as in the H_2O molecule) or induced (as in glass or paper).

6. We can describe the electric forces due to stationary charges by introducing the concept of the electric field, $\vec{E}(\vec{r})$.

Suppose a small positive test charge q is placed at position $\vec{r}$ near a collection of other charges. It will experience a force $\vec{F}(\vec{r})$ which is proportional to q. If q is large, F is large, and vice versa. We define the electric field $\vec{E}(\vec{r})$ by

$$\boxed{\vec{F}(\vec{r}) = q\vec{E}(\vec{r})} \tag{18.3}$$

$\vec{E}$ is a vector quantity with units N/C. Later we will see that we can define a more convenient unit, the volt, such that electric field may also be measured in volts/meter.

The meaning of E is that it is the force which a unit charge will experience in the presence of other charges which create the electric field E. If q is positive F and E are parallel. If q is negative (as for an electron) F and E are antiparallel.

7. We can visualize the electric field by drawing electric field lines. At every point the tangent to a field line points in the direction of E. Where the lines are close together the field is strong. These lines are analogous to the flow lines in a fluid. The flow lines show the path of a little cork floating in a moving stream. Where the flow lines are close together the stream is moving rapidly.

Although E-lines do not represent the "movement" of anything, the fluid flow analogy is useful in understanding them. The "sources" of the E-lines are positive charges and the "sinks" are negative charges. The E-lines sprout out of + charges and end on - charges. Thus the number of E-lines passing out through any closed surface is proportional to the amount of positive charge enclosed by the surface. The constant of proportionality is related simply to the constant in Coulomb's law. This idea, in mathematical form, is called Gauss's law and can be used to figure out E for certain simple charge distributions.

8. Any net charge on a conductor will always move to the outer surface, and the electric field within a perfect conductor is zero.

9. Suppose that a collection of electric charges create an electric field. A charge q placed in this field will experience a force. Suppose we move the charge from point A to point B, pushing against the electric field. It turns out that the work we must do in taking q from A to B does not depend on the path we follow from A to B. This means that E is a conservative field (it conserves the energy of the charge q). For such a field we define the electrostatic potential difference V_{AB} between points A and B such that

$$\boxed{qV_{AB} = qV_A - qV_B = \text{work done in carrying } q \text{ from A to B}} \qquad (18.4)$$

V_A = electrostatic potential at point A.

Potential has units of $\frac{\text{work}}{\text{charge}} = \frac{J}{C}$ = volt.

The magnitude of the force on a charge q is eE. Consider the work done in going a distance d from A to B against a constant electric field E.

Then work done = Force x Distance = qEd = qV_{AB} .

Thus

$$\boxed{V_{AB} = Ed} \qquad \text{(Constant field)} \qquad (18.5)$$

Since V_{AB} is measured in volts, E can be measured in volts/meter.

We can think of qV_A as the electrostatic potential energy of a charge q at point A in an electric field E.

The potential V is thus the potential energy per unit charge in the field E.

Be careful not to confuse the different terms potential (in volts) and potential energy (in Joules).

Potential increases when moving "upstream" against the E-lines.

Since only differences in potential energy are important, the choice of a point at which V = 0 is arbitrary. We usually choose V = 0 far away where R = ∞.

10. The potential at a distance r from an isolated charge Q is

$$\boxed{V(r) = k\frac{Q}{r}} \qquad (18.6)$$

Here k = 9×10^9 volt · m/C.

A positive charge Q causes a positive potential V, and a negative charge Q causes a negative potential.

11. A useful analogy which helps us to understand potential and electric field is that of a topographical map. The potential V is like the elevation, h. The electric field is like the negative of the slope of a hill.

In one dimension we can write

$$E = -\frac{\Delta V}{\Delta x} \quad (18.7)$$

where ΔV is the change in V which occurs in a distance Δx.

The locus of points in three dimensions which are at the same potential form an equipotential surface. In two dimensions these points form an equipotential line. An equipotential line is analogous to a contour line (the locus of points at a constant elevation).

No work is done when a charge is moved along an equipotential surface. This means that electric field lines must be perpendicular to equipotential surfaces. In the map analogy imagine you are looking down from above when a small ball is released on a hillside. The ball is like a positive charge. It moves in the direction of steepest slope, i.e. along an electric field line. Thus electric field lines are like lines of steepest slope on a map.

Imagine a horizontal stretched rubber sheet. Poke a sharp spike up from the bottom and make a pointed mountain. This is the effect a positive charge has on the potential (represented by the elevation of the sheet). Push a spike downward and you have the effect of a negative charge. This analogy isn't exact, but it gives the general idea.

All parts of a conductor are at the same potential. The electric field everywhere within a conductor is zero. Since the surface of a conductor is an equipotential, electric field lines are always perpendicular to the surface at the surface. Further, since the E-lines end on electric charges, the charge density on the surface is proportional to the electric field there.

In our analogy a sphere charged to a potential of 100 volts would look like a flat top mountain at a constant elevation of, say, 100 meters.

12. In our analogy a charge q is analogous to a mass m.

The potential energy of a mass m at elevation h in a gravitational field g is mgh.

The potential energy of a charge q at potential V is qV.

In a constant electric field E the potential at a distance d from the point where $V = 0$ is $V = Ed$. Thus the potential energy may be written $qV = qEd$. This is analogous to mgh, so q is analogous to m

E is analogous to g

d is analogous to h (height above "zero" or sea level).

If a charge q is "lifted" through a potential difference V its potential energy increases by an amount qV. Frequently we are interested in the change in potential energy when an electron or proton of charge e is raised through a potential difference V. For such measurements it is useful to define a new energy unit, the electron volt.

Definition: 1 electron volt is the amount of potential energy gained when a charge $e = 1.6 \times 10^{-19}$ C is lifted through a potential difference of 1 volt.

Thus

$$1 \text{ e.V.} = (1.6 \times 10^{-19}\text{C}) \times (1 \text{ volt}) = 1.6 \times 10^{-19}\text{J} \quad (18.8)$$

Note that one "e.V." is just a very small energy unit which can be used to measure the energy of anything, not just electrons.

When a particle of mass m and charge q falls from rest through a potential difference of V volts it will acquire speed v, where

Gain in KE = Loss in PE

$$\tfrac{1}{2}mv^2 = qV \quad \text{or} \quad v = \sqrt{\frac{2qV}{m}}$$

If an electron falls through a potential difference of 100 volts its kinetic energy will increase by 100 e.V. In calculating velocities be sure to express energy in Joules, not in e.V.

• Qualitative Questions

18M.1 Gravitational forces between ordinary objects are very small and we don't usually notice them. Electrical forces, on the other hand, can be very large. Why don't we then notice an electric force between most objects?

A. They don't contain free electrons.
B. Most objects are electrically neutral.
C. Any electric charge on one's body will immediately drain off to the earth.
D. Gravitational forces cancel the electric forces.
E. Since the universe as a whole is electrically neutral, for every electric force pulling an object in one direction there is a compensating force pulling it in the opposite direction.

18M.2   An entertaining thing to do for a child's birthday party is to
        stick balloons all over the wall of the party room. You can do
        this by first rubbing each balloon against your sweater and then
        pressing it to the wall. Presto Chango, it sticks! Why?

        A. Rubbing the balloon roughens the surface so that it now will
            make very close contact with the wall and be held in place by
            air pressure.
        B. Rubbing the balloon causes moisture to condense on it, and
            surface tension causes the balloon to stick to the wall.
        C. The charge on the balloon induces an opposite charge on the
            surface of the wall.
        D. The charge on the balloon induces charge of the same sign on
            the wall, and these two charges are drawn together by their
            mutual repulsion of other stray charges in the air.
        E. Rubbing the balloon neutralizes the electrical charge on the
            balloon, thereby allowing it to form a weak chemical bond
            with the molecules in the paint on the wall.

18M.3   Suppose that isolated charges Q_1 and Q_2 attract each other with a
        force F. If the separation between these charges were made half
        as great, each charge would then experience a force

        A. F
        B. 2F $F = \dfrac{KQ_1Q_2}{(\frac{1}{2}r)} \quad \dfrac{Ka}{\frac{1}{4}r^2}$
        C. 4F
        D. which cannot be determined unless we know the magnitude of
            Q_1 and Q_2.

18M.4   An electroscope is a simple little gadget which played an
        important role in early investigations of electricity. It is
        still a useful device. It consists of a metal rod connected
        to two very thin metal leaves (usually made of gold foil).
        When the electroscope is uncharged the leaves hang straight
        down. If a charged object is touched to the metal ball on the
        top of the electroscope the leaves suddenly spring apart and
        stay that way.   Why does this happen?

        A. Charge is placed on the leaves
            and attracted to opposite
            induced charge on the glass
            bottle.
        B. A net charge is placed on
            the leaves and spreads out
            over them. Since like
            charges repel, the leaves
            are pushed apart.
        C. Electrons flow from the
            leaves onto the charged
            object which has touched
            the electroscope. This
            makes the leaves lighter   (cont.)
            and thus they rise up.

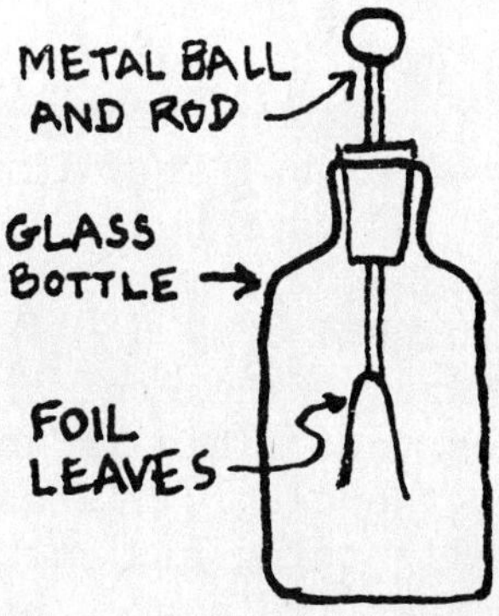

D. The charged object induces charge of one sign on the metal ball on top of the electroscope and, by induction, charge of the opposite sign on the leaves. Since these opposite charges attract, the leaves are pulled up toward the metal ball.

E. The leaves fly apart because electric charge is always conserved.

18M.5 If a charged object is brought near the metal ball on the top of an electroscope (but not allowed to touch it),

A. both leaves will be pulled toward the charged object.
B. both leaves will be pushed away from the charged object.
C. the leaves will not change their position.
D. the leaves will diverge.
E. the leaves will be pulled together more tightly than ever.

18M.6 Not too many years ago Lee Trevino was almost killed when he was zapped by a lightning bolt on the golf course. Recently someone asked him what one should do to avoid injury if he happens to be out on the course in a lightning storm.

"Pull out a one iron," replied SuperMex, "hold it over your head and run like crazy for the clubhouse."

"But Lee," protested his questioner, "isn't that awfully dangerous? Won't that just guarantee that you will get hit by lightning?"

"Hell, no," says Trevino. "Even God can't hit a one iron!"

(You have to be a golfer to appreciate that one.)

What do you think? Would it be particularly dangerous to hold up a metal one iron in a lightning storm?

A. Yes, because electrons flow up to the top of the club and attract the electric charges released in the lightning bolt.
B. Yes, because lightning will only hit a conductor such as a metal club. It will not be attracted to or hit an insulator such as your body.
C. Yes, because metal becomes very hot when electricity flows through it.
D. No, quite the contrary. The lightning will hit the club and miss you. This is the principle of the lightning rod.
E. No, the club will have no effect. Whether or not you are hit by lightning has nothing to do with whether or not you are in an open field holding up a metal club or whether you are sleeping in your cellar. Being hit by lightning is a purely chance event.

Questions 18M.7-18M.11 refer to the drawing here.

18M.7 From the diagram we see that

A. $Q_1 > 0$

B. $Q_1 < 0$

C. Q_1 is a unit charge.

D. There is no force acting on Q_1.

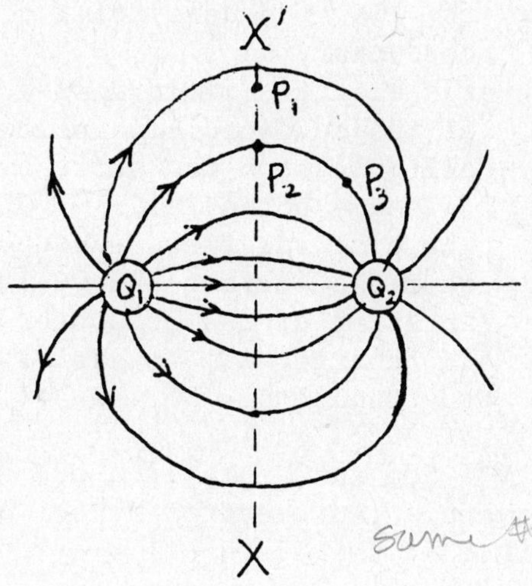

18M.8 From the diagram we also see that

A. $Q_1 = Q_2$

B. $Q_1 = -Q_2$

C. $Q_1 > Q_2$

D. $Q_1 < Q_2$

18M.9 Consider a charge Q placed at P_1 or P_2.

A. There is no force on Q at P_1.
B. The force on Q is the same at P_1 and at P_2.
C. The force on Q is greater at P_2 than at P_1.
D. The force on Q will be directed along the line X-X'.

18M.10 The line X-X' represents

A. a surface of constant force.
B. a surface of constant potential.
C. the only equipotential surface for this array of charge.
D. the locus of points where the electric field vanishes.

18M.11 The work required to move a charge Q from P_3 to P_2 is

A. greater going from P_3 to P_1 to P_2 than it is in going directly from P_3 to P_2 along a straight line.
B. independent of the path followed.
C. independent of the value of Q.
D. independent of the value of Q_1 and Q_2.
E. zero.

18M.12 When there is a net charge present on a perfect conductor, and no other charges are present,

A. the charge density will be greatest where there are sharp protuberances or points on the surface.
B. the charge density will be uniform over the surface of the conductor.
C. the electric field inside the conductor need not be zero if the conductor is hollow.
D. all points on the surface of the conductor need not be at the same potential.
E. the electric field inside the conductor will be greatest near the surface and will gradually decrease in the interior of the conductor.
F. the charge will be distributed throughout the volume of the conductor.

18M.13 A V^{4+} ion and an Mn^{+} ion are separated by 9×10^{-10}m in a crystal. An electron placed between them would experience no force if the distance from the electron to the V^{4+} ion were

A. 1.7 Å
B. 3.0 Å
C. 4.5 Å
D. 5.1 Å
E. 6.0 Å

18M.14 A perfect electric conductor is a material in which charge carriers (usually electrons) can move freely. Because of this property, one would be led to the conclusion that, in the case of electrostatic equilibrium (nothing changing with time),

A. the net charge in a conductor is uniformly distributed throughout its volume.
B. there is a constant non-zero electric field throughout the volume of a conductor.
C. the electric field is zero everywhere throughout the interior volume of a conductor.
D. it is not possible for a conductor to carry a net electric charge.

18M.15 Suppose that a charged cloud induces charge on the surface of the earth. The resulting electric field at the earth's surface will be

A. greatest where the surface charge density (in coulombs/meter^2) is greatest.
B. a result of the charge on the cloud only, and not due to the induced charge on the earth.
C. zero, since the earth is "grounded".
D. the same at all points on the earth's surface.

18M.16 We know that the gravitational force exerted on an object of mass m at the earth's surface is mg. We could think of g as the "force per unit mass" due to all of the mass of the earth. In electricity what quantity is analogous to g, viewed in this way?

A. Electric charge
B. Electric potential
C. Electric field
D. Electric force
E. Electric energy

18M.17 Which, if any, of the following statements are true?

A. If the electric field is zero throughout some region of space, the electric potential must also be zero in that region.
B. If the electric potential is zero throughout some region in space, the electric field must also be zero within that region.
C. If the electric field is zero at a point, the potential must also be zero at that point.
D. If the electric potential is zero at a point, the electric field must also be zero at that point.
E. Lines of electric field point towards regions of higher potential.

18M.18 If the electric potential varies from 100 volts to 150 volts in a distance of 5 meters, the average electric field in this region is

A. 6.25 v/m
B. 5 v/m
C. 7.5 v/m
D. 10 v/m
E. 20 v/m
F. 25 v/m
G. 30 v/m

18M.19 Which of the following best describes the motion of a point charge q of mass m in an electric field?

A. If released from rest it will remain on an equipotential surface.
B. If released from rest it will start to move and will continue to move along one line of the electric field.
C. If released from rest it will start to move along an electric field line, but it may not continue to move along this E-line.
D. It will accelerate with uniform acceleration.
E. It will move with constant velocity.

18M.20 Which of the following statements is true?

A. The potential due to an array of point charges is the sum of the potentials due to each individual charge.
B. The electric field is zero where the potential is zero.
C. The potential due to a single point charge may vary from positive to negative in different regions.
D. The force on a charged particle on an equipotential surface is zero.
E. It is possible for electric field lines to intersect in some cases.

• Multiple Choice: Answers and Comments

18M.1 B

18M.2 C

18M.3 C According to Coulomb's law the force between two charges varies inversely as the square of the distance between them.

$$F = k\frac{Q_1Q_2}{R^2}$$

If R is made half as large, F will increase by a factor of four.

18M.4 B

18M.5 D Suppose the charged object carries a positive charge. Even though it does not touch the electroscope, electrons (with negative charge) will flow up the metal rod, since they are attracted to the positive object. This leaves a net positive charge on the leaves and they thus repel each other, flying apart. If the charged object is moved away, the electrons once again distribute themselves uniformly over the rod and leaves and they are once again uniformly neutral.

18M.6 A Since the force between charged objects increases as the separation decreases, any object which projects up toward a charged cloud (such as a tall tree or a church spire or even a person standing up above her surroundings) will have induced charge appear on the highest points. This is true even for an object which is a relatively good insulator (such as your body). The effect is greatly accentuated for a conductor, tor, for then large amounts of charge can flow to the highest point. This induced charge will attract the charges released when the lightning bolt "strikes". If the metal rod is connected directly to the earth the flow of charge will pass through the metal rod directly to the earth. This is the basis for a lightning rod attached to a building. If the rod is not connected to earth (except via your body, as with the golf club) the electric charge from the lightning will flow through you and fry you to a crisp. You may have seen such burns on lightning struck trees in the forest. In a lightning storm, always keep a low profile!

18M.7 A Electric field lines are sprouting out of Q_1, thus it is a positive charge.

18M.8 B Q_1 must be positive since E-lines sprout out of it and Q_2 must be negative since E-lines end on it. The same number of lines end on each, so the magnitude of Q_1 is the same as that of Q_2.

18M.9 C There is a force on Q at both P_1 and P_2, but the E-lines are closer together at P_2. The force is directed tangent to the E-lines. Even where no line is drawn there is still an electric field.

18M.10 B

18M.11 B

18M.12 A The charge will all be on the outer surface of the conductor. The electric field within the conductor will be zero. This means that the electric field will also be zero inside any hollow cavity. If there is a sharp protuberance charge will tend to be concentrated there. The charges all try to get as far from each other as possible. When they are near a sharp point a charge like Q_1 here tends not to feel the effect of a charge like Q_2 which is "just over the hill". Thus the charges tend to scrunch up near sharp points and cause a big electric field there. When the electric field is strong enough to cause the air molecules to be pulled apart, the resulting spark will occur near a sharp point. For this reason it is important that high voltage apparatus be smoothly polished.

18M.13 E We require

$$k\frac{(4e)(-e)}{R^2} = k\frac{(g)(-e)}{(d-R)^2}$$

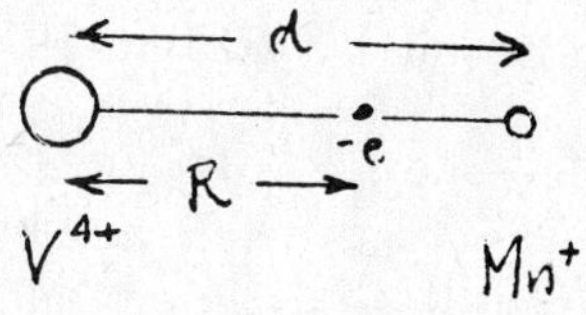

or $4(d - R)^2 = R^2$, $2(d - R) = R$

$$R = \frac{2}{3}d = \frac{2}{3} \times 9 \times 10^{-10}\,\text{m} = 6 \times 10^{-10}\,\text{m}$$

18M.14 C If the electric field were not zero in the conductor the free charges would be pushed away by it. They in turn would set up a "back" electric field which would tend to cancel any electric field which was present. This process would continue until the electric field was zero in the conductor.

18M.15 A Imagine the electric field lines sprouting out of charges. Where there is a high surface charge density there will thus be a strong electric field. The earth is a conductor, and charges will be drawn to the area beneath the charged cloud, and it is here that the electric field will be strongest.

18M.16 C Recall that gravitational potential energy is mgh and electric potential energy is QV or QEd. Q is like mass m and E is like g.

18M.17 B Recall our topographic map analogy. For "potential" think "elevation" and for "~~electric field" think "slope~~". Answer A would then read, "If the slope is zero in a region the elevation there is zero (i.e. sea level)". This need not be true. Apply this analogy of rewording to the other choices and the correct answer will be evident.

18M.18 D

$$E = \frac{\Delta v}{\Delta x} = -\frac{(100 \text{ volts} - 150 \text{ volts})}{5 \text{ meters}} = 10 \text{ volts/m}$$

18M.19 C A charge will start to move along an electric field line, since this is the direction of the force acting. However, after it has picked up speed it may "overshoot" and continue more or less straight ahead even though the electric field line may make a sharp turn to one side. As long as the velocity is low or if the electric field lines don't bend too sharply the particle will move approximately along one line.

18M.20 A

• Problems

18.1 What is the force of attraction between a Na^+ ion and a Cl^- ion separated by 0.5 nm?

18.2 A charge $Q_1 = +9\ \mu C$ is placed at the origin and a second charge $-Q_2 = -4\ \mu C$ is placed at x = 2m. Where on the x axis can a charge Q be placed and experience no force?

18.3 Charges +Q are placed at opposite corners of a square of side a. Charges -Q are placed at the other two corners. What is the magnitude and direction of the force on one of the charges?

18.4 Two identical foam balls, each of mass 0.1 gm, are given equal charges while hung side by side from threads 50 cms long. They are observed to hang motionless with a separation of 2 cms. What is the charge on each ball? (This idea gives a simple way of measuring the charge on an object.)

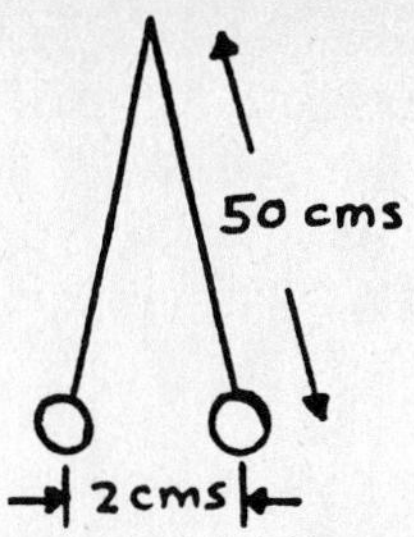

18.5 A proton at rest is released in an electric field of 20,000 N/C (or 20,000 V/m, since 1 N/C = 1 V/m). What is the force on it and what is its acceleration? How far will it move in 2 seconds?

18.6 An electric field of about 3×10^6 V/m will cause an electric breakdown in air. How many electrons would have to be placed on a metal sphere of radius 1 cm to achieve this field at the surface of the sphere?

18.7 Because of the charge on the earth and that accumulated on clouds, it is not unusual for electric fields of a few hundred volts per meter to occur in the atmosphere. Is it likely that such fields could support a charged raindrop? To explore this possibility, calculate the charge necessary to support a 2 mm diameter raindrop in an electric field of 150 V/m.

18.8 Four charges are placed at the corners of a rectangle as shown here. Determine the magnitude and direction of the force which would act on a charge Q placed at the center of the rectangle. What is the potential at this point? Express answer in terms of a, q and k (Coulomb's law constant).

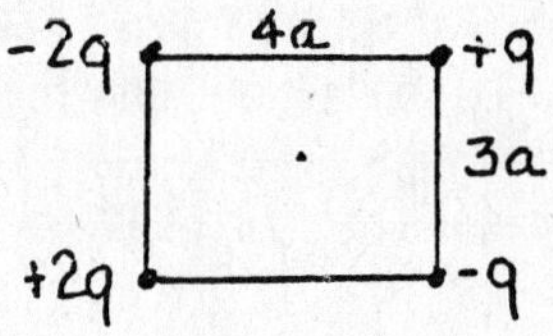

18.9 A charge of -10 μC and another of +20 μC are positioned 4 cm apart. How much work must be done to move a +2 μC charge from point A to point B?

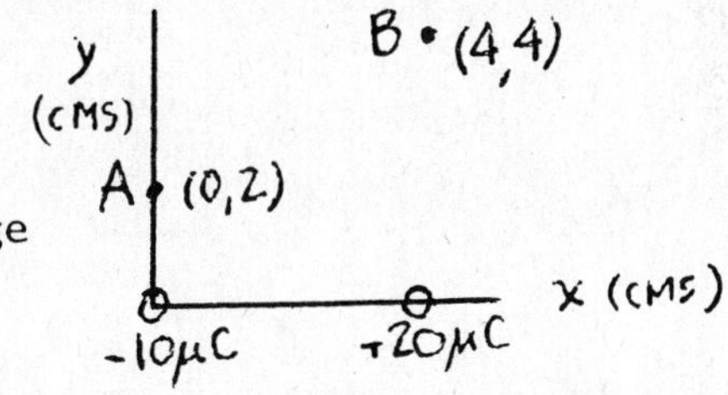

18.10 In an electron gun an electron is emitted from a cathode and accelerated toward a positive anode. A potential difference of 50 volts is applied between the two electrodes, which are separated by 20 mm. Assuming the electric field is uniform, what is the electric field in the region between the electrodes? What is the K.E. of an electron, in Joules and in e.V., when it reaches the anode?

18.11 Through what potential difference must an alpha particle be accelerated in order to reach a speed of 5×10^5 m/s? An alpha particle is a helium nucleus, of charge 2e and mass 6.7×10^{-27} Kg.

18.12 A small Van de Graaff generator has a metal sphere of radius 15 cms which can be raised to a high voltage. If the voltage is too high, sparks will occur which discharge the machine. These occur when the electric field at the surface of the sphere reaches a value of about 3×10^6 V/m, at which point the air breaks down. What is the maximum potential which can be achieved for this machine? Note that in big Van de Graaff machines, the sphere is surrounded with a special gas which doesn't break down as readily as does air.

18.13 Two large parallel plates are separated by distance d. A battery causes a potential difference V between the plates. An electron moving at speed v enters a hole in the positive plate, moving toward a negative plate. What is the minimum value of V which will prevent the electron from striking the negative plate? What time elapses between when the electron enters the hole and when it stops? What is the electric field between the plates?

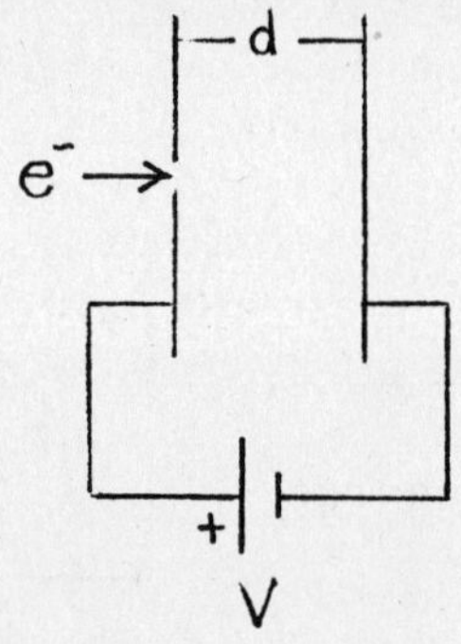

• Problem Solutions

18.1 THE CHARGE ON EACH ION IS OF MAGNITUDE $e = 1.6 \times 10^{-19}\,C$

$$F = k\frac{Q_1 Q_2}{R^2} = \left(9 \times 10^9 \frac{N \cdot m^2}{C^2}\right) \frac{(1.6 \times 10^{-19}\,C)(1.6 \times 10^{-19}\,C)}{(0.5 \times 10^{-9}\,m)^2} = \underline{\underline{9.21 \times 10^{-10}\,N}}$$

18.2 LET US CONSIDER WHAT HAPPENS TO CHARGE Q PLACED IN REGION I. ASSUME Q IS POSITIVE. IT IS PUSHED TO THE LEFT BY THE $+9\mu C$ CHARGE AND PULLED TO THE RIGHT BY THE $-.4\mu C$ CHARGE. THERE WILL THUS ALWAYS BE A NET FORCE TO THE LEFT. SINCE THE POSITIVE CHARGE AT THE ORIGIN HAS THE ADVANTAGE OF BEING BOTH LARGER AND CLOSER THAN THE NEGATIVE CHARGE.

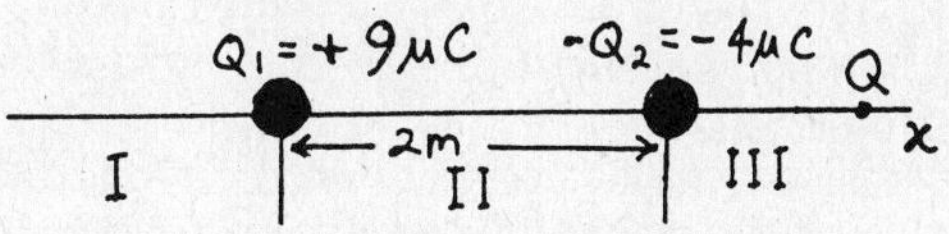

IF THE CHARGE Q IS PLACED IN REGION II IT WILL ALWAYS BE PULLED TO THE RIGHT. SO THE NET FORCE ON IT CAN NEV BE ZERO.

IN REGION III, THE FORCE ON Q CAN BE ZERO. THE CHARGE $+9\mu C$ PUSHES TO THE RIGHT AND THE CHARGE $-4\mu C$ PULLS TO THE LEFT. THE $+9\mu C$ CHARGE IS LARGER, BUT THE $-4\mu C$ IS CLOSER. IF THE CHARGE Q IS AT POINT x, THE FORCE ON IT IS

$$F = k\frac{Q_1 Q}{x^2} - k\frac{Q_2 Q}{(x-2)^2} = 0$$

(NOTE THAT I TAKE Q_1 AND Q_2 TO BE POSITIVE NUMBERS)

DIVIDE OUT kQ : $\frac{Q_1}{x^2} - \frac{Q_2}{(x-2)^2} = 0$

$$\frac{Q_1}{x^2} = \frac{Q_2}{(x-2)^2}$$

$$\frac{\sqrt{Q_1}}{x} = \frac{\sqrt{Q_2}}{x-2}$$

$$\frac{x-2}{x} = \sqrt{\frac{Q_2}{Q_1}} = \sqrt{\frac{4\mu C}{9\mu C}} = \frac{2}{3}$$

$$3(x-2) = 2x$$

$$3x - 6 = 2x$$

$$\underline{\underline{x = 6\,m}}$$

18.3 CONSIDER THE FORCES ACTING ON A CHARGE $-Q$. THE POSITIVE CHARGES EXERT THE TWO FORCES F_1 WHICH ARE EQUIVALENT TO A SINGLE RESULTANT FORCE F_R, WHERE

$$F_R = \sqrt{2}\,F$$

(F_R IS THE DIAGONAL OF A 45° RIGHT TRIANGLE WITH F_1 FOR EACH SHORT SIDE.)

THE NET FORCE ON CHARGE $-Q$ IS THUS DIRECTED TOWARD THE CENTER AND OF MAGNITUDE

$$F = F_R - F_2$$

F_2 IS THE REPULSIVE FORCE BETWEEN THE TWO NEGATIVE CHARGES WHICH ARE SEPARATED BY $\sqrt{2}\,a$.

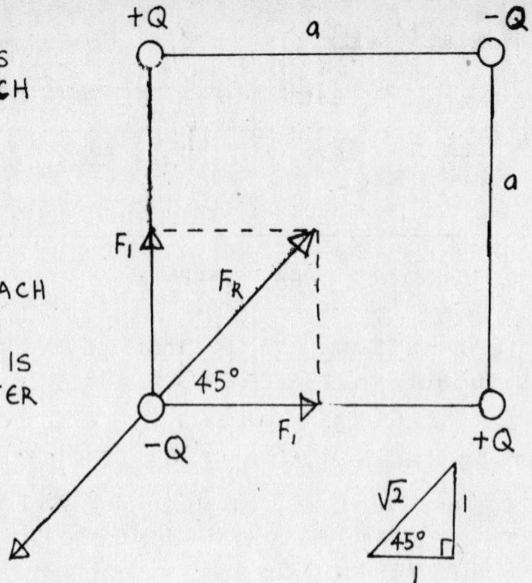

$$F_1 = k\frac{Q^2}{a^2}, \quad F_2 = k\frac{Q^2}{(\sqrt{2}a)^2} = k\frac{Q^2}{2a^2}$$

$$\text{THUS } F = F_R - F_2 = \sqrt{2}\,F_1 - F_2$$

$$= \sqrt{2}\,k\frac{Q^2}{a^2} - \frac{1}{2}k\frac{Q^2}{a^2}$$

$$F = (\sqrt{2} - \tfrac{1}{2})k\frac{Q^2}{a^2}$$

$$F = \left(\frac{2\sqrt{2}-1}{2}\right)\frac{kQ^2}{a^2}$$

18.4

$$F = k\frac{Q^2}{(2d)^2}$$

$$T\cos\theta = mg$$

$$T\sin\theta = F$$

EACH BALL IS IN EQUILIBRIUM, SO

FORCE UP = FORCE DOWN

FORCE LEFT = FORCE RIGHT

DIVIDE,

$$\frac{T\sin\theta}{T\cos\theta} = \frac{F}{mg}$$

$$\tan\theta = \frac{F}{mg}$$

$$\tan\theta = \frac{k\,Q^2}{mg(2d)^2}$$

T = TENSION IN STRING

d = 1 cm

L = 50 cm

m = 0.1 gm

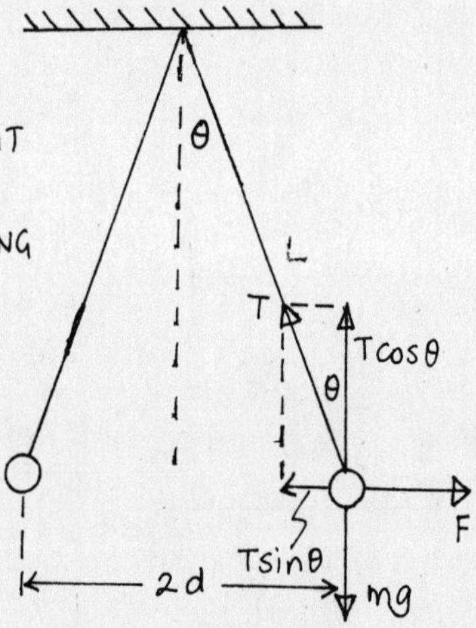

$\theta << 1$ HERE. SO $\tan\theta \simeq \sin\theta = \frac{d}{L}$

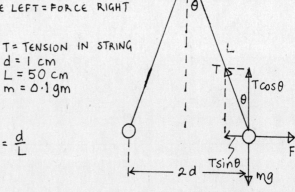

$$\frac{d}{L} = \frac{kQ^2}{4mgd^2}$$

$$Q^2 = \frac{4mg\,d^3}{kL}$$

$$Q = \sqrt{\frac{(4)(10^{-4}\text{kg})(9.8\,\text{m/s}^2)(10^{-2}\text{m})^3}{\left(9\times10^9\,\frac{\text{N}\cdot\text{m}^2}{\text{C}^2}\right)(0.5\,\text{m})}} = 9.3\times10^{-10}\,\text{C}$$

18.5 $F = qE = (1.6 \times 10^{-19}\ C)(2 \times 10^4\ N/c) = \underline{3.2 \times 10^{-15}\ N}$

$$x = \frac{1}{2}at^2 = \frac{1}{2}\left(\frac{F}{m}\right)t^2 = \frac{1}{2}\ \frac{(3.2 \times 10^{-15}\ N)(2s)^2}{(1.67 \times 10^{-31}\ kg)} = 3.8 \times 10^{16}\ m$$

NOTE THAT MASS AND CHARGE OF THE PROTON ARE ASSUMED TO BE KNOWN QUANTITIES. YOU LOOK THEM UP IN A BOOK TO FIND EXACT VALUES.

18.6 THE ELECTRICAL FIELD DUE TO A CHARGED SPHERE IS JUST THAT WHICH WOULD OCCUR IF ALL THE CHARGE WERE CONCENTRATED AT A POINT AT THE CENTER OF THE SPHERE.

THUS $E = k\frac{Q}{R^2}$

$$Q = \frac{ER^2}{k} = \frac{(3 \times 10^6\ N/c)(10^{-2}\ m)^2}{9 \times 10^9} = 3.3 \times 10^{-8}\ C$$

SINCE THE CHARGE ON AN ELECTRON HAS MAGNITUDE e, THE NUMBER OF ELECTRONS IS

$$N = \frac{Q}{e} = \frac{3.3 \times 10^{-8}\ C}{1.6 \times 10^{-19}\ C} = \underline{\underline{2.1 \times 10^{11}\ \text{ELECTRONS}}}$$

18.7 WE REQUIRE $QE = mg$

$m = (\rho)\left(\frac{4}{3}\pi R^3\right)$, ρ = DENSITY OF WATER $= 10^3\ kg/m^3$

$$Q = \frac{mg}{E} = \frac{4}{3}\ \frac{\pi R^3 \rho}{E} g = \frac{(4\pi)(10^3\ kg/m^3)(10^{-3}\ m)^3(9.8\ m/s^2)}{(3)(150\ V/m)} = \underline{\underline{2.74 \times 10^{-7}\ C}}$$

$$Q = \underline{\underline{27.4\ \mu C}}$$

18.8 THE DISTANCE FROM A CORNER TO THE CENTRE IS R,

$$R = \sqrt{(2a)^2 + (3/2\,a)^2} = \frac{5}{2}a$$

FROM THE SKETCH WE SEE THAT X-COMPONENTS OF THE ELECTRIC FIELD CANCEL OUT. THE Y-COMPONENT IS THE RESULTANT FIELD, E.

$$E = E_{1_y} + E_{2_y} + E_{3_y} + E_{4_y}$$

$$= k\frac{2q}{R^2}\cos\theta + k\frac{2q}{R^2}\cos\theta$$

$$- k\frac{q}{R^2}\cos\theta - k\frac{q}{R^2}\cos\theta$$

$$= k\frac{2q}{R^2}\cos\theta = k\frac{2q}{(\frac{5}{2}a)^2}(0.6) = 0.192\frac{k^2q}{a^2}$$

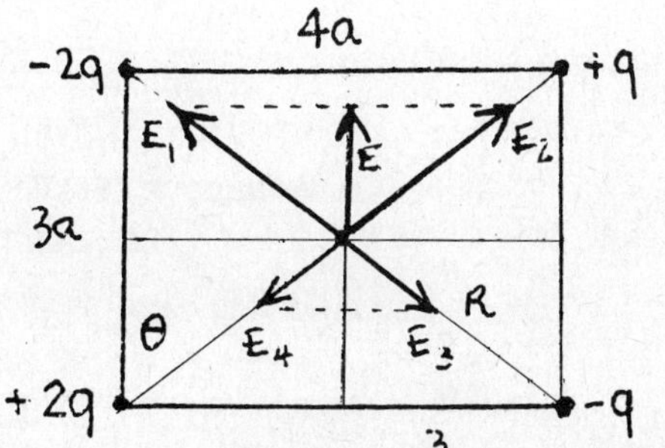

$$\cos\theta = \frac{3}{\sqrt{3^2+4^2}} = 0.6$$

THE POTENTIAL AT THE CENTER OF THE RECTANGLE IS THE SUM OF THE POTENTIALS DUE TO EACH OF THE CHARGES.

$$V = k\frac{(-2q)}{R} + k\frac{(+2q)}{R} + k\frac{(-q)}{R}$$

$$+ k\frac{(+q)}{R} = 0$$

18.9 WORK DONE = GAIN IN P.E. = $QV_2 - QV_1$

WHERE V_1 = POTENTIAL AT POINT (0, 2)

V_2 = POTENTIAL AT POINT (4, 4)

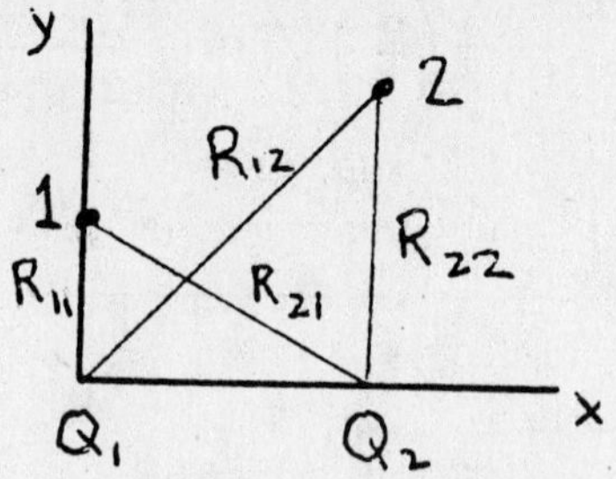

THUS $V_1 = k\dfrac{Q_1}{R_{11}} + k\dfrac{Q_2}{R_{21}}$

$V_2 = k\dfrac{Q_1}{R_{12}} + k\dfrac{Q_2}{R_{22}}$

WORK $W = k\dfrac{QQ_1}{R_{12}} + k\dfrac{QQ_2}{R_{22}} - k\dfrac{QQ_1}{R_{11}} - k\dfrac{QQ_2}{R_{21}}$

$R_{11} = 2$ cms.

$R_{12} = \sqrt{(4)^2 + (4)^2} = 5.66$ cms

$R_{21} = \sqrt{(4)^2 + (2)^2} = 4.47$ cms.

$R_{22} = 4$ cms.

$Q_2 = -2Q_1 = 20\mu C$

SUBSTITUTE $Q_2 = -2Q_1$

$$W = kQQ_1\left[\frac{1}{R_{12}} - \frac{2}{R_{22}} - \frac{1}{R_{11}} + \frac{2}{R_{21}}\right]$$

$$= (9\times10^9\ \text{N}\cdot\text{m}^2/\text{C}^2)(2\times10^{-6}\ \text{C})(-10\times10^{-6}\ \text{C})\left[\frac{1}{0.0566\text{m}} - \frac{2}{0.04\text{m}} - \frac{1}{0.02\text{m}} + \frac{2}{0.0447\text{m}}\right]$$

$W = 6.77$ J

18.10 $E = \dfrac{V}{d} = \dfrac{50\text{ volts}}{20\times10^{-3}\text{m}} = 2500$ V/m

$K.E = qV = (1.6\times10^{-19}\ \text{C})(50\ \text{V}) = 8\times10^{-18}\ \text{J} = 50$ e.V.

BY DEFINITION, AN ELECTRON GAINS 50 e.V. OF KINETIC ENERGY WHEN IT FALLS THROUGH A POTENTIAL DIFFERENCE OF 50 VOLTS.

18.11 $\frac{1}{2}mv^2 = qV$, $V = \dfrac{mv^2}{2q} = \dfrac{mv^2}{2(2e)} = \dfrac{(6.7\times10^{-27}\ \text{kg})(5\times10^5\ \text{m/s})^2}{(4)(1.6\times10^{-19}\ \text{C})} = 2600$ V

18.12 THE ELECTRIC FIELD AND THE POTENTIAL AT THE SURFACE OF THE SPHERE ARE JUST THOSE WHICH WOULD RESULT IF ALL OF THE CHARGE ON THE SPHERE WERE PLACED AT ITS CENTER. THUS

$V = k\dfrac{Q}{R}$ AND $E = k\dfrac{Q}{R^2}$

THUS $kQ = RV$ AND $kQ = ER^2$, SO $RV = R^2E$

AND $V = RE$

$= (0.15\ \text{m})(3.0\times10^6\ \text{V/m})$

$V = 4.5\times10^5$ VOLTS.

18.13 LET THE POTENTIAL OF THE NEGATIVE PLATE BE ZERO AND THAT OF THE POSITIVE PLATE BE V. AT THE NEGATIVE PLATE THE PARTICLE STOPS (VELOCITY = 0).

THUS $KE_1 + PE_1 = KE_2 + PE_2$

$$\frac{1}{2}mv_0^2 + (-e)V = 0 + 0$$

$$\frac{1}{2}mv^2 = eV, \quad v_0^2 = \frac{2eV}{m}, \quad v_0 = \sqrt{\frac{2eV}{m}}, \quad V = \frac{mv_0^2}{2e}$$

THE ELECTRIC FIELD IS $E = \frac{V}{d}$

THE FORCE ON THE ELECTRON IS $F = eE$

ACCELERATION $a = \frac{F}{m} = \frac{eE}{m}$

$$v = v_0 - at = 0, \quad t = \frac{v_0}{a} = \frac{v_0}{eE/m} = \frac{mv_0}{eE} = \frac{m}{eE}\sqrt{\frac{2eV}{m}}$$

$$E = \frac{V}{d}, \text{ so } t = \frac{m}{eV/d}\sqrt{\frac{2eV}{m}} \qquad t = \left(\sqrt{\frac{2m}{eV}}\right) d$$

19 Capacitors

• Summary of Important Ideas, Principles, and Equations

1. Two conducting bodies separated by an insulator are called a capacitor. Most often the two bodies are metal plates. If charge +Q is placed on one plate and charge −Q on the other, there will be a potential difference V created between them. This potential difference will be proportional to Q and will also depend on the size, shape and separation of the plates and on the nature of the insulating material between the plates. The relationship between Q and V may thus be written

$$\boxed{Q = CV} \qquad (19.1)$$

   This equation defines the capacitance C of the two conductors.

   Q is in Coulombs, V in volts and C in farads (F). Frequently used units are microfarads (μF, 10^{-6} F) and picofarads (pF, 10^{-12} F).

   The term "capacitance" refers to the ability of the two conductors to store electric charge. For a given potential difference V a large capacitor can store more charge than a small one. Note that a capacitor has capacitance independent of whether or not is is charged up. The capacitance is analogous to the capacity of a bucket. A two gallon bucket has a given capacity for holding water whether or not there is any water in the bucket. Q is called the charge on the capacitor if there is charge +Q on one plate and charge −Q on the other. We are interested in cases where the charges on the two plates have equal magnitudes because this is the situation which occurs in virtually all real situations.

2. Two flat plates of area A and separation d are called a parallel plate capacitor and have capacitance

$$\boxed{C = \kappa C_o = \kappa \frac{\varepsilon_o A}{d}} \qquad (19.2)$$

Here κ is the dielectric constant of the insulator placed between the plates. For air or vacuum $\kappa = 1$. $\varepsilon_o = 8.85 \times 10^{-12}$ m/F. ε_o, called the permittivity of free space, is simply related to the constant k in Coulomb's law, $k = 1/4\pi\varepsilon_o$.

A dielectric insulator, like oil or glass or mica, increases the capacitance significantly because it is polarizable. The free charge on the plates distorts the insulator slightly, causing polarization charges to appear on the insulator faces next to the plates. These polarization charges tend to cancel the effect of the free charges.

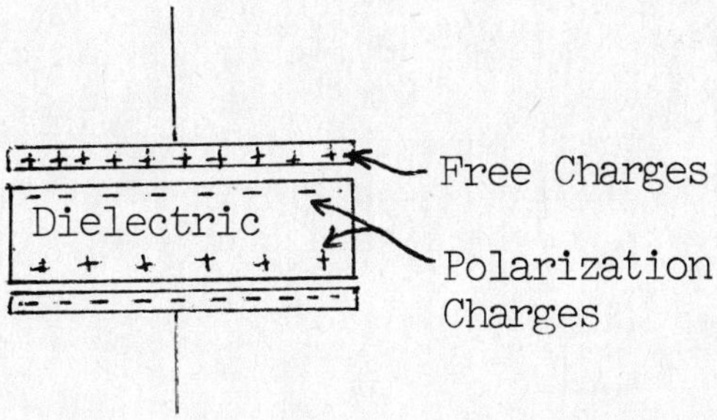

Thus in order to create a given potential difference V a bigger free charge Q must be placed on the plates.

3. A capacitor <u>stores electric energy</u>. A capacitance C charged to a potential difference V by placing charge Q on the plates has a stored potential energy

$$PE = \frac{1}{2}QV = \frac{1}{2}CV^2 = \frac{1}{2}\frac{Q^2}{C} \qquad (19.3)$$

PE is measured in Joules here.

We can also think of the energy as stored in the electric field between the plates. The energy stored per unit volume will be

$$\text{Electrostatic energy per unit volume} = \frac{1}{2}\kappa\varepsilon_o E^2 \qquad (19.4)$$

Here E = electric field in volts/meter (V/m).

Since the potential energy of a charge Q at a point where the potential is V is QV you might expect the stored energy of the capacitor to be QV instead of $\frac{1}{2}QV$. We can understand the factor of $\frac{1}{2}$ by recognizing that as charge is added little by little to the plate the potential gradually builds up from zero to V. Consider the analogy of adding water to a bucket with a final depth h. The mass of water m is like the charge Q. The potential difference is like the water pressure ρgh, i.e. proportional to h. If you fill a bucket to a height h the <u>average</u> height water had to be lifted is only $\frac{1}{2}h$ (since at first the bucket is empty). Thus the gravitational potential energy of the water in the bucket is $mgh_{ave} = \frac{1}{2}mgh$. Similarly, the potential energy of the charged capacitor is QV_{ave}, where V_{ave} is the average potential difference through which the charge had to be lifted to place it on the plates. $V_{ave} = \frac{1}{2}V$, so $PE = \frac{1}{2}QV$.

4. Combinations of capacitors are equivalent to a single capacitor. Two important cases occur.

When the top plates and bottom plates of two capacitors are connected together by a wire, the two capacitors are connected in parallel. When connected like this $V_1 = V_2$. These two capacitors are equivalent to a single capacitor C_p, with potential difference $V = V_1 = V_2$ and charge $Q = Q_1 + Q_2$.

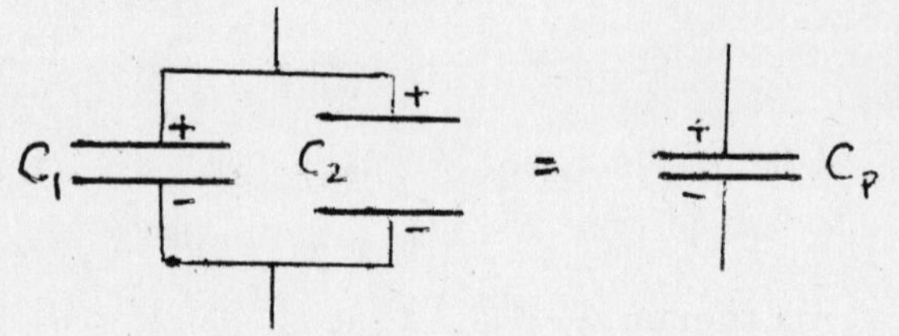

$$\boxed{C_p = C_1 + C_2} \qquad (19.5)$$

We can understand this equation by recognizing that adding C_2 to C_1 in parallel is just like increasing the effective area of C_1, where we saw C was proportional to plate area.

When only one plate from each of two capacitors is connected together the two capacitors are connected in series and are equivalent to a single capacitor C_s,

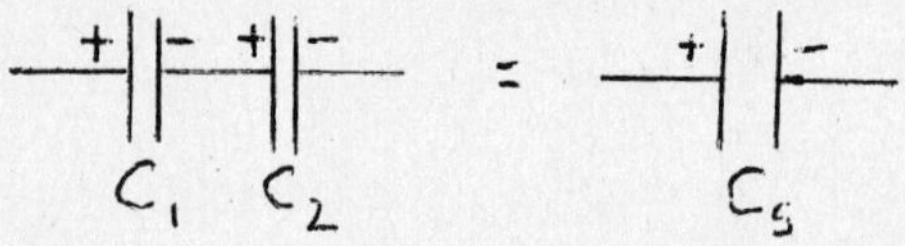

$$\frac{1}{C_s} = \frac{1}{C_1} + \frac{1}{C_2}$$

$$\text{or } C_s = \frac{C_1C_2}{C_1 + C_2} \qquad (19.6)$$

We can understand this equation by imagining the two capacitors to be connected by such a short wire that their common plates merge into a single plate. This plate is an equipotential surface. Its presence does not affect the capacitance between the two outer plates, which form a single equivalent capacitor with separation $d_1 + d_2$. Since $C = \kappa\varepsilon_0 A/d$. Increasing d decreases C, so $C_s < C_1$ and $C_s < C_2$, whereas $C_p > C_1$ and $C_p > C_2$.

Any number of capacitors may be replaced by an equivalent capacitor by combining them two at a time. To reduce a complex capacitor network first combine those which are obviously in parallel. Work from the "inside out" until you have only one capacitor left.

For N capacitors in parallel or series we find

$$C_p = C_1 + C_2 + \ldots + C_N$$

$$\text{and } \frac{1}{C_s} = \frac{1}{C_1} + \frac{1}{C_2} + \ldots + \frac{1}{C_N}$$

Note that an ideal battery provides a fixed potential difference between its terminals.

• Qualitative Questions

19M.1 The capacitance between two conductors depends on

A. the charge on them.
B. the potential difference between them.
C. the electric field between them.
D. the energy stored between them.
E. None of the above.

19M.2 Suppose that a capacitor is charged and then disconnected from the battery. When a dielectric material is now inserted between the plates of the capacitor the capacitance is increased. Which of the following best explains why this happens?

A. The induced charge on the surface of the dielectric adds to the free charge on the adjacent plate, thereby effectively increasing the net charge stored on the capacitor.
B. The induced charge on the surface of the dielectric effectively neutralizes the effect of some of the free charge on the adjacent plate, thereby reducing the potential difference between the plates.
C. The dielectric slab has a capacitance of its own, and this capacitance effectively adds to the capacitance of the parallel plate capacitor.
D. The dielectric slab is a good insulator in general, and its presence allows the presence of higher potential differences and hence greater charge storage.
E. The induced charge in the dielectric is spread throughout the volume of the dielectric. This results in greater capacitance as compared to the case where all of the charge is localized on two-dimensional plates.

19M.3 Suppose that while a capacitor is connected to a battery a dielectric slab is placed between the plates.

A. The capacitance will not be affected, since the battery holds the potential difference fixed.
B. More charge will flow onto the plates from the battery.
C. Charge will flow from the capacitor plates back into the battery.
D. The effective capacitance will be reduced.
E. The potential difference between the plates will increase because of the presence of induced charges on the surface of the dielectric.

The following questions refer to the two identical capacitors shown in the circuit here. They remain connected to the battery while a dielectric slab is inserted into the lower capacitor. The battery maintains a constant potential difference between points A and B. The next five questions refer to the changes which occur when the slab is inserted.

19M.4 The charge on the upper capacitor will

A. increase.
B. decrease.
C. remain unchanged.

19M.5 The charge on the lower capacitor will

A. increase.
B. decrease.
C. remain unchanged.

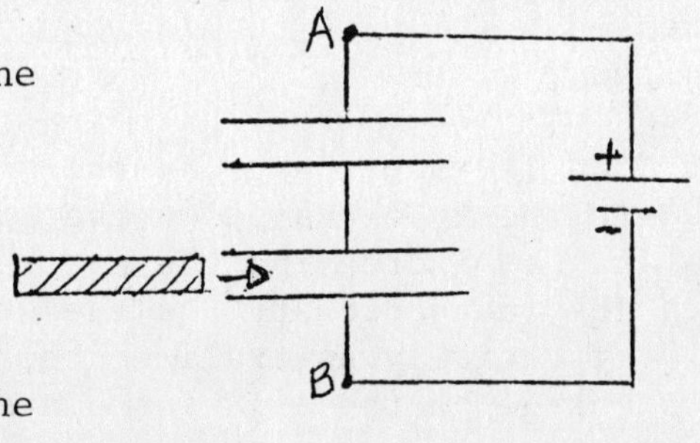

Battery

19M.6 The potential difference across the upper capacitor will

A. increase.
B. decrease.
C. remain unchanged.

19M.7 The potential difference across the lower capacitor will

A. increase.
B. decrease.
C. remain unchanged.

19M.8 The total energy stored in the two capacitors will

A. increase.
B. decrease.
C. remain unchanged.

19M.9 You are designing a capacitor to be used in an experiment where a particular value, C, of capacitance is required. The capacitor must be capable of operating up to a given maximum voltage, V_m. You decide to use a pair of parallel metal plates immersed in an oil bath. Unfortunately, you find that on your first test run the plates arc over. The boss is mad and you are embarrassed. What should you do to redesign the capacitor?

A. Simply drain out the oil and use the capacitor as air-filled.
B. Increase the area of the plates and move them closer together.
C. Increase the area of the plates and move them farther apart.
D. Decrease the area of the plates and move them farther apart.
E. Decrease the area of the plates and move them closer together.

19M.10 A sheet of aluminum foil is placed between the plates of a parallel plate capacitor, as shown here. What effect will this have on the capacitance?

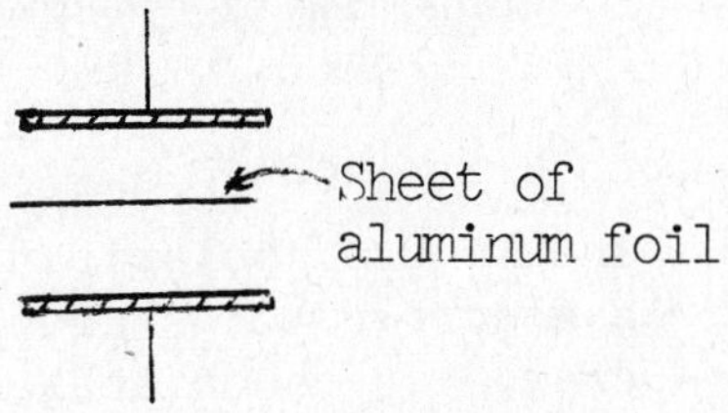

A. It will increase the capacitance.
B. It will decrease the capacitance.
C. It will have no effect on the capacitance, assuming the sheet of negligible thickness.

19M.11 A sheet of aluminum foil of negligible thickness is inserted between the plates of a capacitor, as shown here. It is connected to the upper plate. If this is done, the capacitance will

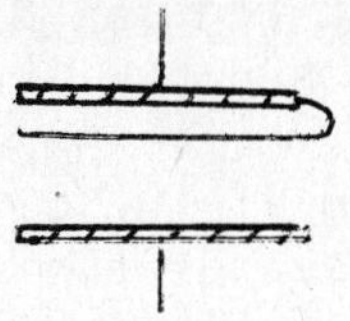

A. increase.
B. decrease.
C. remain unchanged.

19M.12 Suppose that a capacitor is connected to a battery. Is it true that each plate will receive a charge of the same magnitude? If so, why? If not, why not?

A. Each plate will receive equal amounts of charge only if they are of the same area.
B. A capacitor is designed so that all of the electric field lines emanating from one plate terminate on the other plate. From Gauss's law we then know that there is an equal amount of charge on each plate.
C. No, each plate will not receive an equal amount of charge, whether or not they have equal areas. The charge on a plate depends on the shape of the plate as well as on its area.
D. Yes, each plate receives the same charge because each is at the same potential.

19M.13 A parallel plate capacitor is charged by a battery and then disconnected. If now the plates are pulled apart,

A. the electric field between the plates will decrease.
B. the potential difference between the plates will decrease.
C. the charge on the plates will decrease.
D. the energy stored in the capacitor will increase.

19M.14 Suppose the energy stored in a capacitor charged to a potential difference V is U_o. If this capacitor were charged to a potential difference 2V the stored energy would be

A. $\frac{1}{2}U_o$ B. $\frac{1}{2}U_o$ C. U_o D. $2U_o$ E. $4U_o$

• Multiple Choice: Answers and Comments

19M.1 E Capacitance depends just on the size and shape of the conductors, their separation and the material between them. It has nothing to do with charge, potential, electric field or stored energy.

19M.2 B

19M.3 B The induced polarization charge on the surface of the dielectric tends to cancel the effect of some of the free charge on the plates. This would tend to cause a drop in potential between the plates, but the connected battery holds the potential difference between the plates at a fixed value. To do this it makes more charge flow on to the plates. Thus the total free charge on the plates for a given potential difference (and hence the capacitance) is increased by insertion of the dielectric.

19M.4 A Insertion of the dielectric increases the effective capacitance of the two capacitors connected in series. Since V is fixed by the battery, and Q = CV, increasing C increases Q also.

19M.5 A Same reasoning as in 19M.3.

19M.6 A Q increases and C remains the same for the upper capacitor, so since Q = CV, V must increase.

19M.7 B The sum of the potential differences across the two capacitors is fixed, so since the potential difference across the upper one increases, the difference across the lower one decreases. Note that from Q = CV it is hard to tell at a glance what is happening, since both Q and C increase for the lower capacitor. From this equation it is not easy to see what V is doing. However, C increases by a bigger factor than does Q, so V decreases.

19M.8 A Stored energy = $\frac{1}{2}CV^2$. V is constant and C increases.

19M.9 C The capacitor arced over because the electric field was too great. The electric field is related to the potential difference V and the plate separation d by E = V/d, thus increasing d will decrease E as needed. However, this will also decrease capacitance C. This can be compensated for by increasing the plate area.

19M.10 C The sheet has been placed along an equipotential surface. A conductor can always be placed on an equipotential without changing the electric fields. Opposite charges will be induced on the top and bottom surfaces of the aluminum and it will appear that the electric field passes right through just as if the aluminum were not there.

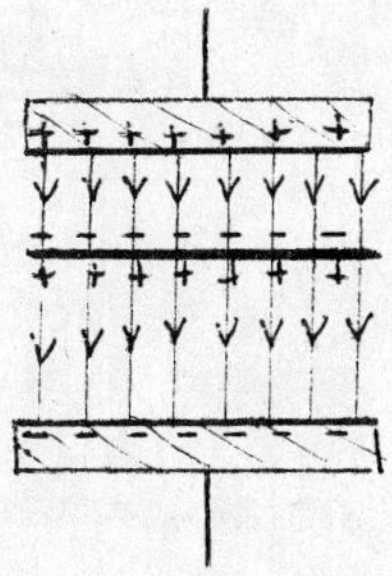

19M.11 A This is equivalent to moving the plates closer together, hence the capacitance is increased, since $C = \varepsilon_0 A/d$.

19M.12 B

19M.13 D It is assumed that the dimensions of the plates are large compared to their separation, even after pulled apart. The charge on the plates stays constant once they are disconnected from the battery. The electric field outside a plane conductor depends only on the surface charge density, thus it stays constant. The potential difference will increase when the plates are pulled apart, since V = Ed. The stored energy is $\frac{1}{2}QV$, and thus it will increase. One must do work to pull the plates apart, because the opposite charges on the two plates attract each other. This work goes into stored energy.

19M.14 E Energy $U = \frac{1}{2}CV^2$. If V is doubled, V^2 increases by a factor of 4.

• Problems

19.1 A parallel plate capacitor has a plate area of 200 cm^2 and a separation of 0.1 mm. What is its capacitance?

19.2 A parallel plate capacitor is made by rolling up three sheets of aluminum foil, with a 0.05 mm thick sheet of paper between each layer of foil. The two outer sheets of aluminum are joined by a wire. What area of foil is needed if the paper has a dielectric constant of 3.5 and the capacitance is to be 20 μF? What maximum voltage can be applied to the capacitor if the paper breaks down under an electric field of 14 kV/mm?

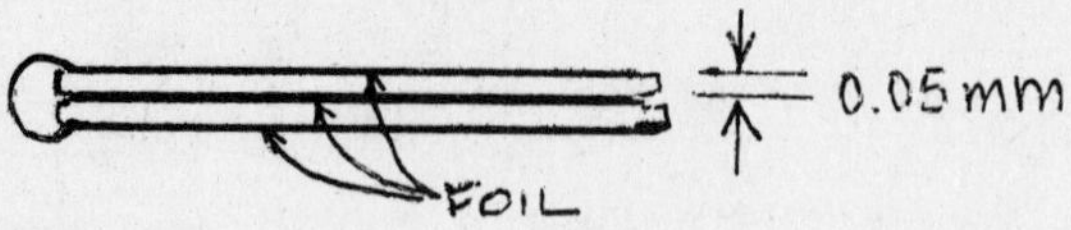

19.3 What is the capacitance of a parallel plate capacitor with plate area 2 m^2 and plate separation 1 mm if half the space between the plates is filled with a material of dielectric constant 5 using the two arrangements sketched.

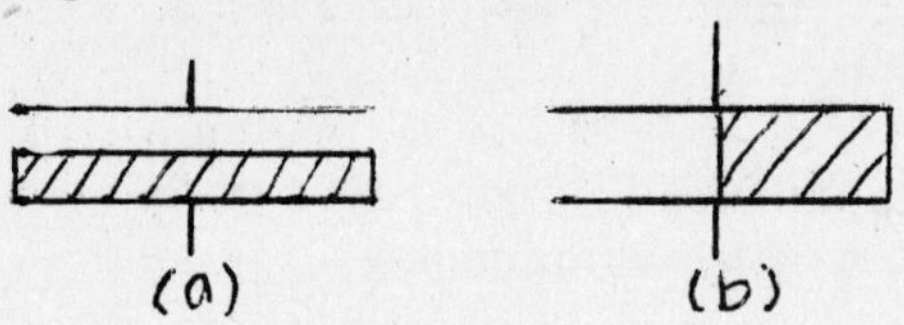

19.4 A 3 μF capacitor and a 1 μF capacitor are connected in series to a 12 volt battery. A material of dielectric constant 5 is then inserted in the 1 μF capacitor. Calculate the charge and potential difference and the stored energy for each capacitor before and after insertion of the dielectric, and the total stored energy before and after.

19.5 Determine the equivalent capacitance of each of the arrangements shown here.

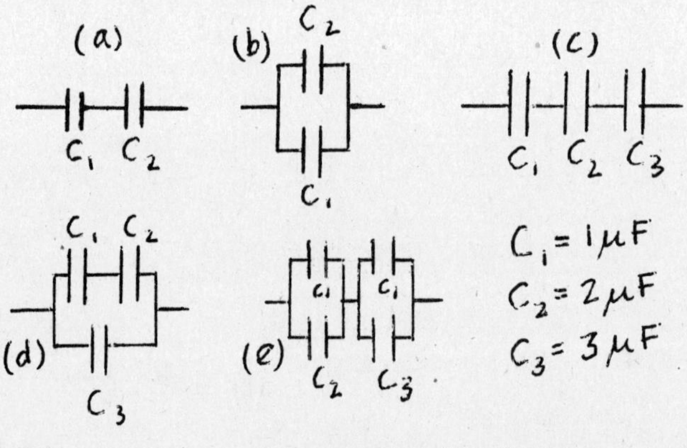

19.6 A photoflash unit operates by discharging a capacitor across a Xenon flash lamp. What capacitance is needed if a charging voltage of 800 volts is used and the lamp requires 20 Joules of energy?

• Problem Solutions

19.1 $C = \frac{\epsilon_0 A}{d} = \frac{(8.85 \times 10^{-12}\ F/m)(200)(10^{-2}\ m)^2}{(0.1 \times 10^{-3}\ m)} = \underline{\underline{1.77 \times 10^{-9}\ F}}$

BE CAREFUL ALWAYS TO CHANGE DIMENSIONS TO METERS.

19.2 THIS ARRANGEMENT IS EQUIVALENT TO TWO CAPACITORS IN PARALLEL. TO SEE THIS. IMAGINE MIDDLE SHEET OF ALUMINIUM IS A "DOUBLE" FOIL.

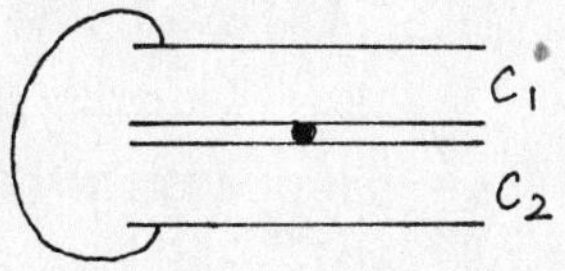

NOW IMAGINE THE DOUBLE FOIL SEPARATED.

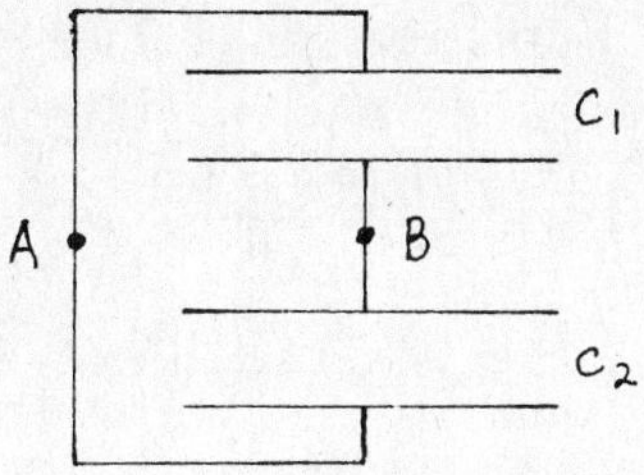

NOW REARRANGE AND CONNECT POINTS A AND B TO THE CHARGING BATTERY.

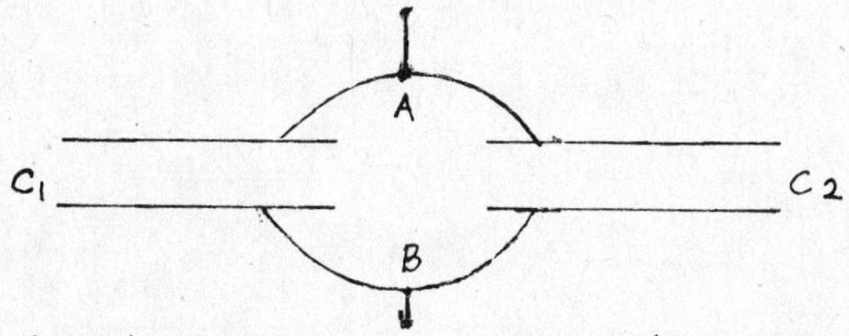

C_1 AND C_2 ARE EQUIVALENT TO A SINGLE CAPACITOR C_P,

$$C_p = C_1 + C_2 = 2C_1 \quad \text{SINCE } C_1 = C_2$$

$$C_p = 2\frac{K\epsilon_0 A}{d} = \frac{(2)(3.5)(8.85 \times 10^{-12}\ F/m)(A)}{(0.05 \times 10^{-3}\ m)}$$

GIVEN $C_p = 20 \times 10^{-6}\ F$, $\underline{\underline{A = 16.1\ m^2}}$ THREE SHEETS EACH OF THIS AREA ARE NEEDED, SO

$$V = Ed = \left(\frac{14000\ V}{10^{-3}\ m}\right)(0.05 \times 10^{-3}\ m) = \underline{\underline{700\ \text{VOLTS}}}$$

$$A_{TOT} = (3)(16.1) = \underline{\underline{48.3\ m^2}}$$

19.3 (a) IS TWO CAPACITORS IN SERIES, SO

$$C_S = \frac{C_1 C_2}{C_1 + C_2}, \quad \text{WHERE } C_1 = \frac{\epsilon_0 A}{d/2} = \frac{2\epsilon_0 A}{d}$$

$$\text{AND} \quad C_2 = \frac{K\epsilon_0 A}{d/2} = \frac{2K\epsilon_0 A}{d}$$

$$C_S = \frac{\left(\frac{2\epsilon_0 A}{d}\right)\left(\frac{2K\epsilon_0 A}{d}\right)}{\frac{2\epsilon_0 A}{d} + \frac{2K\epsilon_0 A}{d}} = \left(\frac{K}{1+K}\right)\left(\frac{\epsilon_0 A}{d}\right)$$

$$C_S = \left(\frac{5}{1+5}\right)\left(\frac{(8.85 \times 10^{-12}\ F/m)(2\ m^2)}{10^{-3}\ m}\right) = \underline{\underline{1.48 \times 10^{-8}\ F}}$$

(b) IS TWO CAPACITORS IN PARALLEL, $C_P = C_1 + C_2$

$$= \frac{\epsilon_0 A}{d} + \frac{K\epsilon_0 A}{d}$$

$$C_P = (1+K)\frac{\epsilon_0 A}{d} = \underline{\underline{1.06 \times 10^{-7}\,F}}$$

19.4 FIRST CONSIDER SITUATION BEFORE DIELECTRIC IS INSERTED, WITH $C_1 = 3\mu F$ AND $C_2 = 1\mu F$

EQUIVALENT CAPACITANCE IN SERIES IS $C_S = \frac{C_1 C_2}{C_1 + C_2} = \frac{(3)(1)}{(3+1)} = 0.75\mu F$

CHARGE IS $Q = C_S V = (0.75 \times 10^{-6} F)(12\text{ VOLTS}) = 9\mu C$

THUS $\underline{\underline{Q = Q_1 = Q_2 = 9\mu C}}$ (IN SERIES EACH CAPACITOR HAS SAME CHARGE)

POTENTIAL DIFFERENCES ARE $V_1 = \frac{Q_1}{C_1} = \frac{(9 \times 10^{-6} C)}{(3 \times 10^{-6} F)} = \underline{\underline{3\text{ VOLTS}}}$

$$V_2 = \frac{Q_2}{C_2} = \frac{9 \times 10^{-6} C}{1 \times 10^{-6} F} = \underline{\underline{9\text{ VOLTS}}}$$

OBSERVE THAT $V_1 + V_2 = 12$ VOLTS, AS IT MUST.

TOTAL STORED ENERGY IS $P.E. = \frac{1}{2} CV^2 = \frac{1}{2}(0.75 \times 10^{-6} F)(12\text{ VOLTS})^2$

$$= \underline{\underline{5.4 \times 10^{-5} J}}$$

STORED ENERGY IN EACH IS $P.E._1 = \frac{1}{2} C_1 V_1^2 = \frac{1}{2}(3 \times 10^{-6} F)(3\text{ VOLTS})^2$

$$= \underline{\underline{1.35 \times 10^{-5} J}}$$

$$P.E._2 = \frac{1}{2} C_2 V_2^2 = \frac{1}{2}(1 \times 10^{-6} F)(9\text{ VOLTS})^2$$

$$= \underline{\underline{4.05 \times 10^{-5} J}}$$

NOTE THAT $PE_1 + PE_2 = PE$, AS IT MUST.

AFTER DIELECTRIC IS INSERTED C_1 IS UNCHANGED BUT C_2 IS INCREASED, $C_2' = KC_2 = 5C_2 = 5\mu F$

NOW EQUIVALENT CAPACITANCE IS $C_S = \frac{C_1 C_2'}{C_1 + C_2'} = \frac{(3)(5)}{3+5} = 1.875\,\mu F$

$Q = C_S V = (1.875 \times 10^{-6} F)(12 V) = \underline{\underline{22.5\mu C}}$

$Q = Q_1 = Q_2 = \underline{\underline{22.5\mu C}}$

$V_1 = \frac{Q_1}{C_1} = \frac{(22.5 \times 10^{-6} C)}{(3 \times 10^{-6} F)} = \underline{\underline{7.5\text{ VOLTS}}}$

$V_2 = \frac{Q_2}{C_2'} = \frac{(22.5 \times 10^{-6} C)}{(5 \times 10^{-6} F)} = \underline{\underline{4.5\text{ VOLTS}}}$

$PE_1 = \frac{1}{2} C_1 V_1^2 = (\frac{1}{2})(3 \times 10^{-6} F)(7.5 V)^2 = \underline{\underline{8.44 \times 10^{-5} J}}$

$PE_2 = \frac{1}{2} C_2' V_2^2 = (\frac{1}{2})(5 \times 10^{-6} F)(4.5 V)^2 = \underline{\underline{5.06 \times 10^{-5} J}}$

$PE = \frac{1}{2} C_S V^2 = (\frac{1}{2})(1.875 \times 10^{-6} F)(12 V)^2 = \underline{\underline{13.5 \times 10^{-5} J}}$

NOTE CAREFULLY WHAT HAPPENED HERE. TOTAL ENERGY INCREASED. VOLTAGE ACROSS C_2 DECREASED, ALTHOUGH CHARGE ON EACH CAPACITOR INCREASED BY FACTOR OF 2.5.

19.5 (a) $C_S = \frac{C_1C_2}{C_1+C_2} = \frac{(1)(2)}{1+2}\mu F = \underline{\underline{0.67\mu F}}$

(b) $C_p = C_1 + C_2 = (1+2)\mu F = \underline{\underline{3\mu F}}$

(c) $\frac{1}{C_S} = \frac{1}{C_1} + \frac{1}{C_2} + \frac{1}{C_3} = \frac{1}{1} + \frac{1}{2} + \frac{1}{3} = \frac{6}{6} + \frac{3}{6} + \frac{2}{6} = \frac{11}{6}$

$C_S = \frac{6}{11} = \underline{\underline{0.55\mu F}}$

(d) FIRST COMBINE C_1 AND C_2 IN SERIES $C_S = \frac{C_1C_2}{C_1+C_2} = \frac{(1)(2)}{(1+2)} = \frac{2}{3}\mu F$

NOW COMBINE C_S AND C_3 IN PARALLEL,

$C = C_S + C_3 = \left(\frac{2}{3} + 3\right)\mu F = \underline{\underline{3.67\mu F}}$

(e) C_1 AND C_2 IN PARALLEL YIELD $C_p = C_1 + C_2 = (1+2)\mu F = 3\mu F$

C_1 AND C_3 IN PARALLEL YIELD $C_p' = C_1 + C_3 = (1+3)\mu F = 4\mu F$

C AND C' IN SERIES YIELD $C_S = \frac{C_pC_p'}{C_p + C_p'} = \frac{(3)(4)}{(3+4)}\mu F = \underline{\underline{1.71\mu F}}$

19.6 $PE = CV^2$, $C = \frac{2\,PE}{V^2}$ $\frac{(2)(20J)}{(800V)^2} = 62.5 \times 10^{-6} F = \underline{\underline{62.5\mu F}}$

20 Steady Electric Currents

• Summary of Important Ideas, Principles, and Equations

1. A battery is a device which, by chemical means, is able to produce a potential difference across its terminal. In olden times this potential difference was called an "electromotive force", and we still use the symbol $\mathcal{E}$ for it. In a diagram we indicate a battery with this symbol, 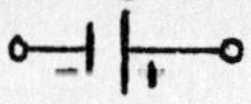 where the "+" indicates the positive terminal.

2. The amount of charge ΔQ that passes through a given area in time Δt is called the current, I, through that area.

$$I = \frac{\Delta Q}{\Delta t} \tag{20.1}$$

   The unit of current is the ampere, 1 Amp = Coulomb/second. Current is positive in the direction positive charge moves.

3. If A is an area perpendicular to the flow of current I, the current density through A is defined as

$$j = \frac{I}{A} \text{ (Amp/m}^2\text{)} \tag{20.2}$$

   Suppose the current is carried by a drifting cloud of charged particles, each of charge q and density n per meter^3. If the average drift velocity of the cloud is v_d, the current and current density are

$$I = nqv_dA \quad \text{and} \quad j = \frac{I}{A} = nqv_d \tag{20.3}$$

4. In real conductors (except superconductors) the flow of current is impeded because the charge carriers (usually electrons) keep bumping into things. Because of this resistance it is necessary that

a potential difference be maintained across a conductor in order to maintain a current flow. In many common materials the current flow through an object (such as a wire) is proportional to the potential difference maintained between the ends of the wire. This relationship is called Ohm's Law, and is written

$$\boxed{V = IR} \qquad (20.4)$$

Here R is the resistance, measured in Ohms (Ω), where $1\Omega = 1$ volt/ampere.

5. The resistance of a conductor, such as a length of wire, will depend on its length and cross-sectional area and on the kind of material from which it is made. The resistance of a wire is kind of like the resistance of a pipe to fluid flow. Resistance is proportional to length and inversely proportional to cross-sectional area. Thus

$$\boxed{R = \rho\frac{\ell}{A}} \qquad (20.5)$$

ρ is the resistivity, in Ohm-meters.

Be careful not to confuse the terms resistance and resistivity. Resistivity depends just on the nature of the material and its temperature. It is tabulated in books. Resistance depends on resistivity and on size and shape of the conductor.

In many materials, such as metals, resistivity increases with increasing temperature, because then the atoms vibrate more vigorously and collide with the drifting electrons more often. The resistivity at temperature T (^{o}C) is

$$\rho = \rho_{oC}(1 + \alpha T)$$

α = Temperature coefficient of resistivity ($^{o}C^{-1}$)

T = Temperature (^{o}C)

ρ_{oC} = Resistivity at 0 ^{o}C (20.6)

Materials in which the resistivity decreases with increasing temperature are called semiconductors (e.g. silicon, germanium, gallium arsenide). In these substances more conduction electrons are freed as the temperature increases, and this effect outweighs the increased lattice scattering.

Materials with relatively large resistivity are called insulators, but even they conduct slightly.

Some materials (mostly metals) when cooled to temperatures below about 20 ^oK have zero resistance. They are called superconductors and have great promise for practical application.

6. When current flows through a conductor the charge carriers bump into atoms and cause heating. The rate at which thermal energy is delivered to a resistance R when current I flows is I^2R. Combining this with Ohm's Law, V = IR, allows us to write

$$\boxed{P = I^2R = IV = \frac{V^2}{R}} \qquad (20.7)$$

Here P is the power (in watts) dissipated as heat.

7. A network of resistors is equivalent to a single resistor in which the same current will flow for a given applied potential difference.

When one end only of two resistors is connected together the two resistors are connected in series. The same current flows in each in this case. The equivalent resistance is R_s.

$$\boxed{R_s = R_1 + R_2} \qquad (20.8)$$

This is represented in a circuit diagram like this: 

If N resistors are connected in series, $R_s = R_1 + R_2 + \ldots + R_N$.

When two ends of two resistors are connected together the resistors are connected in parallel. The same potential difference is found across each of the resistors in this case. The equivalent resistance is R_p.

$$\boxed{\frac{1}{R_p} = \frac{1}{R_1} + \frac{1}{R_2}} \quad \text{or} \quad \boxed{R_p = \frac{R_1R_2}{R_1 + R_2}} \qquad (20.9)$$

Parallel resistors are indicated like this: 

To find the resistance equivalent to a complicated network proceed step by step, first combining pairs of resistors which are either connected in series or parallel. Notice that some resistors are not in series or parallel connection, so they cannot be immediately replaced by an equivalent resistor. For example, in this circuit R_1 is connected in series with R_2, but it is not in series or in parallel with R_3. Thus we first combine R_1 and R_2 to obtain an equivalent resistor R_s which can then be combined in parallel with R_3.

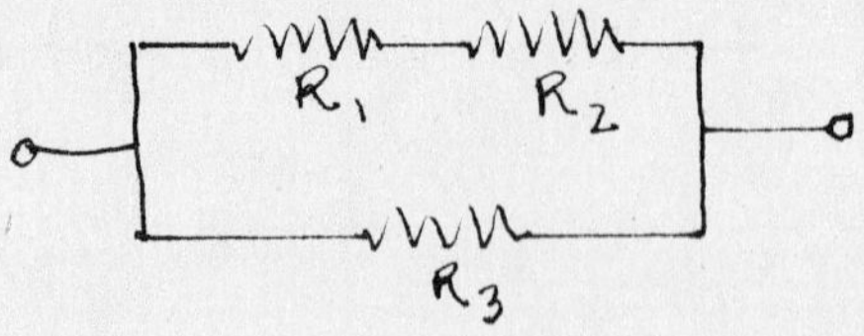

We can summarize the relations for series and parallel resistors.

Series	Parallel
$V_s = V_1 + V_2 + \ldots$	$V_p = V_1 = V_2 = \ldots$
$I_s = I_1 = I_2 = \ldots$	$I_p = I_1 + I_2 + \ldots$
$R_s = R_1 + R_2 + \ldots$	$\frac{1}{R_p} = \frac{1}{R_1} + \frac{1}{R_2} + \ldots$
$R_s > R_1, R_2, R_3, \ldots$	$R_p < R_1, R_2, R_3, \ldots$

8. A real battery is a seat of EMF plus an internal resistance. To see what this means, suppose no load is connected to the terminals of a car battery. If you measure the potential difference between the terminals you will find a 12 volt difference. If now you connect a load resistance R_L, such as the resistance of your headlamp filaments, to the battery, you will find that the potential difference between the terminals has dropped slightly, perhaps to 11.9 volts. This happens because a real battery has some internal resistance, R_i. The same current flows in R_i and in the load, R_L. The potential difference across the load is $V_L = IR_L$ and the potential difference across the internal resistance is $V_i = IR_i$. The sum of these is the EMF produced by the battery, $\mathcal{E} = V_i + V_L$.

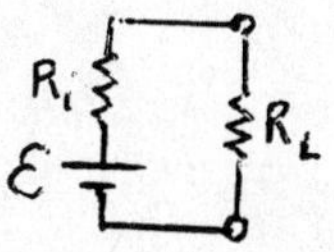

The total power delivered by the battery will thus be divided between the load resistance and the internal resistance. Usually we want as much power as possible delivered to the load. This will occur when the load is matched to the internal resistance.

Thus for maximum load power we require

$$R_L = R_i \tag{20.10}$$

The power delivered to the load is

$$P_L = \frac{R_L \mathcal{E}^2}{(R_L + R_i)^2} \tag{20.11}$$

The maximum load power is

$$P_{MAX} = \frac{1}{4} \frac{\mathcal{E}^2}{R_L} \tag{20.12}$$

9. The analogy between the flow of electric current in a circuit and the flow of water in an irrigation system is very helpful. The mass (or volume) of water is analogous to the amount of charge, Q. The current flow in kilograms per second (or in liters per second) is analogous to the electric current I (in Coulombs per second, i.e. in amperes). The resistance of a pipe is like the resistance of an electrical conductor. The pressure difference between the ends of a pipe is analogous to the difference in potential (in volts) between the ends of a resistor. A pump can create a pressure difference which will cause fluid flow. A battery (or an electronic power supply) can create a potential difference (called an EMF) which can cause an electric current to flow. Note that in both cases one must have a complete closed circuit in order for current to flow.

Recall that mgh is the gravitational potential energy of a mass m at elevation h. Similarly, QV is the potential energy of a charge Q at a point where the electric potential is V. Q is like m and V is like gh.

Consider an irrigation system which works like this: A pump (like a battery) lifts water up to a tank at the top of a hill. The pressure created causes water to flow downhill through some pipes (just like charge flowing "downhill" from high potential to low potential through a resistance R). The water may then come to another pump which lifts the water back up another hill, from which it again flows on downhill. This process can be repeated until finally the water is brought back to the original pump. Here I am imagining that no water is taken out of the irrigation system (not too realistic). In my analogy I assume that there is enough resistance in the pipes so that the water does not gain kinetic energy in flowing downhill. Instead, its loss in potential energy does work in heating the resistive pipes. An electric circuit works just like this. Batteries (seats of EMF) create potential differences by chemically separating charges. Positive charge then flows "downhill" to lower potential, with the potential drop across a resistor R given by Ohm's Law, $V = IR$. When the current comes to a junction it splits, some going each way, just the way water would do if the pipes branch out. Additional batteries may be inserted in the circuit at various points. These will each give a little boost in potential (or a drop if they are hooked in backwards), just as a pump can change the pressure in a water line.

Notice that a current will flow only if there is a complete closed circuit. A given battery tends to make positive charges flow <u>away</u> from its positive terminal (like charges repel). Current flow is positive in the direction positive charge moves. In real metals electrons (with negative charge) are the actual charge carriers, and thus they physically move opposite the direction of the current. This is of no concern for most purposes, however, so don't worry about it. Just treat all circuits as if the current were carried by positive charges.

The main points in our analogy are worth remembering:

Fluid System	Electric Circuit
Volume of water $\rightarrow$	Amount of charge, Q (coulombs)
Flow of water, liters/sec $\rightarrow$	Flow of current, I (in C/sec or amps)
Resistance of pipe $\rightarrow$	Resistance of conductor (ohms)
Pressure difference $\rightarrow$	Potential difference (volts)
(Pressure) = (Resistance) x (Flow) $\rightarrow$	$V = IR$ for each resistor

A closed circuit is required for current flow in both cases.

In a circuit diagram a straight line represents a copper wire which we approximate as having zero resistance. Thus there is no potential drop along it, and all points connected by such a wire are at the same potential.

When two points in a circuit are connected by a "zero resistance" wire (particularly when this is done accidentally) they are said to be "short circuited" (or just "shorted").

When one point in the circuit is "grounded" (connected to a large conductor like the earth or the body of a car) we take the potential at that point to be zero. In a circuit diagram "grounding" is indicated like this: 

10. In a typical electric circuit problem one is given a network of resistors containing some batteries. We wish to find the current in each resistor and the potential at various points in the circuit. This can be done by systematically applying the ideas above. They are expressed as <u>Kirchoff's Rules</u>.

Current flowing into a junction	=	Current flowing out	(KR-I)
Sum of EMF's around a circuit loop	=	Sum of "IR" potential drops around that loop	(KR-II)

A "<u>junction</u>" is a point where three or more wires come together.

Always use the rules for combining resistors in series or in parallel first. Use Kirchoff's Rules only when this can't be done.

In some cases it isn't obvious in which direction the current is flowing. In this case simply make a guess in drawing the arrow showing current flow. If you are wrong you will find a negative value for that current.

To illustrate this method I will find the current in each resistor in Figure 21.1 and the potential at a labelled point.

First I note that the 2 Ω and 4 Ω resistors are connected in series. The presence of the 120 V battery in between them doesn't matter. The same current flows in each. Thus I start by replacing them by a single 6 Ω resistor.

The 6 Ω and 12 Ω resistors on the left are also in series. Replace them by a single 18 Ω resistor. This 18 Ω is in parallel with the 36 Ω, so use the parallel resistance formula to find the resistance R_p equivalent to them. $R_p = (18)(36)/(18 + 36) = 12$ Ω. Note that the fact that the 36 Ω is slanting downward doesn't matter. The wires can be pushed around to different orientations without changing the circuit connections.

After making all of the possible series and parallel reductions the circuit looks like Figure 21.2.

Now apply Kirchoff's Rules. I label the currents in the three branches I_1, I_2 and I_3. Since the batteries are pushing current in opposite directions in the two top resistances it isn't clear to me which way it actually flows. In drawing the arrows as I have I am just making a guess. If I am wrong, some of the currents will turn out negative, showing that current is flowing opposite to what I guessed.

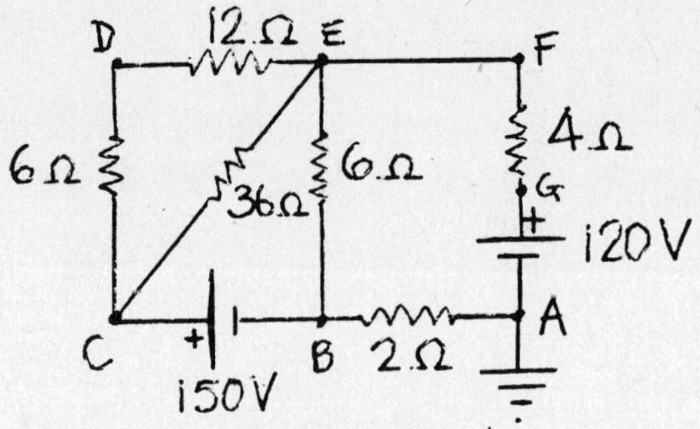

FIGURE 21.1

Kirchoff's Rules yield:

KR-1: $I_1 = I_2 + I_3$ (1)

KR-2: $150 = 12I_1 + 6I_3$ (2)

KR-3: $120 = -6I_2 + 6I_3$ (3)

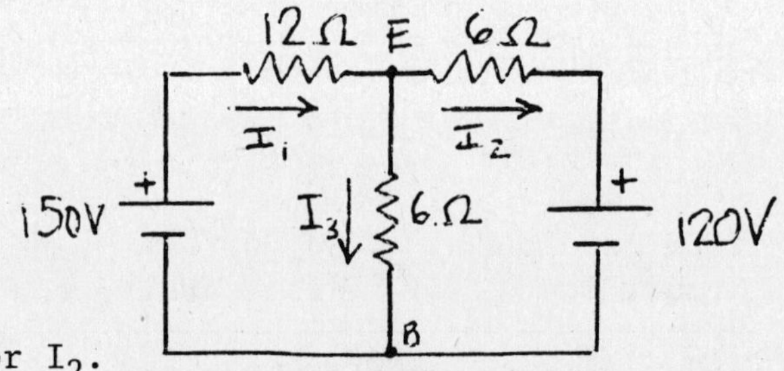

FIGURE 21.2

Substitute (1) in (2) and solve for I_2.

$$150 = 12(I_2 + I_3) + 6I_3$$

$$I_2 = \frac{150 - 18\ I_3}{12} \quad (4)$$

Substitute I_2 in (3) and solve for I_3.

$$120 = (-6)\left(\frac{150 - 18\ I_3}{12}\right) + 6I_3, \quad I_3 = \underline{13A}$$

Substitute this back in (4) to find I_2.

$$I_2 = \frac{150 - 18I_3}{12} = \frac{150 - 18(13)}{12} = \underline{-7A}$$

$$I_1 = I_2 + I_3 = 13 - 7 = \underline{6A}$$

Notice that I_2 is negative, so the actual current flows opposite the direction assumed initially.

The potential difference (PD) across each resistor is obtained by applying V = IR to each.

Thus across the 12Ω resistor in Fig. 21.2 is a PD of $12I_1 = 72$ volts.

Across the upper 6Ω resistor in Fig. 21.2 is a PD of $6I_2 = 42$ volts.

Across the lower 6Ω resistor is a PD of $6I_3 = 78$ volts.

Looking at Fig. 21.1 we see that the PD across the 36Ω is 72 volts, so the current through it is 72 volts/36 ohms = 2A. This is drawn in as an arrow in Figure 21.3, where the actual currents and their directions are shown. (Figure 21.3 on next page.)

The PD across the 6Ω and 12Ω in series in Fig. 21.1 is also 72 volts, so the current through them is 72 volts/18 ohms = 4A.

The PD between points C and D, across the 6Ω resistor, is given by V = IR = (4A)(6 ohms) = 24 volts.

The PD between D and E across the 12 resistor is (4A)(12 ohms) = 48 volts.

The sum of these two is 72 volts, just the PD across the 36 volts, as it must be.

The current in each resistor is thus as shown in Fig. 21.3.

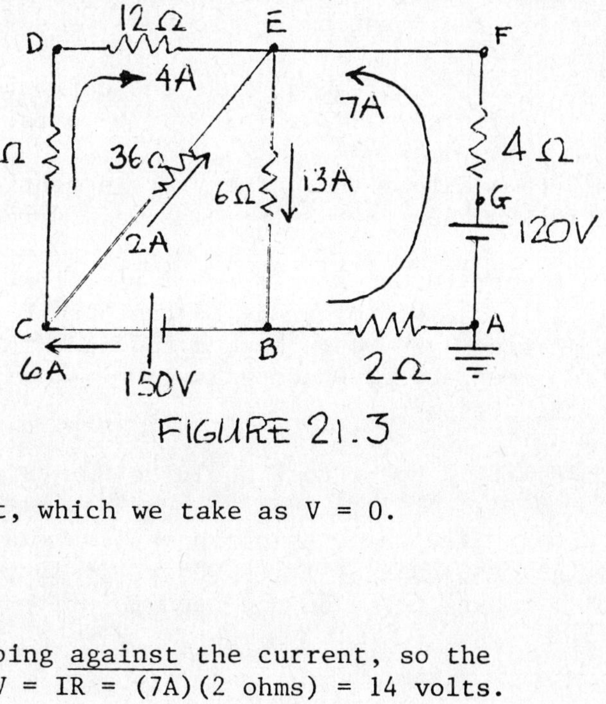

FIGURE 21.3

By inspection we can write down the potential at each labelled point. Start at A, the grounded point, which we take as V = 0.

$$V_A = 0$$

In going from A to B we are going <u>against</u> the current, so the potential rises by an amount V = IR = (7A)(2 ohms) = 14 volts. Thus $V_B = +14$ volts.

When we go from B to C the potential jumps up by +150 volts in crossing the battery, since we are going from the negative terminal to the positive terminal. Thus $V_C = V_B + 150 = 14$ volts + 150 volts = 164 volts.

From C to D the potential drops an amount V = (4A)(6 ohms) = 24 volts.

The potential is dropping here because we are going with the flow of current, i.e. like going downstream to lower elevation (potential).

From D to E the potential again drops, this time by an amount (4A)(12 ohms) = 48 volts.

Note that the total potential drop from C to D to E is thus 24 volts + 48 volts = 72 volts.

If we had gone directly from C to E across the 36 ohms resistor we would have seen the potential drop to be (2A)(36 ohms) = 72 volts. Naturally we find the same result using either of the two parallel paths.

The potential at D is thus $V_D = V_C - 24 = 164 - 24 = 140$ volts.

The potential at E is $V_E = V_D - 48 = 140 - 48 = 92$ volts.

As a check we might note that the potential drops by (13A)(6 ohms) going from E to B, so $V_B = V_E - 78 = 92 - 78 = 14$ volts, just as we found before.

There is no resistance between E and F, since they are connected by a wire of essentially zero resistance, so $V_F = V_E = 92$ volts.

From F to G we are going upstream against the current, so potential rises. $V_G = V_F + (7A)(4 \text{ ohms}) = 92 + 28 = 120$ volts.

When we go from G to A across the battery the potential drops by 120 volts, bringing us back to $V_A = 0$ where we started.

If one used a voltmeter to check voltage differences between various points in the circuit, as you might do in trouble shooting to find the source of a problem, you would expect to find, for example, that the voltage difference between D and B is 140 - 14 = 126 volts, with D positive with respect to B.

The potential of F with respect to C would be 92 - 164 = -72 volts.

Written out in detail like this, this process looks a little complicated. Once you get the hang of it, however, it is really much simpler than this appears. You do not need to write out every step as I have.

11. A galvanometer is a meter movement which can be used to make an ammeter or voltmeter with which to measure currents and voltages. The galvanometer has a small internal resistance R_g (usually less than 100 ohms). When a small fixed current I_g flows through the galvanometer the needle on the meter will deflect full scale. The amount of current needed is very small and depends on the design of the particular galvanometer.

To make an ammeter, connect a "shunt" (i.e. parallel) resistor R_{sh} parallel with the galvanometer.

Suppose that a current I, flowing in a circuit into which the ammeter has been connected, causes a current I_g to flow in the galvanometer, causing full scale deflection of the meter. We can find the relation between I and I_g in terms of R_g and R_{sh} by noting that the potential difference across R_g is equal to the potential difference across R_{sh}.

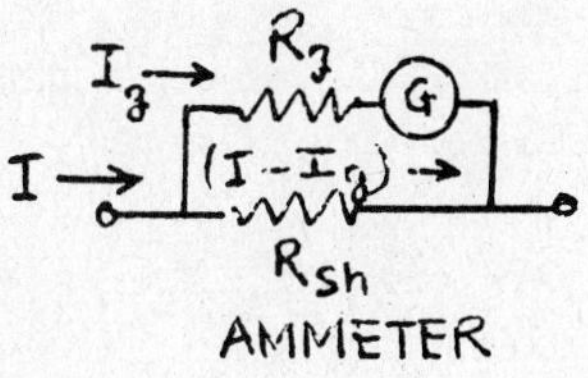

Thus $V = I_g R_g = (I - I_g)R_{sh}$

or

$$R_{sh} = \frac{I_g}{(I - I_g)} R_g \qquad (20.13)$$

Thus suppose we have a galvanometer movement which has a resistance of 10 ohms and which deflects full scale when 0.1 mA flows through it. What shunt resistance is needed so that a current of 10A in the circuit will cause full scale deflection?

Answer: $R_{sh} = \dfrac{10^{-4}\text{A}}{(10\text{A} - 10^{-4}\text{A})}$ (10 ohms) $\simeq 10^{-4}$ ohms.

We can use a galvanometer to make a voltmeter by connecting a resistance R_s in series with it. If a voltage V applied to the voltmeter is to cause full scale deflection of the meter, then current I_g must flow in the meter and the resistance R_s needed is given by

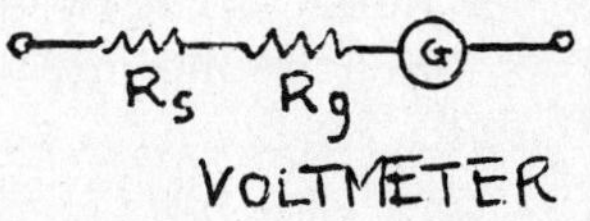

$$V = I_g (R_g + R_s) \qquad (20.14)$$

Both an ammeter and a voltmeter will cause a disturbance in the circuits they are measuring because they introduce new resistances into the circuits.

12. A Wheatstone Bridge is a useful circuit which illustrates the use of a null measurement. This technique is important and widely used because it enables one to make a measurement with a meter which is reading zero, and hence the calibration of the meter is not critical. The Wheatstone Bridge drawn here can be used to measure an unknown resistance R_x in terms of known resistances R_1, R_2 and R_3. R_1 is a variable resistor which is varied until there is no potential difference between points B and D. When this is true no current will flow between B and D, and the detecting meter will read zero. This is the condition of balance.

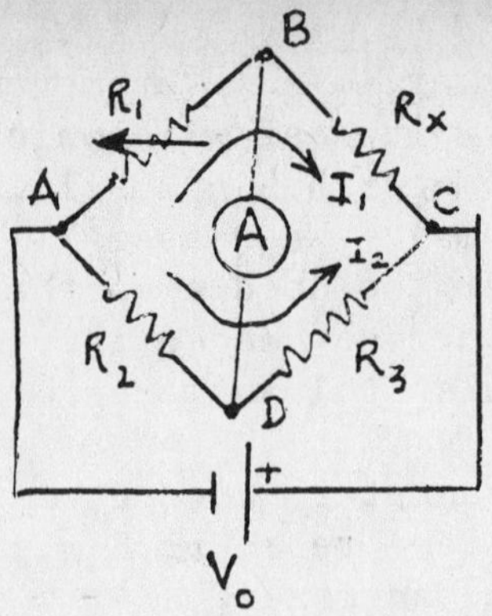

This occurs when $V_{AB} = V_{AD}$ or $I_1R_1 = I_2R_2$.   (5)

But the potential difference between A and C is the battery voltage V_o, whether we go through R_1 and R_2 or through R_3 and R_4.

Thus $V_o = I_1(R_1 + R_x) = I_2(R_2 + R_3)$.   (6)

From (5), $\frac{I_1}{I_2} = \frac{R_2}{R_1}$ and from (6), $\frac{I_1}{I_2} \quad \frac{R_2 + R_3}{R_1 + R_x}$

Equate these and simplify.

$$\frac{R_2}{R_1} = \frac{R_2 + R_3}{R_1 + R_x} \qquad R_2(R_1 + R_x) = R_1(R_2 + R_3)$$

$$R_2R_x = R_1R_3$$

$$\boxed{\frac{R_x}{R_1} = \frac{R_3}{R_2}}$$

Balance condition for a Wheatstone Bridge.   (20.15)

• Qualitative Questions

20M.1 In the Stanford linear accelerator a beam of electrons is accelerated down a pipe two miles long. Does this beam constitute an electric current?

A. Yes.
B. No, because the electrons are not in a conductor.
C. No, because the beam is not electrically neutral.
D. No, because the beam is electrically neutral.

20M.2 If you put your fingers across the terminals of a 1½ volt flashlight battery, the current that would flow through your body would be about 1 microampere (1 millionth of an ampere). How many electrons per second would this current represent?

A. 1,000,000 electrons/sec
B. 1,500,000 electrons/sec
C. 10^{-6} electrons/sec
D. 1.5×10^{-6} electrons/sec
E. 6.2×10^{12} electrons/sec
F. You can't tell without knowing the body's resistance.

20M.3 Suppose I have two spools of copper wire, one with a diameter somewhat larger than the diameter of the other. Suppose I cut off two short pieces of the same length, one from each spool. Then I do the same with two longer pieces.

Which piece of wire would have the largest resistance between its ends?

A. The short, fat wire.
B. The short, skinny wire.
C. The long, fat wire.
D. The long, skinny wire.
E. They would all have the same resistance since they are all made of the same material.

20M.4 In buying a battery you may find that one of the specifications provided is the "ampere-hour" rating. This is a measure of

A. the maximum current the battery can provide.
B. the maximum potential difference the battery can provide.
C. the maximum electric charge the battery can deliver.
D. the maximum power the battery can deliver.
E. the maximum electrical energy the battery can deliver.

20M.5 When two resistors are connected in parallel

A. the power dissipated in each is the same.
B. the current through each is the same.
C. the potential difference across each is the same.
D. one will usually contain more net electric charge than the other.

20M.6 Below is shown an oscilloscope trace used to plot current (vertical) vs. voltage (horizontal) for a carbon resistor. When the voltage is increased, the trace will look like:

A. A
B. B
C. C
D. D
E. E

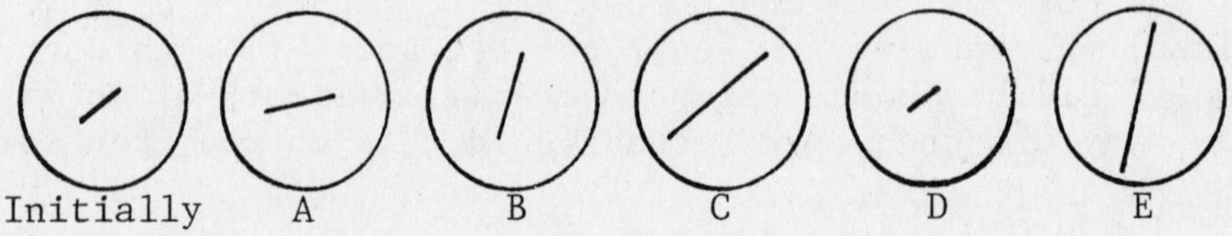

20M.7 Which of the following is the most accurate statement?

A. "Resistance" and "resistivity" have the same meaning.
B. The resistivity of a given conductor depends on its size and shape and on the material of which it is composed.
C. The resistivity of a piece of copper is less than the resistivity of a piece of iron.
D. The resistance of a thick copper wire is greater than the resistance of a thin copper wire of the same length.
E. Resistance in a metal conductor is due to the "back" internal electric fields which the conduction electrons exert on each other.

20M.8 If the length and diameter of a wire of circular cross-section are both doubled, and if the resistance was initially R, the new resistance will be

A. $\frac{1}{4}$ R
B. $\frac{1}{2}$ R
C. R
D. 2 R
E. 4 R

20M.9 Suppose that the radio and the heater in an automobile are both on. If the radio draws 2 Amps and if the heater has a resistance of 1.5 Ohms, how much current do these two accessories draw from the 12-volt electrical system when they operate simultaneously?

A. 2 A
B. 8 A
C. 10 A
D. 12 A
E. 18 A
F. 20 A

20M.10 Which of the following units, or combinations of units, is a measure of energy?

A. Kilowatt
B. Coulomb/farad
C. Kilowatt-Hour
D. Ampere-Hour
E. Coulomb/second
F. None of the above.
G. More than one of the above.

20M.11 I heat my house with natural gas, and during the winter I use an average of 400 Therms per month. This costs me approximately $100 per month. How much would it cost if I used electric heating instead?

A. $ 66
B. $ 72
C. $ 97
D. $122
E. $144
F. $152
G. $188
H. $202
I. $888

1 Therm = 100 cu. ft.
1 cu. ft. of gas yields 1,000 BTU
1 BTU = 1,000 Joules
1 Therm costs $0.25 in our area.
1 kW-Hr costs $0.013 in our area (very cheap when compared to the rest of the U.S.)
Line voltage = 120 VAC

20M.12 Suppose you have a portable radio which operates on 12 V. What value resistance should you put in series with the radio in order to have it operate from a 6-volt tractor battery?

A. 12 Ohms
B. 6 Ohms
C. 2 Ohms
D. 0.5 Ohm
E. It won't work. Putting a resistance in series will not increase the voltage.
F. It won't work. A resistance must be placed <u>in parallel</u> with the radio to increase the voltage.
G. One can't answer this question without knowing how much current the radio uses.

20M.13 Suppose that a load resistor R_L is connected across the terminals of a real battery. The power dissipated in the load resistor

A. will be independent of the value of R_L.
B. will increase when R_L is increased.
C. will decrease when R_L is increased.
D. may increase or decrease when R_L is increased, depending on the initial value of R_L.
E. may decrease or stay the same, depending on the initial value of R_L.

20M.14 A load resistance R_L is connected to a 6 volt battery whose internal resistance is 0.1 ohms. The battery is rated at 120 ampere-hours. Which of the following is the most accurate statement?

A. The maximum current the battery can deliver is 30 amps.
B. The potential difference across the load will be a maximum when R_L = 0.1 ohm.
C. The potential difference between the battery terminals will be a maximum when R_L = 0.
D. The maximum power the battery can deliver to the load is 90 watts.
E. The maximum time for which the battery can deliver current without being recharged is 2 hours.

Questions 20M.15 - 20M.19 refer to the analogy between an electric circuit and water flowing in an irrigation system.

20M.15 The quantity of charge Q is analagous to

A. the elevation above sea level.
B. the volume of a storage tank.
C. the pressure generated by a pump.
D. the flow in gallons per minute.
E. a mass (or volume) of water.

20M.16 A difference in electric potential between two points in a circuit is analogous to

A. a difference in pressure caused either by a pump or a difference in elevation.
B. a difference in the flow of water at two different points.
C. a difference in the resistance of two pipes.
D. a difference in the amount of water stored in two tanks.

20M.17 A flow of water, in liters/second, is analogous to

A. a flow of electric potential.
B. an electric current (i.e. a flow of charge in Coulombs per second).
C. the steady heating of an electrical resistor.
D. the potential difference created by a battery.

20M.18 A battery is analogous to

A. a very large diameter pipe.
B. a pump.
C. a storage tank.
D. the total amount of water in the system.

20M.19 The resistance of a pipe carrying water is analogous to

A. the EMF generated by a battery.
B. the tendency of current to flow from high to low potential.
C. the resistance of an electrical conductor.
D. the resistivity of an electrical conductor.

20M.20 Shown here are four possible ways of connecting a lamp to a flashlight battery. In which of the ways could the lamp light?

A.
B.
C.
D.
E. More than one of the arrangements shown will light.
F. None of the arrangements shown will light.

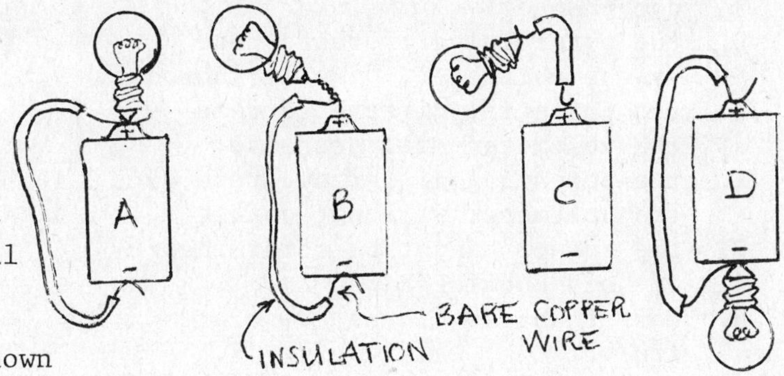

20M.21 If the circuit shown here were broken at point X,

A. the current in R_1 would not change.
B. the current in R_3 would increase.
C. the current in R_6 would decrease.
D. the potential difference between X and Z would decrease.
E. the potential difference between Y and Z would change.

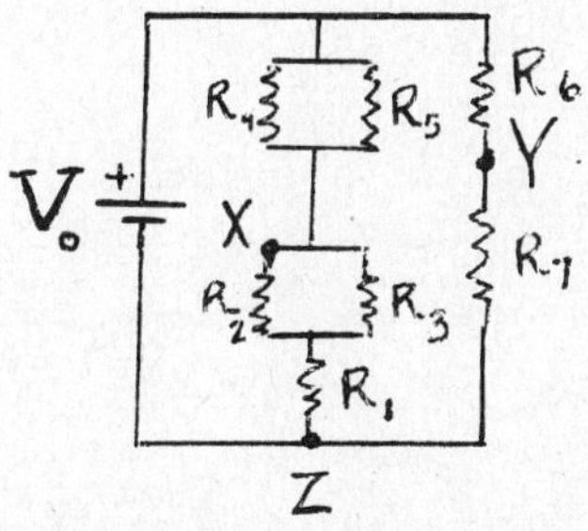

20M.22 When the switch S in the circuit shown here is closed, what change will occur in the current passing through the resistor R?

A. There will be no change.
B. The current will increase.
C. The current will decrease.
D. Whether the current will increase or decrease depends on the value of R.
E. None of the above is correct.

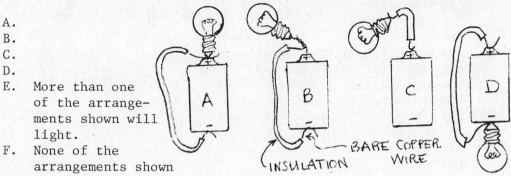

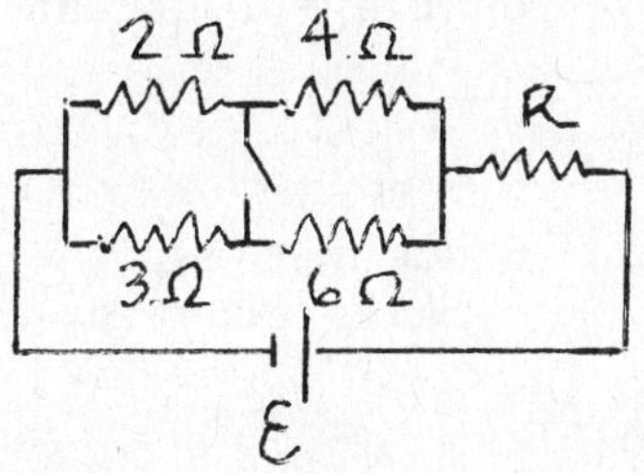

20M.23 For any network in which current flows, it is always true that

A. the sum of the currents around a closed loop is zero.
B. the sum of the currents entering a junction is equal to the sum of the currents leaving the junction.
C. the sum of the EMF's around any closed loop is zero.
D. the sum of the IR drops around any closed loop is zero.
E. the total current, including all branches, is zero.

20M.24 Negligible current flows in the voltmeter here. If the circuit is broken at point P,

A. the power delivered by the battery will increase.
B. the current in the ammeter, I, will increase.
C. the current in the ammeter will not change.
D. the potential difference read by the voltmeter, V, will increase.
E. the potential difference read by the voltmeter will decrease.
F. the potential difference read by the voltmeter will not change.
G. the answer cannot be determined without knowing R_x.
H. More than one of the above are true.

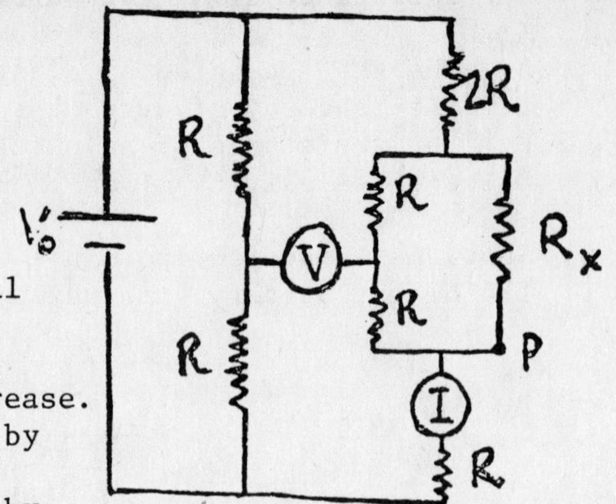

• Multiple Choice: Answers and Comments

20M.1 A The beam of pure electrons has a net charge. Because it is moving, it constitutes a current. A current need not be in a solid such as metal.

20M.2 E $I = \frac{Q}{t} = \frac{Ne}{t}$, $N = \frac{It}{e} = \frac{(10^{-6}A)(1s)}{(1.6 \times 10^{-19} C)} = 6.25 \times 10^{12}$ Electrons

20M.3 D $R = \rho\frac{\ell}{A}$ Wire with small diameter and large length will thus have greatest resistance, since resistivity ρ is the same in all cases.

20M.4 C Charge = (Current) x (time)

20M.5 C

20M.6 C Since V = IR, the slope of the I vs. V curve is R, which remains constant in an ordinary resistor.

20M.7 C

20M.8 B $R = \rho\frac{\ell}{A} = \rho\frac{\ell}{\pi r^2}$ If $\ell_1 = 2\ell$ and $r_1 = 2r$,

$$R_1 = \rho\frac{(2\ell)}{\pi(2r)^2} = \frac{1}{2} R$$

20M.9 C The heater draws current $I_H = \frac{V}{R_H} = \frac{12 \text{ volts}}{1.5\ \Omega} = 8$ A

Radio and heater are in parallel, so I = 2 A + 8 A = 10 A.

20M.10 C Energy = (Power) x (Time).

20M.11 E

GAS USED = (400)(100) = 4×10^4 CU. FT.

THIS YIELDS ENERGY $\mathcal{E}$, $\mathcal{E} = (4 \times 10^4 \text{ CU.FT.})(1000 \text{ BTU./CU. FT.})(1000 \text{ J/BTU})$

$= 4 \times 10^{10}$ JOULES.

1 KW-HR = (1000 JOULE/SEC)(3600 SEC) = $3 \cdot 6 \times 10^6$ JOULES

THUS $\mathcal{E} = 4 \times 10^{10}$ JOULES $= (4 \times 10^{10})(3.6 \times 10^6)^{-1}$ KW-HR $= 1.11 \times 10^4$ KW-HR

COST = $(1.11 \times 10^4$ KW-HR$)(\$0.013$ /KW-HR$) = \$144$

20M.12 E The potential drop across the resistor plus the potential drop across the radio must be equal to the potential difference created by the battery, which is only 6 volts in this case. Thus there is no way to get 12 volts across the radio by inserting a resistance.

20M.13 D The power delivered to the load is a maximum when $R_L = R_i$. Thus whether or not the power in the load increases when R_L is increased depends on whether R_L was initially greater or less than R_i.

20M.14 D A is false, since the maximum current will flow when $R_L = 0$, so $I = \mathcal{E}/R_i$ = 6 volts/0.1 ohms = 60 A.

The potential difference V_L will be a maximum when there is no potential drop across the internal resistance. This means there is no current flow, which results when $R_L = \infty$.

$$P_{MAX} = \frac{\mathcal{E}^2}{4R_L} = \frac{(6 \text{ V})^2}{(4)(.1 \ \Omega)} = 90 \text{ W}$$

If the battery provides current at a very low rate it can operate for an indefinite time.

20M.15 E

20M.16 A

20M.17 B

20M.18 B

20M.19 C

20M.20 F In order for current to flow one must have a closed circuit, and that is the case for none of the arrangements shown.

20M.21 B Observe first that the current in R_6 and R_7, and thus the potential at point Y, is unaffected by breaking the circuit at X, since the battery is connected directly to R_6 and R_7 in series. Thus E is false.

When the circuit is broken at X the resistance in the left branch of the circuit will decrease. This means that the PD across R_4 will drop, as will the PD across R_1. This means that the drop across R_3 must increase, since the sum of the drop across R_1, R_3, and R_4 must still add up to the fixed battery voltage V_o. Notice that this effect is a little tricky. Although breaking the circuit at X causes the current in R_1, say, to drop, the current in R_3 increases, because now all of the current passes through R_3 instead of dividing between R_2 and R_3.

20M.22 A Two ohms is to 4 ohms as 3 ohms is to 6 ohms, thus the two ends of the switch are at the same potential, and no current will flow between them when the switch is closed.

20M.23 B

20M.24 E To analyze this circuit first consider two limiting values for R_x. If $R_x = 0$ the right side of the voltmeter will be a potential of $V_o/3$ and the left side will be at $\frac{1}{2}V_o$. If $R_x = \infty$, the right side of the voltmeter is at $(2/5)V_o$ and the left side is a gain at $\frac{1}{2}V_o$. Breaking the circuit at P corresponds to having $R_x = \infty$, which means that the meter will read $\frac{1}{2}V_o - 2V_o/5$. Initially it was reading something between this value and a maximum of $\frac{1}{2}V_o - V_o/3$, so the voltmeter reading decreases when the circuit is broken at P.

• Problems

20.1 There are about 8.4×10^{28} electrons/m^3 in copper. What is the drift velocity of electrons in a copper wire of diameter 2 mm in which a current of 10 A flows? You'll be surprised at the result.

20.2 An object like a metal ring can be electroplated with gold with the set-up shown here. Gold is dissolved from one electrode and flows to the other electrode (the ring) when a potential difference is applied by the power supply. The gold ions in solution carry a charge of +e. One mole of gold (6×10^{23} atoms) has a mass of 197 (the atomic weight). How long would you have to let a current of 2 A flow in order to deposit 0.5 gm of gold?

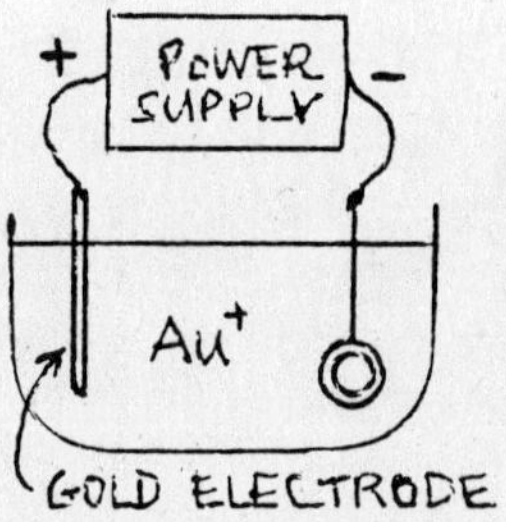

20.3 The two batteries in an ordinary flashlight cause a current of 0.2 A to flow in a lamp filament by causing a potential difference of 3 volts across the filament. What is the resistance of the filament?

20.4 The resistivity of a hot tungsten filament in an incandescent lamp is about 5 x 10^{-7} Ohm-meter. What would be the required diameter of such a filament 5 cms long if it is to have a resistance of 1.2 ohms?

20.5 Nichrome wire is often used as a heating element in an electric furnace. The temperature coefficient of resistivity of nichrome is 4 x 10^{-4} per ^{o}C. If the resistance of a heating element at 20 ^{o}C is 2 ohms, what would be the resistance at the operating temperature of 500 ^{o}C?

20.6 An 800 W hair dryer operates from a 120 volt outlet. What current will it draw?

20.7 A 4 ohm resistor is to be used as a temperature sensor (its resistance varies with temperature). Current is made to flow through it by connecting it to a 10 volt power supply. What resistance should be placed in series with the resistor if the power dissipated in the sensor resistor is not to exceed 10 mW?

20.8 Determine the resistance between terminals X and Y for the arrangements of resistors shown here.

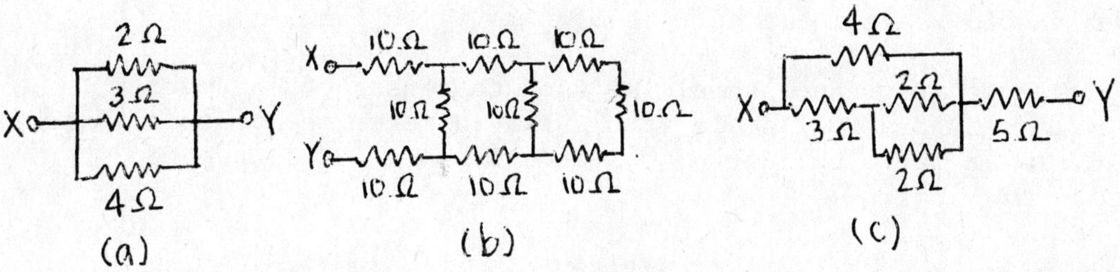

20.9 When no load is connected to a car battery the measured voltage between the terminals is found to be 12 volts. When the terminals are shorted (i.e. connected with a conductor of essentially zero resistance) a current of 120 A flows. What load resistance should be used with the battery in order to obtain maximum power in the load? In engineer's jargon one says that the open circuit voltage is 12 volts and the short circuit current is 120 A.

20.10 What is the power dissipated in the 4 ohm resistor here?

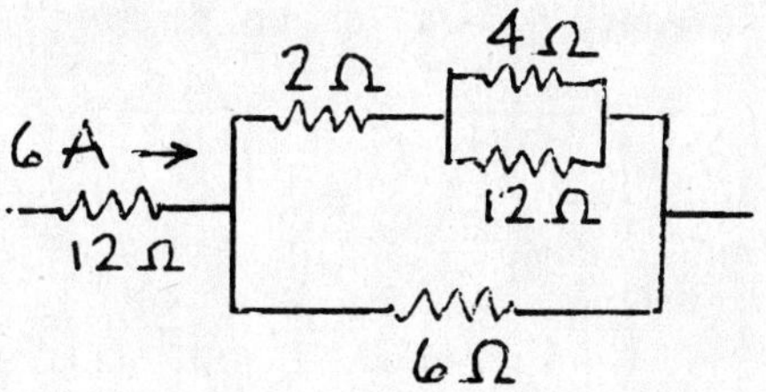

20.11 What current flows in the battery here? This problem requires some thinking, but not much calculating. Ask yourself what the potential would be at points X, Y and Z if the R_1 and $2R_1$ resistors were removed. Look at the symmetry of the problem. Now ask what would happen if these resistors were inserted. The circuit with these resistors absent is easy to reduce using series and parallel combinations.

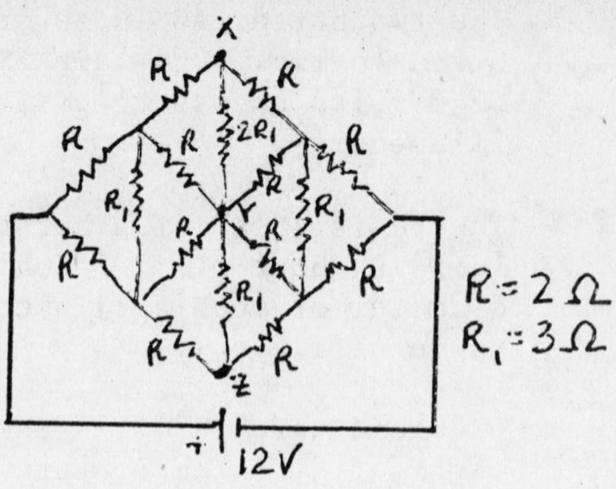

20.12 What is the current through each resistor and potential at point X for the circuit shown?

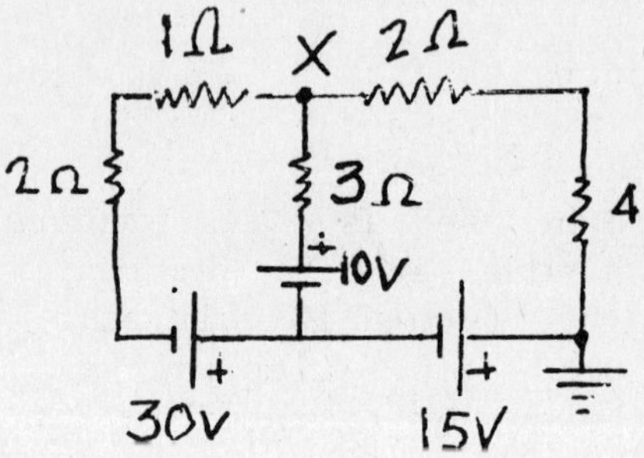

20.13 How could you construct a voltmeter which reads 100 volts full scale by using a galvanometer with a resistance of 120 ohms which reads full scale for a current of 2 mA?

20.14 What shunt resistance should be used to construct an ammeter which reads 2 A full scale if the galvanometer movement deflects full scale for a current of 0.2 mA and has a resistance of 12 ohms?

• Problem Solutions

20.1 FROM EQN. (20.3) $I = nqv_dA$

$$A = \pi r^2, \quad v_d = \frac{I}{nq\pi r^2} = \frac{(10\,A)}{(8.4\times10^{28}\,m^{-3})(1.6\times10^{-19}\,C)(\pi)(10^{-3}\,m)^2}$$

$$v_d = 2.4\times10^{-4}\ m/s = 0.24\ mm/s$$

THUS IN ONE MINUTE THE ELECTRONS DRIFT ONLY 14 mm.

20.2 SUPPOSE 0.5 gm OF GOLD CONTAIN N ATOMS.

$$\frac{N}{0.5} = \frac{6\times10^{23}}{197}, \quad N = \frac{0.5}{197}\times 6\times10^{23} = 1.52\times10^{21}\ \text{GOLD ATOMS.}$$

WHEN A GOLD ATOM GOES INTO SOLUTION IT LOSES ONE ELECTRON AND BECOME A GOLD ION, Au^+. N SUCH ATOMS CARRY CHARGE $Ne = Q$.

$$I = \frac{Q}{t}, \quad \text{SO} \quad t = \frac{Q}{I} = \frac{Ne}{I} = \frac{(1.52\times10^{21})(1.6\times10^{-19}\,C)}{(2A)} = \underline{\underline{122\ s}}$$

20.3 $V = IR$, $R = \frac{V}{I} = \frac{3V}{0.2A} = \underline{\underline{15\ \Omega}}$

20.4 $R = \rho \frac{l}{A} = \rho \frac{l}{\pi r^2}$, $r = \sqrt{\frac{\rho l}{\pi R}} = \sqrt{\frac{(5\times10^{-7}\ \Omega\text{-m})(5\times10^{-2}\ m)}{\pi(1.2\ \Omega)}}$

$r = 8.1 \times 10^{-5}$ m

$d = 2r = \underline{\underline{0.16\ mm}}$

20.5 $R_1 = \rho_1 \frac{l}{A}$, $R_2 = \rho_2 \frac{l}{A}$, so $\frac{R_2}{R_1} = \frac{\rho_2 l/A}{\rho_2 l/A} = \frac{\rho_2}{\rho_1}$

$R_2 = \frac{\rho_2}{\rho_1} R_1 = \frac{\rho_1(1+\alpha\Delta T)}{\rho_1} R_1$. ρ_1 = RESISTIVITY AT 20°C

$\Delta T = T - 20°C = 500°C - 20°C = 480°C$

$R_2 = [1 + (4\times10^{-4}\ °C^{-1})(480°C)][2\ \Omega]$

$\underline{\underline{R_2 = 2.4\ \Omega}}$

20.6 $P = IV$, $I = \frac{P}{V} = \frac{800W}{120V} = \underline{\underline{6.7\ A}}$

20.7

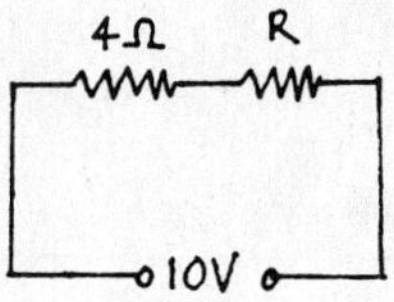

$I = \frac{V}{R_S} = \frac{10}{4+R}$

POWER DISSIPATED IN 4 Ω RESISTOR IS

$P = I^2 R_T = \left(\frac{10}{4+R}\right)^2 (4) = 0.01\ W$

$(4+R)^2 = \frac{400}{0.01} = 4\times10^4$

$4 + R = 2\times10^2 = 200$

$\underline{\underline{R = 196\ \Omega}}$

20.8 (a) $\frac{1}{R_p} = \frac{1}{R_1} + \frac{1}{R_2} + \frac{1}{R_3} = \frac{1}{2} + \frac{1}{3} + \frac{1}{4} = \frac{6}{12} + \frac{4}{12} + \frac{3}{12} = \frac{13}{12}$

$\underline{\underline{R_p = \frac{12}{13} = 0.92\ \Omega}}$

(b)

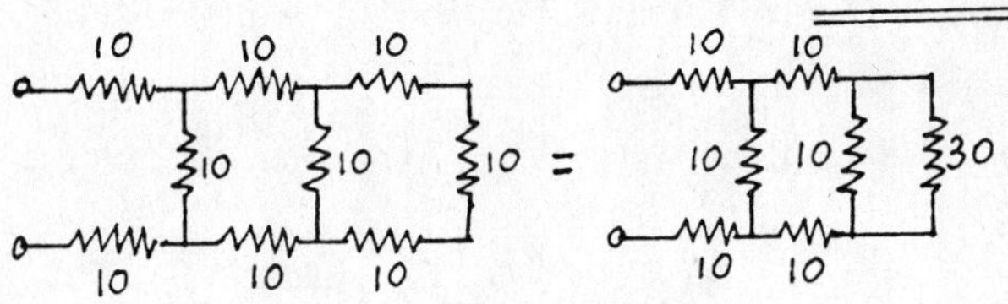

$\frac{1}{R_p} = \frac{1}{10} + \frac{1}{30} = \frac{3}{30} + \frac{1}{30} = \frac{4}{30}$

$R_p = \frac{30}{4} = 7.5\ \Omega$

(ALL RESISTANCES IN OHMS)

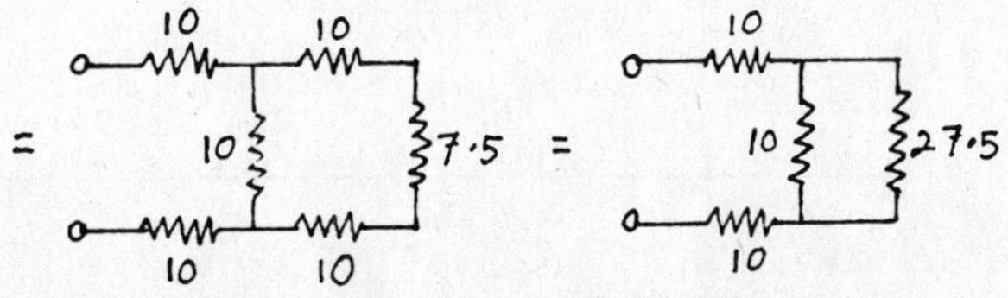

$\frac{1}{R_p'} = \frac{1}{10} + \frac{1}{27.5}$

$R_p' = \frac{(10)(27.5)}{10 + 27.5} = 7.33\ \Omega$

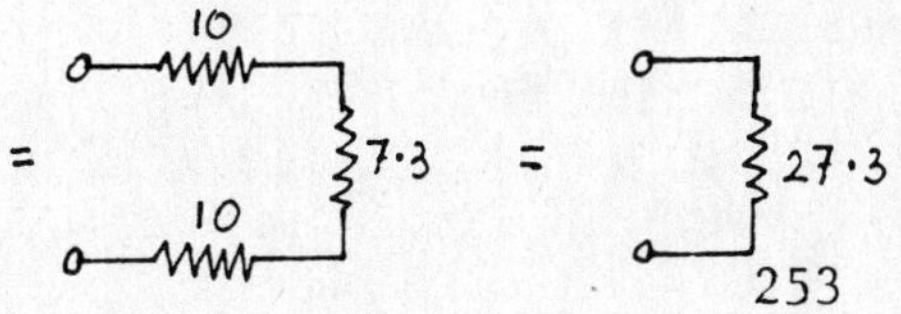

$\underline{\underline{R = 27.3\ \Omega}}$

(c)

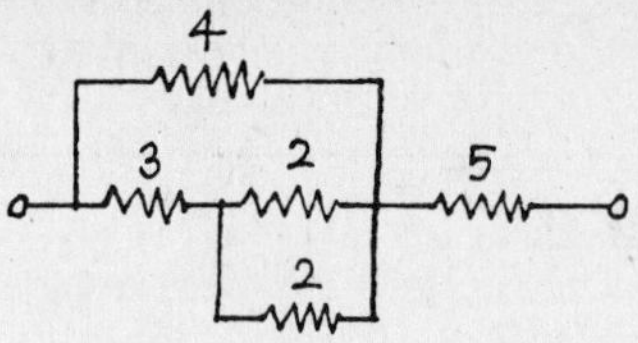

$$\frac{1}{R_p} = \frac{1}{2} + \frac{1}{2} = 1, \quad R_p = 1$$

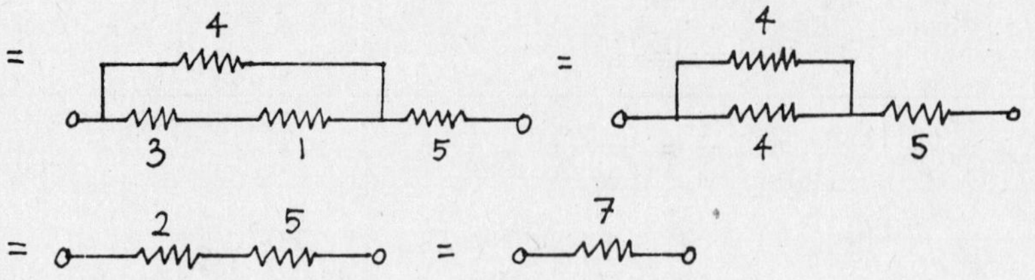

$\underline{\underline{R = 7\,\Omega}}$

20.9

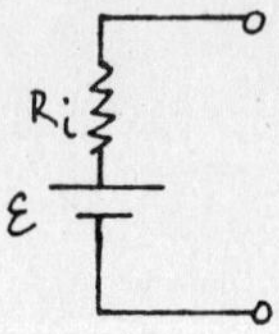

OPEN CIRCUIT VOLTAGE IS $\mathcal{E} = 12\text{V}$

SHORT CIRCUIT IS $I_{sc} = \dfrac{\mathcal{E}}{R_i} = 120\text{A}$

$$\text{SO } R_i = \frac{\mathcal{E}}{I_{sc}} = \frac{12\text{V}}{120\text{A}} = 0{\cdot}1\,\Omega$$

FOR MAXIMUM POWER OUTPUT $\underline{\underline{R_L = R_i = 0{\cdot}1\,\Omega}}$

21·1 FIRST REDUCE RESISTANCES USING SERIES AND PARALLEL COMBINATIONS.

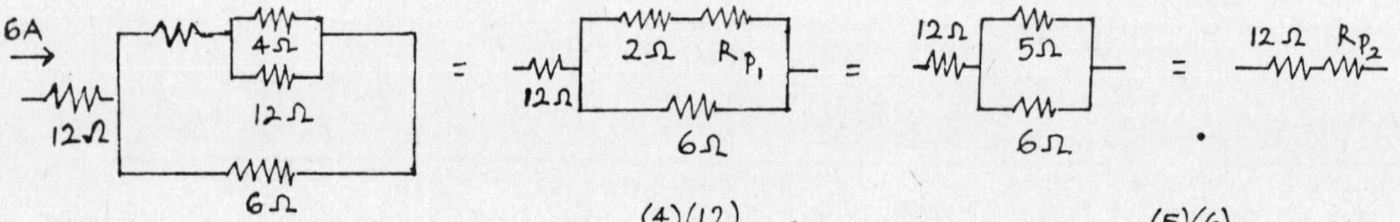

$$R_{p_1} = \frac{(4)(12)}{4+12} = 3\,\Omega \qquad R_{p_2} = \frac{(5)(6)}{5+6} = 2{\cdot}73\,\Omega$$

THE POTENTIAL DROP ACROSS R_{p_2} IS $IR_{p_2} = (6\text{A})(2{\cdot}73\,\Omega) = 16{\cdot}4$ VOLTS

THUS CURRENT THROUGH R_{p_1} AND $2\,\Omega$ IN SERIES IS

$$I' = \frac{V}{R} = \frac{16{\cdot}4 \text{ VOLTS}}{(2+3)\,\Omega} = 3{\cdot}27\text{A}$$

POTENTIAL DROP ACROSS R_{p_1} IS THUS $V = I'R_{p_1} = (3{\cdot}27\text{A})(3\,\Omega) = 9{\cdot}82\text{ V}$

POWER DISSIPATED IN $4\,\Omega$ IS THUS

$$P = \frac{V^2}{R} = \frac{(9{\cdot}82\text{V})^2}{4\,\Omega} = \underline{\underline{24{\cdot}1\text{ W}}}$$

21·2 REMOVE RESISTORS R_1 AND $2R_1$ AND OBSERVE THAT BECAUSE OF THE SYMMETRY OF THE CIRCUIT WITH RESPECT TO REFLECTION IN A HORIZONTAL LINE THROUGH THE CENTER POINTS X, Y AND Z ARE ALL AT THE SAME POTENTIAL. THE SAME IS TRUE FOR OTHER POINTS ON THE SAME VERTICAL LINE. THUS NO CURRENT WILL FLOW BETWEEN THESE POINTS. WE MAY THEREFORE PLACE ANY RESISTORS WE WISH THERE. FOR EXAMPLE, WE COULD REPLACE THE R_1 AND $2R_1$ RESISTORS WITH "SHORTS", R = 0. THEN CIRCUIT WOULD LOOK LIKE THIS:

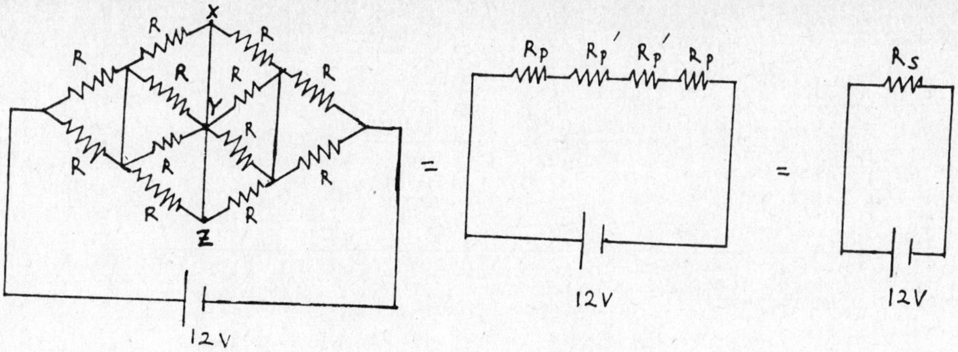

$R_p = \frac{R}{2}$, $\frac{1}{R_p'} = \frac{1}{R} + \frac{1}{R} + \frac{1}{R} + \frac{1}{R}$, $R_p' = \frac{R}{4}$, $R_s = 2R_p + 2R_p'$

$R_s = \frac{3}{2} R = \frac{3}{2}(2\,\Omega) = 3\,\Omega$

$I = \frac{V}{R_s} = \frac{12V}{3\Omega} = \underline{\underline{4A}}$

21·3

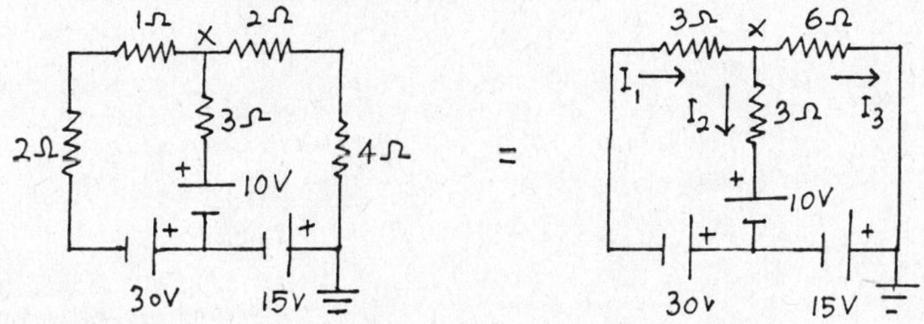

$I_1 = I_2 + I_3$ (1)

$30 + 10 = -3I_1 - 3I_2$ (2)

$15 - 10 = 3I_2 - 6I_3$ (3)

SUBSTITUTE (1) IN (2): $40 = -3(I_2 + I_3) - 3I_2$

$40 = -6I_2 - 3I_3$

$20 = -3I_2 - \frac{3}{2}I_3$ (4)

ADD (4) TO (3):

$25 = -\frac{15}{2}I_3$, $I_3 = -\frac{50}{15} = \underline{\underline{-3.33A}}$

SUBSTITUTE THIS VALUE OF I_3 IN (4):

$20 = -3I_2 - \frac{3}{2}(-3.33A)$, $I_2 = \underline{\underline{-5.0A}}$

FROM (1), $I_1 = I_2 + I_3 = -5.0 - 3.33 = \underline{\underline{-8.33A}}$

THUS THE ACTUAL CURRENTS FLOW OPPOSITE THE DIRECTIONS SHOWN BY THE ARROWS.

FROM GROUND TO POINT X THE POTENTIAL DROPS (CURRENT IS FLOWING "DOWNHILL" THROUGH THE 6Ω RESISTOR)

THUS $V_x = -IR = -(3.33A)(6\Omega) = \underline{\underline{-20.0 \text{ VOLTS}}}$

21.4

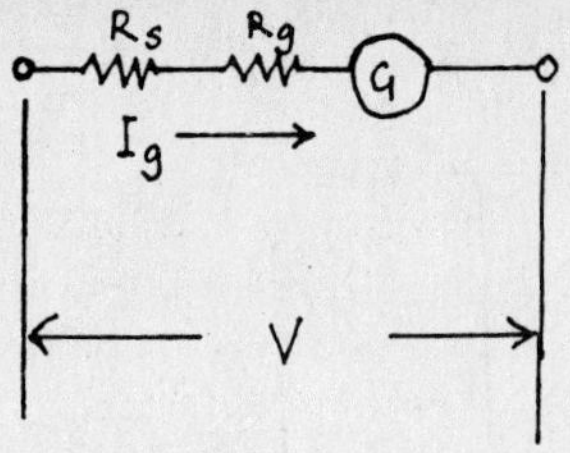

$$V = I\,(R_s + R_g)$$

$$R_s = \frac{V - I_g R_g}{I_g}$$

$$R_s = \frac{(100\,\text{V}) - (2\times10^{-3}\,\text{A})(120\,\Omega)}{(2\times10^{-3}\,\text{A})} = 49{,}880\,\Omega$$

$$\simeq \underline{\underline{50{,}000\,\Omega}}$$

21.5 FROM EQ. (21.1), $R_{sh} = \left(\dfrac{I_g}{I - I_g}\right) R_g$

$$= \left(\frac{0.2\times10^{-3}\,\text{A}}{2\,\text{A} - 0.2\times10^{-3}\,\text{A}}\right)(12\,\Omega)$$

$$\underline{\underline{R_{sh} = 0.0012\,\Omega}}$$

Magnetism

21

• Summary of Important Ideas, Principles, and Equations

1. A magnetic field can be created by moving electric charges (i.e., electric currents) or by magnetic particles (such as individual electrons or protons, or certain magnetic atoms, such as iron or gadolinium). We may envision the magnetic field by using the concept of magnetic field lines, much as we did for the electric field. Where the lines are close together the field is strong; where they are far apart the field is weak. The field at any point is directed tangent to the magnetic field line at that point. The magnetic field, B, is measured in Tesla in the SI system. Another unit of magnetic field is the Gauss, $1G = 10^{-4}T$. The earth's magnetic field is about 1G.

2. A moving charge will experience a force in a magnetic field given by

$$F = Bqv \sin\theta \tag{21.1}$$

where B = magnetic field in Tesla
q = charge of particle
v = velocity of particle
θ = angle between B and v

The direction of the force on a positive charge is obtained by Right Hand Rule No. 1 (RHR-I). Use your right hand. Point fingers along the magnetic field B with your thumb in the direction of the velocity v. The force F will point out perpendicular to your palm. For a negative particle the force will point in the opposite direction.

3. A force acts on a current in a magnetic field. The direction of the force is given by RHR-I, with your thumb pointing along the current I. The force on a length of wire carrying current I in a magnetic field B which makes an angle θ with the wire is

$$F = BI\ell\sin\theta \qquad (21.2)$$

4. A charge moving perpendicular to a uniform magnetic field B will move in a circle of radius R, where the magnetic field provides the needed centripetal force. R is called the cyclotron radius.

$$Bqv = \frac{mv^2}{R}, \quad R = \frac{mv}{qB} \qquad (21.3)$$

The number of revolutions made per second is f_c, the cyclotron frequency.

$$\omega_c = \frac{v}{R} = \frac{qB}{m}$$

$$f_c = \frac{\omega_c}{2\pi} = \frac{1}{2\pi} = \frac{qB}{2\pi m} \qquad (21.4)$$

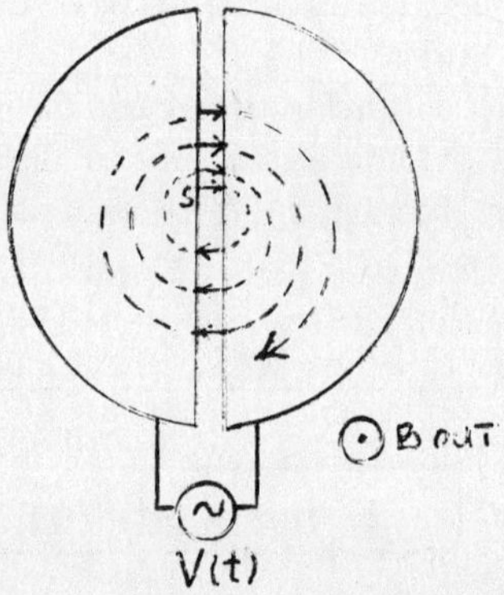

A cyclotron is a particle accelerator which uses a magnetic field to bend charged particles into a circular path. Imagine a magnetic field coming out of the paper. Particles are emitted from source S with low velocity. They are accelerated by a voltage applied between the two "D's" drawn here. There is no electric field inside the D (an evacuated metal can that looks like the letter D). There the particle is bent in a semi-circle by the magnetic field. When it comes back to the gap the alternating voltage has changed sign and again accelerates the particle across the gap. Again the particle travels a semi-circle, but a bigger one now because it is going faster. This process is repeated over and over as the particle spirals out, finally to be removed from the machine at very high energy. Protons are accelerated in this way and allowed to smash into a target which they make radioactive. These radioactive materials are used, for example, in hospitals for diagnosis and therapy.

5. The Hall Effect occurs when a current passes through a conductor placed in a magnetic field oriented perpendicular to the current. The magnetic force on the carriers causes some of them to move to the top side of the conductor. Charge piles up here, setting up a "back" electric field which eventually causes an electric force which cancels out the effect of the magnetic field. The field necessary to do this is the Hall field, E_H.

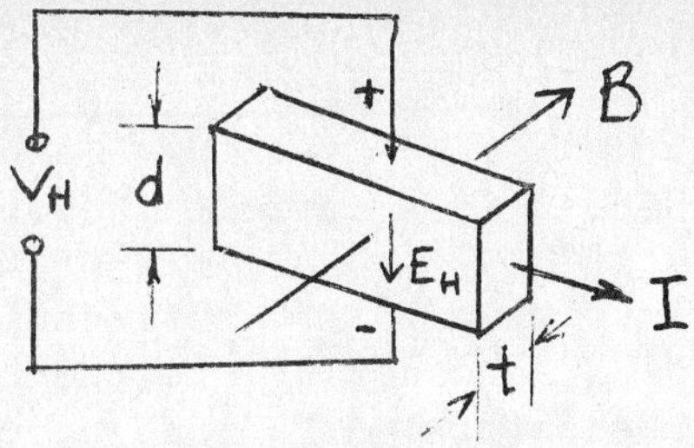

$$\boxed{qE_H = qvB} \quad \text{or} \quad \boxed{E_H = vB} \qquad (21.5)$$

A potential difference V_H (the Hall voltage) is set up between the top and bottom of the conductor. A measurement of this voltage for known magnetic field yields the sign of the carriers and their density, n. Conversely, the Hall Effect can be used to measure carrier velocity. Blood flowmeters have been constructed on this principle.

A magnetohydrodynamic (MHD) generator is essentially a giant Hall Effect device in which a huge jet engine blasts charged gas particles (i.e., ions in a plasma) through strong magnetic fields created by superconducting coils. The Hall Voltage is then used as a power source for domestic use. Such devices are like generators without any wires.

$$I = (tdv)(nq)$$

$$\boxed{V_H = E_H d = vBd = \frac{I}{tdnq} Bd = \frac{BI}{tnq}} \qquad (21.6)$$

n = number of carriers/m^3 q = carrier charge
t = conductor thickness   B = magnetic field

6. Magnetic fields can be created by currents. Recall that electric and magnetic field lines look like the flow lines of a fluid. Static electric field lines sprout out of positive charges (sources) and end on negative charges (sinks). The field lines are like the flow lines in a tub of water where water flows in from a faucet (the source) and out through a drain (the sink). Similarly, magnetic field lines are analogous to a whirlpool flow, stirred up by a paddlewheel. Electric currents are like the paddlewheels which stir up the magnetic field lines. Magnetic fields have no sources and sinks, i.e., there are no known magnetic monopoles (but I have a hunch they will be discovered somewhere, someday, in the universe). Thus magnetic field lines always end on themselves.

The magnetic field due to a current can be found using an experimental result, Ampere's law.

$$\Sigma B_{||} \Delta \ell = \mu_o I \tag{21.7}$$

Here I is a current flowing, say, in a wire.

Imagine a path C surrounding I. Let $\Delta \ell$ be a little length of this path, and let $B_{||}$ be the component of the magnetic field parallel to the path at $\Delta \ell$. Add up the contributions $B_{||}\Delta \ell$ for the closed path and you can find B in terms of the current I and the experimental constant μ_o. This formula is only practical for rather symmetric current configurations.

Application of Ampere's law to a long straight wire yields the following formula for the magnetic field surrounding the wire at a distance r from the wire.

$$B = \frac{\mu_o}{2\pi} \frac{I}{R} \tag{21.8}$$

Here $\mu_o = 4\pi \times 10^{-7}$

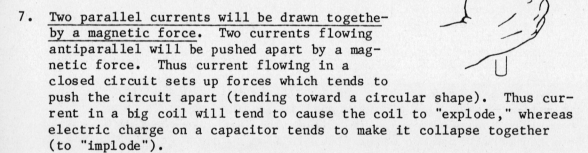

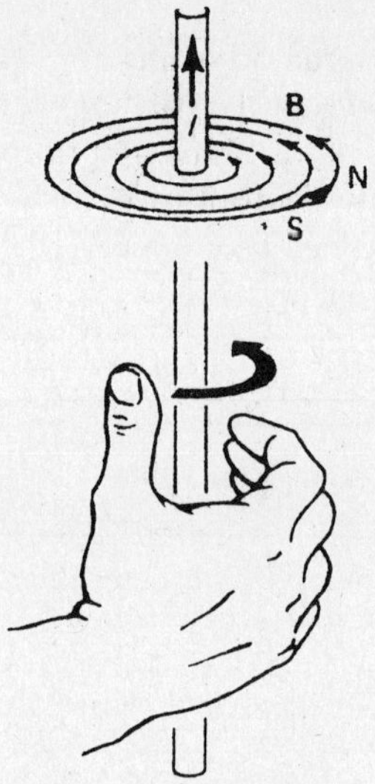

The B-lines are concentric circles centered on the wire. In general, the direction of B is found by using Right Hand Rule No. 2 (RHR-II). Point your right thumb along I and your fingers will curl around the current like the lines of B. This rule can be used for currents not flowing in straight wires in order to get the general direction of B.

7. Two parallel currents will be drawn togethe- by a magnetic force. Two currents flowing antiparallel will be pushed apart by a magnetic force. Thus current flowing in a closed circuit sets up forces which tends to push the circuit apart (tending toward a circular shape). Thus current in a big coil will tend to cause the coil to "explode," whereas electric charge on a capacitor tends to make it collapse together (to "implode").

8. A single current loop acts just like a little bar magnet. It will create a "magnetic dipole" field (shaped like the earth's field), and it will try to line up with its axis parallel to any magnetic field applied to it. Its magnetic moment vector μ points along the axis of the loop in the direction that the current would advance a right hand screw (RHR-III). The magnitude of μ is

$$\mu = IA \tag{21.9}$$

where current I flows around a small loop of area A. A magnetic moment μ placed in a magnetic field B will experience a torque

$$\boxed{\tau = \mu B \sin\theta} \qquad (21.10)$$

where θ = angle between μ and B.

9. A solenoid is a long cylindrical coil made with many turns of wire. The magnetic field inside it is almost constant, with

$$\boxed{B = \mu_o nI} \qquad (21.11)$$

where n = number of turns per meter length.

10. The field inside a toroid (a doughnut) is

$$\boxed{B = \frac{\mu_o NI}{2\pi R}} \qquad (21.12)$$

R is the distance from the center and N is the total number of turns.

• Qualitative Questions

21M.1 Suppose that a charged particle moves in a magnetic field. Such a particle may experience a magnetic force, and the

A. force and velocity vectors can have any angle between them.
B. force and magnetic field vectors can have any angle between them.
C. velocity and magnetic field vectors can have any angle between them.
D. none of the above is true.
E. more than one of the above are true.

21M.2 An electron moving in the uniform magnetic field indicated here will experience a force in the direction

A. toward A.
B. toward B.
C. toward C.
D. toward D.
E. out of the paper.
F. into the paper.

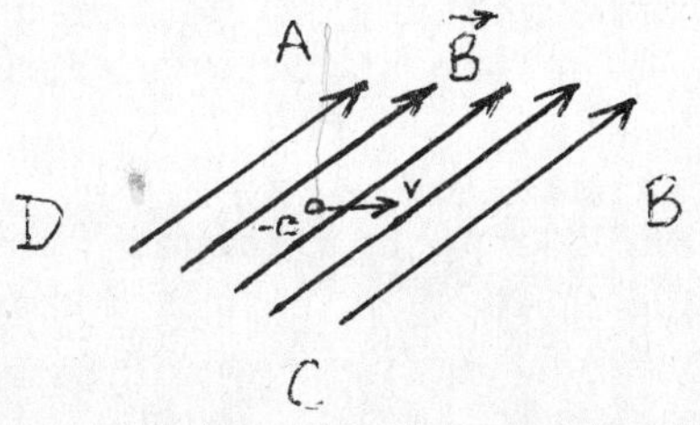

21M.3 An electron shot into a uniform magnetic field in such a way that its velocity vector makes an angle of 45° with the field will move

A. in a straight line.
B. in an ellipse.
C. in a circle.
D. on a helix (i.e., a corkscrew path).
E. on a parabolic path.

21M.4 If an electron is not deflected in moving through a certain region of space, then we can be certain

A. there is no magnetic field there.
B. there is no electric field there.
C. there is no magnetic or electric field there.
D. that the velocity of the electron remains constant.
E. none of the above is true.

21M.5 Which of the following is the most accurate statement concerning the operation of a cyclotron?

A. The speed of the particles remains constant.
B. The energy of the particles is increased by the magnetic field.
C. The energy of the particles is increased by the electric field.
D. The energy of the particles is increased by both the electric and magnetic fields.
E. The particles travel in circular orbits of fixed radius.

21M.6 Which of the following is an accurate statement concerning the Hall Effect?

A. A magnetic field must be applied parallel to the current flow.
B. The Hall field is set up parallel to the current flow.
C. The Hall voltage is measured in a direction perpendicular to the direction of current flow.
D. The polarity of the Hall voltage is independent of the sign of the carriers.
E. We would expect a larger Hall voltage in a good conductor like copper than in a semiconductor like germanium.

21M.7 If the magnetic field due to a long straight wire carrying a current is 12×10^{-4} T at a distance of 2 cm from the wire, the field at a distance of 4 cm will be

A. 24×10^{-4} T.
B. 12×10^{-4} T.
C. 6×10^{-4} T.
D. 3×10^{-4} T.

$B = 12 \times 10^{-4}$ T

d = 2 cm
= .02 m

21M.8 Current I flows in each of four parallel wires positioned in a square array. The current flows in the same sense in each wire, i.e., out of the paper in the diagram. Toward which of the labelled points will the force on wire X point?

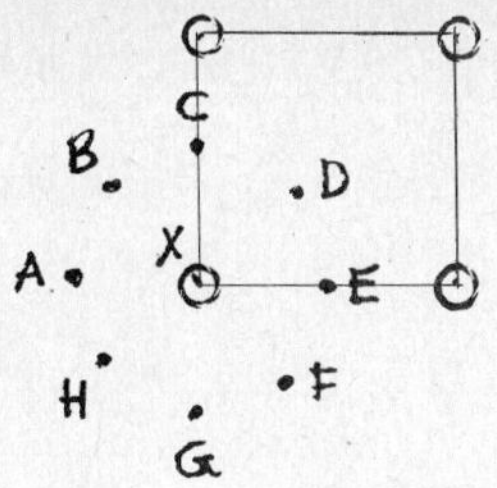

21M.9 A beam of <u>electrons</u> is shot through a hole bored through the center of a long solenoid carrying a current i, as shown here. When the current is turned off the beam strikes the screen at point P. When the current is turned on the beam will

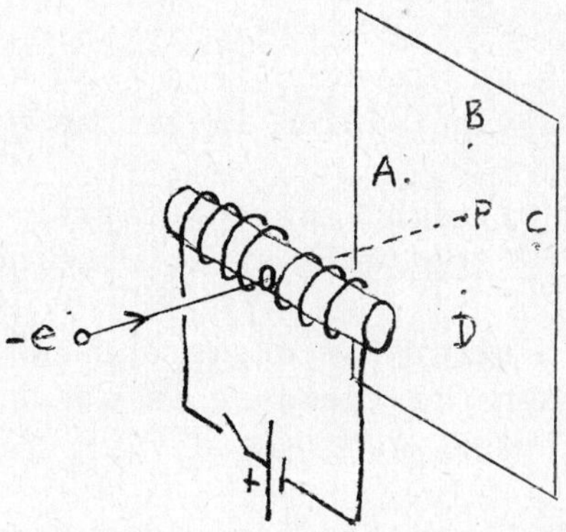

A. be deflected toward point A.
B. be deflected toward point B.
C. be deflected toward point C.
D. be deflected toward point D.
E. still hit the screen at point P, but with greater speed than when the current was off.
F. still hit the screen at point P, but with less speed than when the current was turned off.
G. still hit the screen at point P, with the same speed as when the current was turned off.

21M.10 When a current carrying coil of wire is place in a uniform magnetic field,

A. it will experience a net force parallel to the plane of the loop.
B. it will experience a net force perpendicular to the plane of the coil.
C. it will tend to orient itself with its plane parallel to the magnetic field.
D. it will tend to orient itself with its plane perpendicular to the magnetic field.
E. More than one of the above is true.
F. None of the above is true.

• True–False Questions

21TF.1 The force due to a magnetic field on a moving charge does not change the energy of the charge.

21TF.2 The force experienced by a moving charge in a magnetic field is directed tangential to the magnetic field lines.

21TF.3 Magnetic charges are deflected by magnetic fields, but electric charges are not.

21TF.4 If an electron is curved to the left by a magnetic field, a curving proton in the same field will curve to the right.

21TF.5 A straight current carrying wire alligned parallel to a magnetic field will experience no force.

21TF.6 In order for a current carrying wire to experience a force in a magnetic field it is necessary that the wire carry a net charge.

21TF.7 The magnetic field due to a current configuration of any shape is always proportional to the current.

21TF.8 Magnetic field lines have no beginning or end.

21TF.9 A current is like a paddlewheel which stirs up the swirling magnetic field lines.

21TF.10 The magnetic field in a solenoid decreases as one moves away radially from the center.

ANS: 1.T 2.F 3.F 4.T 5.T 6.F 7.T 8.T 9.T 10.F

• Multiple Choice: Answers and Comments

21M.1 C Given a magnetic field direction, a particle can be projected in any direction. However, the force is always perpendicular to both velocity and magnetic field.

21M.2 F Use RHR-I.

21M.3 D The motion due to the component of velocity perpendicular to B will be a circle. The velocity component parallel to the field is unaffected, so the electron drifts at constant speed parallel to the field while trvelling a circle around it, resulting in a helical path.

21M.4 E There could be both a magnetic field and an electric field oriented such that their effects cancel, or else there could be no field whatever, or there could be an electric field oriented parallel to the electron's velocity.

21M.5 C The particles spiral out, accelerated by the electric field each time they cross the gap between the D's.

21M.6 C

21M.7 C The magnetic field due to a straight wire varies as 1/R, so doubling R decreases B by a factor of 1/2.

21M.3 D Label the wires 1,2,3. The magnetic field due to each is found from RHR-II and indicated here. The resultant field B is thus directed toward point F. Using RHR-I the force on wire X is directed toward D. Parallel currents tend to be drawn together.

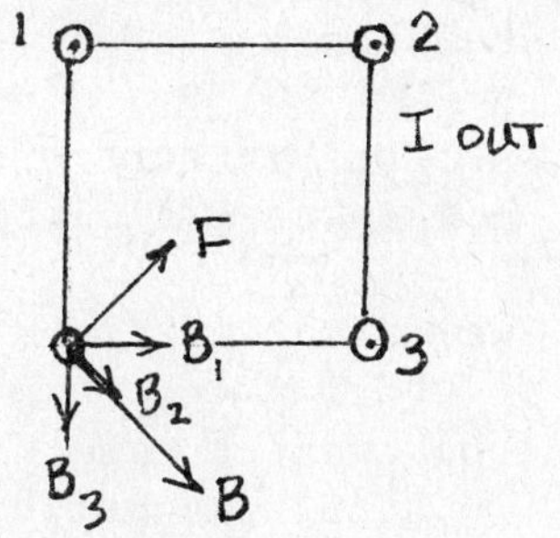

21M.9 D Using RHR-II one sees that the magnetic field in the solenoid points to the left. Using RHR-I I see that a positive charge will be deflected toward B, so a negative electron will be deflected toward D.

21M.10 D A small coil acts just like a compass needle and will line up with its axis parallel to the magnetic field.

• Problems

21.1 At a point where the earth's field is inclined toward the north at an angle of 30° below the horizontal it has a magnitude of 5×10^{-5} T. What is the magnitude and direction of the force exerted on an electron travelling due north with a velocity of 4×10^{6} m/s?

21.2 A mass spectrometer is a device which sorts out ions of different masses by curving them in a magnetic field. Particles with given charge and different masses are bent in circles of different radii by a perpendicular magnetic field. Suppose that singly ionized chlorine atoms of atomic weights 35 and 37 are accelerated through 8 kV and then bent in circles by a field of 0.48T. (An atom of atomic weight 35 has a mass 35u, where u = 1.67×10^{-27} kg).

(a) What is the velocity of each particle as it enters the magnetic field?

(b) What is the diameter of the circle in which each is bent?

21.3 Two parallel wires each carry currents of 10A in opposite directions. The wires are separated by 2 cms. What is the magnitude and direction of the magnetic field at points A, B and C?

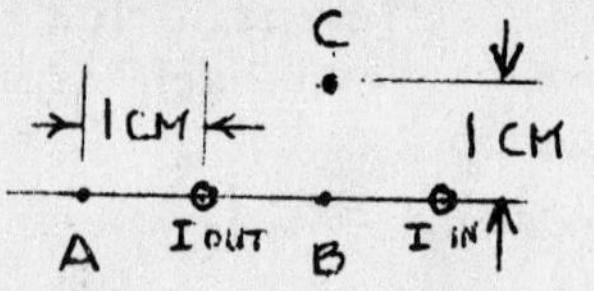

21.4 Suppose that you wish to construct a solenoid of 2 cms radius in which the magnetic field will be 0.01T when a current of 2A flows in the coil. How many turns of wire per centimeter are required?

21.5 A circular coil of wire of radius 1 cm has 100 turns of wire and carries a current of 0.5A.

(a) What is the magnetic moment of the coil?

(b) What torque will be exerted on the coil when it is placed in a magnetic field of 5 x 10-3T which makes an angle of 60° with the plane of the coil?

• Problem Solutions

21.1 $F = qvB\sin\theta = (1.6\times10^{-19}\,C)(4\times10^{6}\,m/s)(5\times10^{-5}\,T)(\sin 30°)$

$F = 1.6\times10^{-17}\,N$

USING RHR-I I SEE THAT FORCE IS DIRECTED TO THE EAST. (REMEMBER THAT ELECTRON HAS NEGATIVE CHARGE)

21.2 (a) GAIN IN KE = LOSS IN PE

$\frac{1}{2}mv^2 = qV$

$v = \sqrt{\frac{2qV}{m}}$

$m_1 = 35\times1.67\times10^{-27}\,kg = 5.85\times10^{-26}\,kg$

$m_2 = 37\times1.67\times10^{-27}\,kg = 6.18\times10^{-26}\,kg$

$v_1 = \sqrt{\frac{(2)(1.6\times10^{-19})(8000)}{5.85\times10^{-26}}} = 2.09\times10^{5}\,m/s$

$v_2 = \sqrt{\frac{(2)(1.6\times10^{-19})(8000)}{6.18\times10^{-26}}} = 2.04\times10^{5}\,m/s$

(b) $d_1 = 2R_1 = \frac{2m_1v_1}{qB} = \frac{(2)(5.85\times10^{-26})(2.09\times10^{5})}{(1.6\times10^{-19})(0.48)} = 0.318\,m$

$d = 2R = \frac{2m_2v_2}{qB} = \frac{(2)(6.18\times10^{-26})(2.04\times10^{5})}{(1.6\times10^{-19})(0.48)} = 0.328\,m$

21.3 USE RHR-II TO FIND DIRECTION OF B AT EACH POINT. ADD VECTORIALLY B DUE TO EACH WIRE.

(a) AT A, $B_1 = \frac{\mu_0 I}{2\pi R_1}$, $B_2 = \frac{\mu_0 I}{2\pi R_2}$

$R_2 = 3R_1 = 0.03\,m$

$B_A = B_1 - B_2 = \frac{\mu_0 I}{2\pi}\left(\frac{1}{R_1} - \frac{1}{R_2}\right)$

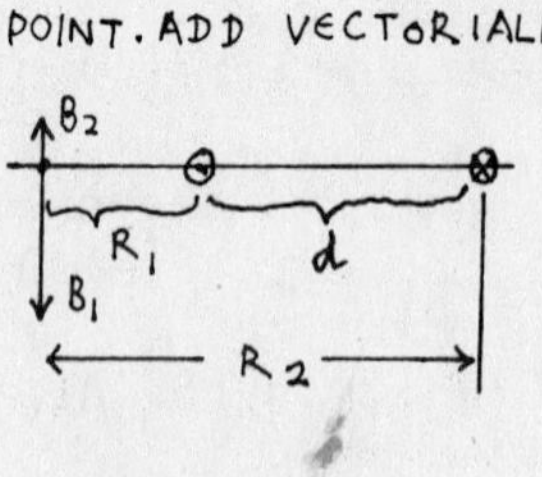

$$B = \frac{(4\pi \times 10^{-7})(10)}{2\pi}\left(\frac{1}{0.01} - \frac{1}{0.03}\right) = 1.33 \times 10^{-4}\ \text{T DIRECTED DOWN}$$

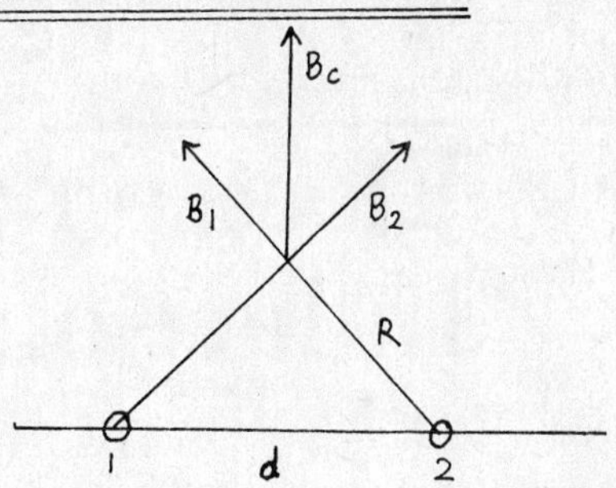

(b) AT B FIELDS ADD,

$$B_B = B_1 + B_2 = 2B_1 = 2\frac{\mu_0 I}{2\pi(\frac{d}{2})} = \frac{8\mu_0 I}{4\pi d}$$

$$B_B = \frac{(8)(4\pi \times 10^{-7})(10)}{(4\pi)(0.02)} = 4 \times 10^{-4}\ \text{T}$$

(c) AT C, $B_1 = \frac{\mu_0 I}{2\pi R} = B_2$

$$B_c = 2B\cos 45°$$

$$R = \sqrt{2}\ \text{cms} = \sqrt{2} \times 10^{-2}\ \text{m}$$

$$B = \frac{2\mu_0 I}{2\pi R}\cos 45° = \frac{(2)(4\pi \times 10^{-7})(10)(\cos 45°)}{(2\pi)(\sqrt{2} \times 10^{-2})} = 2 \times 10^{-4}\ \text{T DIRECTED UP}$$

21.4 $B = \mu_0 n I$, $n = \frac{B}{\mu_0 I} = \frac{0.01}{(4\pi \times 10^{-7})(2)} = 3979$ PER METER

OR $n \cong 40$ PER CM.

21.5 (a) 100 TURNS CARRYING 0.5 A IS EQUIVALENT TO ONE TURN CARRYING 50 A.

$$\mu = IA = NI_1\pi R^2 = (100)(0.5)(\pi)(10^{-2})^2 -$$

$$\mu = 1.57 \times 10^{-2}\ \text{A.m}^2$$

(b) $\tau = \mu B \sin\theta$

μ IS ALONG THE AXIS OF THE COIL, SO IT MAKES AN ANGLE OF 30° WITH THE MAGNETIC FIELD B.

$$\tau = (1.57 \times 10^{-2})(5 \times 10^{-3})(\sin 30°) = 3.93 \times 10^{-5}\ \text{N.m}$$

22 Electromagnetic Induction

• Summary of Important Ideas, Principles, and Equations

1. The <u>magnetic flux</u> Φ through an area A is proportional to the number of lines of the magnetic field B which pass through the area. In the simple case of a plane area A placed in a uniform field B,

$$\boxed{\Phi = B\ A = BA\cos\theta} \qquad (22.1)$$

Here B = $B\cos\theta$ = component of B perpendicular to A
θ = angle between B and a line drawn perpendicular to A. Φ is called flux (which means "flow") because if we were considering the flow lines in a fluid (instead of the B field) it would measure the flow in m^3/s through A.

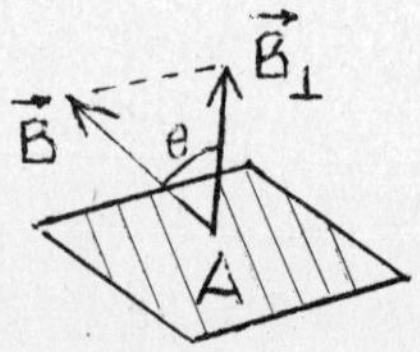

2. <u>Faraday's law</u> is based on experimental observation. It may be stated as follows: Consider an area A bounded by a curve C. <u>If the magnetic flux through area A changes, an electromotive force will be induced around the path C which bounds A.</u> This result can be expressed more exactly by the equation

$$\boxed{\varepsilon = -\frac{\Delta\Phi}{\Delta t}} \qquad (22.2)$$

Here $\Delta\Phi$ is the change in flux which occurs in the short time Δt, and ε is the induced EMF in volts.

If a conductor of resistance R forms the boundary C, an induced current will flow in it when the flux changes. If there is a gap in the conductor, the EMF will appear across the gap.

If the conductor consists of a coil of N loops, the induced EMF is N times as large (like N induced EMF's in series).

$$\boxed{\varepsilon = -N\frac{\Delta\phi}{\Delta t}} \qquad (22.3)$$

Let us consider the case where the flux through a circuit loop is changing. This flux change can occur in the following ways:

(1) The magnetic field can be changed by moving the source farther away or, if the magnetic field is due to an electromagnet, by reducing the current in the electromagnet.

(2) By moving the loop in which the EMF is to be induced into a region of stronger or weaker magnetic field.

(3) By rotating the loop so that the number of lines of magnetic field passing through the loop is changed.

(4) By deforming the shape of the loop so that its effective area is changed.

3. The direction of the induced EMF or induced current is indicated by the minus sign in Faraday's law. Consider a circular loop of wire placed in a magnetic field B, as shown here. Suppose we increase B by turning up the current in the electromagnet generating B. This will make a current flow in the loop. Which way will the current flow, clockwise or counterclockwise? The answer (of which the minus sign is supposed to remind us) is called Lenz/s law No. 1 (LL-I).

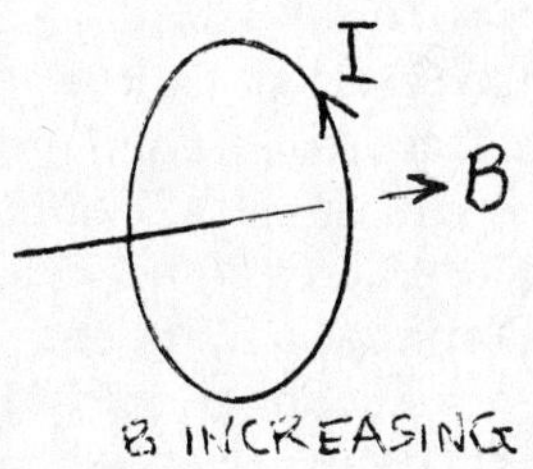

LL-I	The sense of the EMF induced in a loop is such that the current that would flow due to such an EMF would itself create magnetic flux that will oppose the change in flux being impressed on the loop.

By using RHR-I for magnetic fields we see that an induced current flowing as I have drawn it will itself cause a field which will oppose the increase in field impressed from outside. That this must be so follows from the conservation of energy. If the induced field added to the increasing field applied, the result would induce an even bigger current, which would in turn induce more current, and so on. Infinite amounts of energy would be generated in this way.

If a conductor is moved in a magnetic field, currents will be induced and the magnetic field will then exert a force on the current carrying conductor. This force on the induced current will always be in a direction to oppose the motion. This is called Lenz's law No. 2 and follows from energy reasoning also.

All of these ideas can be summed up in Browne's law No. 57. Induced electromagnetic effects are like conservative Idaho Republicans. They always oppose all change in all cases.

4. If a conductor of length ℓ is moved with velocity v perpendicular to a magnetic field B a motional EMF will be induced between the ends of the conductor.

$$\varepsilon = B\ell v \tag{22.4}$$

This result follows from Faraday's law if we imagine the straight conductor to form one side of a rectangular loop. As the conductor is moved the area of the loop increases, thereby increasing the magnetic flux through the loop. This causes an induced EMF to be created in the loop. The loop need not be a conductor in order for an EMF to be induced (it does have to be a conductor for an induced current to flow, however). Thus the three fixed sides of the loop might be made of wood or even air. In fact, this part of the "loop" can be purely in our minds. An EMF will still be induced between the ends of the moving conductor.

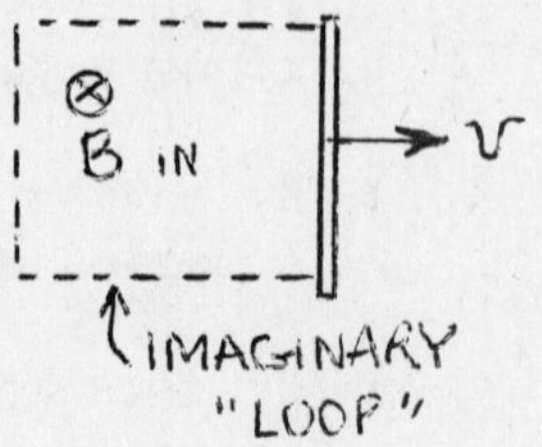

5. Suppose that current I_1 flowing in coil #1 causes a magnetic field which passes through coil #2. If I_1 is changed the magnetic flux Φ_2 through coil #2 will also change. The magnitude of the change in flux will depend on the size of the change in current, ΔI_1, and on the shape and separation of the coils and on the number of turns in each. These geometrical factors are described by the mutual inductance M_{21} for the two circuits, defined as the flux in circuit 2 due to a current of 1 Ampere in circuit 1.

$$M_{21} = \frac{\Phi_2}{I_1} \tag{22.5}$$

The EMF induced in circuit 2 by a changing current in circuit 1 is thus

$$\varepsilon_2 = -\frac{\Delta\phi_2}{\Delta t} = -M_{21}\frac{\Delta I_1}{\Delta t} \tag{22.6}$$

The unit of M_{21} is the henry ($1H = 1\ \frac{\text{volt}\cdot\text{sec}}{\text{amp}}$)

6. When current flows in a circuit a magnetic field is created, and these magnetic field lines pass through the circuit which causes them, constituting a magnetic flux through the circuit. If the current changes the flux will change, and this will induce an EMF in the circuit. The self-inductance L of a circuit is the magnetic flux in the circuit due to a current of 1 ampere in the circuit.

$$\boxed{L = \frac{\Phi}{I}} \qquad (22.7)$$

The induced EMF is $\boxed{\varepsilon = -\frac{\Delta\phi}{\Delta t} = -L\frac{\Delta I}{\Delta t}}$ (22.8)

L is measured in henries (H).

All circuits have self-inductance, but the self-inductance is largest when the circuit is in the form of a coil, such as a solenoid. Inductors (coils) are often introduced purposely into a circuit, and in a diagram they are represented like this:  The resistor represents the resistance of the wire of which the coil is wound.

The minus sign in eq. (23.8) reminds us that the sense of the induced EMF is such that the induced current flows in a direction which tends to oppose the change in flux through the circuit. For this reason this self-induced EMF is called a "back" EMF. If one tries to increase the current in a circuit, the back EMF will act to oppose the increase. If you try to decrease the current, the back EMF will try to keep it going as before.

7. Suppose you try to make a current flow in a circuit with appreciable self inductance. In trying to build the current up from zero to some final value it is necessary to do work against the induced back EMF which opposes the build up of current. The work done is stored in the magnetic field created in the circuit. The magnetic energy stored in an inductance L carrying current I is

$$\boxed{U = \tfrac{1}{2}LI^2} \qquad (22.9)$$

This is analogous to the electric energy $\frac{1}{2}CV^2$ stored in a capacitor.

8. A superconductor is a material whose electrical resistance vanishes below a critical temperature (usually below 20°K). Such materials are very useful for creating intense magnetic fields in large volumes. The superconductivity is quenched, however, if the magnetic field exceeds a certain critical value.

9. When a coil is rotated in a fixed magnetic field (such as that due to a permanent magnet) an oscillating induced EMF is induced in the coil. The coil can be attached to an external load and an alternating current driven through the load. Such a device is called a generator.

• Qualitative Questions

22M.1 One aspect of electromagnetic induction

A. refers to the observation that an uncharged conductor can be charged by placing it close to a charged object and then cutting it in half.
B. refers to the observation that when there is relative motion between a wire and a magnetic field, a potential difference is impressed between the ends of the wire.
C. takes place whenever a conductor is in a magnetic field.
D. causes charged particles to move in circular paths.

22M.2 Two wire loops are oriented with their axes parallel, as shown here. If a current is suddenly made to flow in one of the loops in the direction shown, what will be the force which acts on the second loop?

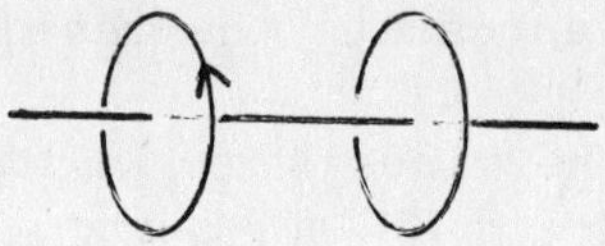

A. No net force will act on the loop.
B. The loop will experience a force to the right.
C. The loop will experience a force to the left.

22M.3 If you hold a sheet of copper in a strong permanent magnet, with the plane of the sheet perpendicular to the magnetic field, and quickly jerk it out,

A. you will experience a magnetic force opposing your action.
B. you will experience a magnetic force assisting your action.
C. you will feel no magnetic force.
D. any force you feel will be due mainly to iron impurities in the copper, since copper itself is not magnetic.
E. None of the above is true.

22M.4 Immediately after the switch S is closed, in what direction does current flow in R?

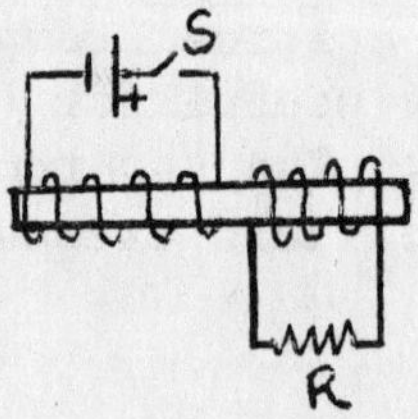

A. Left.
B. Right.
C. No current flows in R.

22M.5 Quite a long time after switch S in 23M.4 is closed, in what direction does current flow in R?

A. Left.
B. Right.
C. No current flows.

22M.6 Immediately after switch S in 23M.4 is opened, in what direction does current flow in R?

A. Left.
B. Right.
C. No current flows in R.

22M.7 A bar magnet is positioned inside a coil, as shown here. In the following "work" refers to any work done as a consequence of the fact that the bar is magnetic. If the bar is suddenly pulled out of the coil,

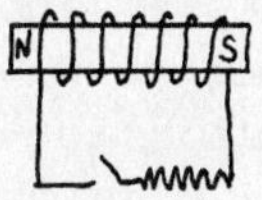

A. no work will be done, independent of whether or not the switch is closed.
B. more work will be done if the switch is open.
C. more work will be done if the switch is closed.
D. equal (non-zero) amounts of work will be done whether or not the switch is open.
E. whether the work done is positive or negative depends on whether the magnet is pulled out of the right end or the left end of the coil.

22M.8 Lenz's law tells us that induced currents always flow in a direction such that they oppose any change in magnetic flux through a conductor. If they flowed in the opposite direction, this would lead to a violation of

A. the law of Biot Savart.
B. Ampere's law.
C. Coulomb's law.
D. the basic speed law.
E. the law of conservation of energy.

22M.9 Suppose that a bar magnet is dropped down a long copper tube. Neglecting air friction,

A. the magnet will fall at the same rate as in the absence of the copper tube.
B. the magnet will fall with constant acceleration less than g.
C. the magnet will fall at constant acceleration always greater than g.
D. the magnet will fall at non-constant, non-zero acceleration.
E. the magnet will reach a constant terminal velocity.

22M.10 If a cylinder of metal, such as aluminum, is rapidly withdrawn from a coil of n turns of wire, as sketched here, which of the following best describes what will happen?

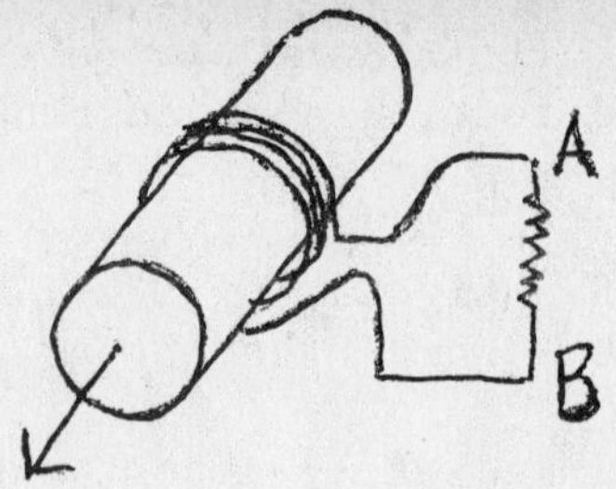

A. The coil will tend to rotate.
B. The coil will experience a net force.
C. An induced current will flow from A to B.
D. An induced current will flow from B to A.
E. No current will be induced in the circuit.

22M.11 A bar magnet is suspended as a pendulum, as shown here. A coil of N turns lies on the table below the point of suspension. The pendulum is pulled to one side and released. It swings back and forth with frequency f. Which of the following is the most accurate statement?

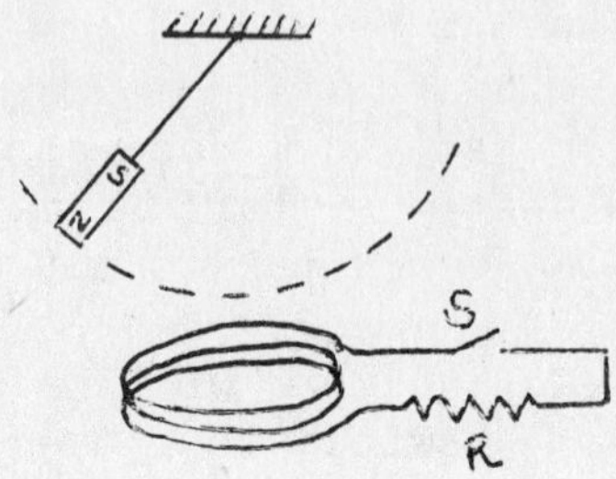

A. An EMF oscillating at frequency f will be induced in the coil.
B. An EMF oscillating at frequency 2f will be induced in the coil.
C. Whether or not an EMF will be induced in the coil depends upon whether or not the switch is closed.
D. The magnitude of the induced EMF will depend on the value of the resistance R.
E. The induced EMF will be independent of the number of turns in the coil.

22M.12 A rectangular loop is caused to move with constant velocity through a uniform magnetic field which is confined to the region indicated here. The coil is initially outside the field on the left, and it is finally outside the field on the right. Which of the following curves best illustrates the EMF generated between the terminals of the coil as a function of time?

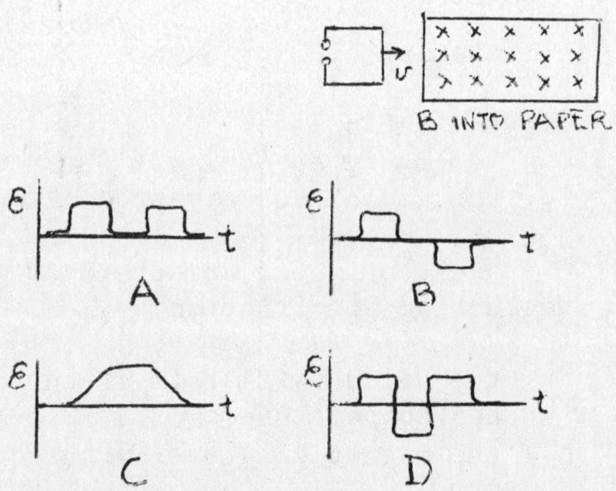

E. None of the above, since no EMF will be generated in an open circuit.

22M.13 Suppose that an electromagnet with a very large inductance and a fairly low resistance are connected in series to a 100 volt DC power supply, as shown here. If the switch S were suddenly opened, the potential difference between the contacts of the switch

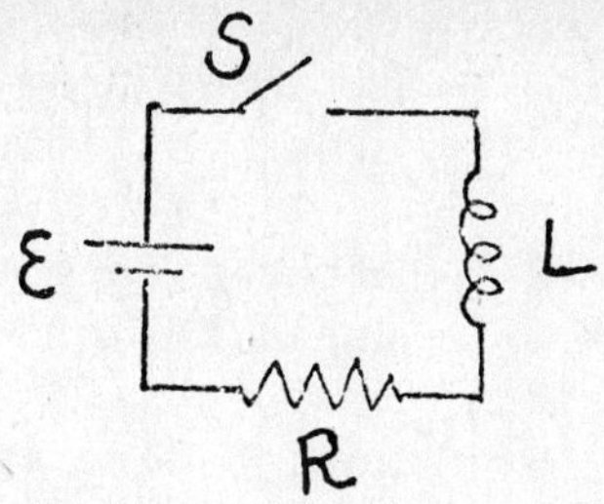

A. would be zero initially.
B. would be less than 100 volts.
C. would be equal to 100 volts.
D. could be much greater than 100 volts.
E. none of the above is true.

22M.14 A copper ring and a wooden ring of the same dimensions are placed so that there is the same changing magnetic flux through each. How do the induced electric fields in each compare?

A. The induced electric fields are the same in both.
B. The induced fields are greatest in the copper.
C. The induced fields are greatest in the wood.

22M.15 The self inductance of a coil depends on

A. the current flowing in it.
B. the flux in it.
C. the EMF applied to it.
D. geometrical factors such as size, shape and number of turns.
E. more than one of the above.
F. none of the above.

• Multiple Choice: Answers and Comments

22M.1 B

22M.2 B Using RHR-II we see that the field due to the current in the left loop passes through the right loop from right to left. Thus the induced current will flow as shown to try to oppose this field build up. The magnetic moment of each loop is thus as shown for the two small bar magnets here. The loops act just like little magnets with like poles together, and they will repel each other. You can also deduce this result using RHR-I if you observe that the magnetic field at the right loop due to the left one has a radial component.

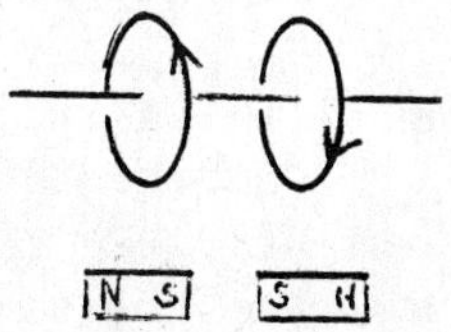

22M.3 A Induced effects always oppose whatever you try to do (LL-II).

22M.4 A Current due to the battery creates a magnetic field directed from right to left, so induced current tries to create a field directed from left to right.

22M.5 C After a long time the flux is not changing and the induced EMF and current are zero.

22M.6 B Just before switch is opened the field is directed from right to left. If the switch is opened, so that field tries to decrease to zero, an induced current will flow which tries to maintain this field.

22M.7 C If the switch is closed an induced current will flow when the magnet is pulled out, since pulling the magnet out will tend to reduce the flux in the coil due to the magnet. This induced current will set up a magnetic field which will try to pull the bar magnet back into the coil, i.e., the force will oppose pulling it out. This will make it necessary to do more work in pulling the magnet out.

22M.8 E

22M.9 E If no copper tube were present the magnet would fall faster and faster due to the force of gravity. With the copper tube present currents are induced in the walls of the tube as the magnet moves along, carrying its magnetic field with it. Since the field of the magnet is strong near it and weaker farther away, the flux through a given section of the ring will change as the magnet passes by. As the magnet gains speed the rate of change of flux keeps increasing. Thus the induced currents get larger and larger and set up larger and larger induced magnetic fields. These induced fields exert a force to oppose the continued acceleration of the magnet downward. Finally this retarding magnetic force equals the force of gravity on the magnet and no more acceleration occurs. Then the magnet falls with a constant terminal velocity.

22M.10 E The aluminum isn't magnetic, so pulling it out doesn't change the flux in the coil and no EMF is induced.

22M.11 B As the magnet swings across the flux builds up and then decreases in the coil. When the magnet swings back on the second half of its cycle the same thing happens. Thus the induced flux, and hence the induced EMF, has gone through two cycles while the pendulum has gone through one, so its frequency is 2f.

22M.12 B No EMF is induced when the coil is completely in the uniform field, for then the flux in the coil isn't changing.

22M.13 D Before the switch is opened a current is flowing and a significant magnetic field may exist in the inductance. When the switch is opened the flux in the circuit will drop to zero, and this decrease depends on how rapidly the circuit is broken. If one jerks the switch open quickly a hugh EMF can be induced (big enough to electrocute you!). Typically the EMF appears across the contacts of the switch and will cause arching. The large induced EMF's which can occur in circuits with large inductance (like electromagnets) make it necessary to take special safety precautions.

22M.14 A The induced EMF depends just on the rate of change of flux, not on any material which may be present. In a Betatron (a kind of particle accelerator) an EMF is induced around a circular path in a vacuum, and electrons are then accelerated by it.

22M.15 D Inductance, like capacitance, depends only on goemetric factors and on the material filling the coil. The inductance of a coil filled with iron is much induced because the iron can be magnetized and thus add greatly to the B field created in the coil (and hence to the flux).

• Problems

22.1 A circular coil of 50 turns and radius 5 cms is placed with its plane oriented at 90° to a magnetic field of 0.1T. The field is now increased at a steady rate, reaching a value of 0.5T after 2 seconds. What EMF is induced in the coil?

22.2 One can create very large induced currents, and hence large magnetic fields, by rapidly collapsing a loop which is placed in a magnetic field. This is done by setting off a high explosive surrounding the conductor. Suppose that a single copper loop of resistance 6×10^{-5} ohms and area 30 cms^2 is placed in a magnetic field of 0.8T and collapsed to essentially zero area in 10^{-3} second. What current will be induced to flow? Incidentally, melting of the conductors often sets a limit on the fields obtainable. Fields up to 1000T have been obtained in this way.

22.3 When an airplane flies through the earth's magnetic field EMF's are induced in the electronic circuits on board. Is this likely to be a serious problem? To check out the magnitude of the effect calculate the EMF induced in a single horizontal conductor of length 60 cms positioned east-west in an airplane flying due north at 800 km/hr at a point where the earth's magnetic field has a magnitude of 0.8 gauss inclinded at 30° below the horizontal.

22.4 A battery and a resistor are connected to two parallel metal rails along which a metal bar can slide. The bar and rails have negligible resistance and friction. A uniform magnetic field of 0.08T acts out of the plane of the paper. If the mass of the bar is 0.5 gm and its length is 4 cms, what is its initial acceleration when the switch is closed? What is its final terminal velocity? Sketch the velocity as a function of time. The "rail gun" is a device based on this principle, and its potential uses include creating thermonuclear fusion, launching satellites, and hurling raw materials from the surface of the moon out to an orbiting space factory.

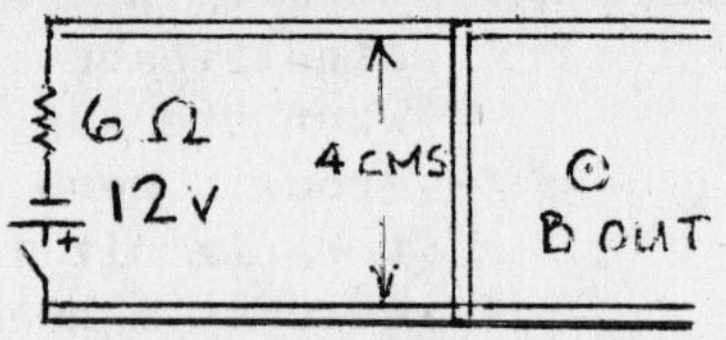

22.5 A long solenoid of square cross section (2 cms x 2 cms x 30 cms) has 200 turns per cm. When a current of 2A flows in it an average field of 0.08T is created over the cross section of the coil. What is the self-inductance of such a coil? What EMF will be induced if the current is reduced to zero in 1 ms?

22.6 What energy is stored in an inductance of 0.4H when a current of 3 A flows in it?

22.7 What EMF is generated as a function of time by a generator consisting of a coil of 120 turns and area 160 cms^2 which rotates at 1800 RPM in a magnetic field of 0.5T oriented perpendicular to the axis of rotation?

• Problem Solutions

22.1

$$\mathcal{E} = N\frac{\Delta\phi}{\Delta t} = \frac{N\Delta B A}{t} = \frac{N\Delta B\pi R^2}{t} = \frac{(50)(.5T - .1T)(\pi)(0.05)^2}{2}$$

$$\mathcal{E} = 0.078V$$

I DON'T WORRY HERE ABOUT THE MINUS SIGN IN OHM'S LAW BECAUSE I WANT ONLY THE MAGNITUDE OF $\mathcal{E}$.

22.2

$$\mathcal{E} = \frac{\Delta\phi}{\Delta t}$$

$$\mathcal{E} = \frac{BA}{\Delta t} = \frac{(.8T)(30\times10^{-4}m^2)}{10^{-3}s} = 2.4 \text{ VOLTS}$$

$$I = \frac{\mathcal{E}}{R} = \frac{2.4V}{6\times10^{-5}\Omega} = 4\times10^4 A$$

22.3 ONLY THE COMPONENT $B_\perp$ OF MAGNETIC FIELD PERPENDICULAR TO BOTH THE VELOCITY AND THE LENGTH OF THE CONDUCTOR WILL CONTRIBUTE TO THE MOTIONAL EMF. THUS

$$\mathcal{E} = B_\perp \ell v = (B\sin\theta)(\ell v)$$

$$= (0.8\times10^{-4}T)(\sin 30°)(0.6m)\left(800\times\frac{1}{3.6}\times\frac{m}{s}\right)$$

$$\mathcal{E} = 5.3\times10^{-3} V = 5.3 mV$$

22.4 WHEN SWITCH IS CLOSED A CURRENT I WILL FLOW,

$$I = \frac{\mathcal{E}}{R} = \frac{12V}{6\Omega} = 2A$$

THE MAGNETIC FIELD WILL EXERT A FORCE F ON THE MOVABLE BAR,

$$F = I\ell B = (2A)(0.04m)(0.08T) = 6.4 \times 10^{-3} N$$

THIS WILL CAUSE AN INITIAL ACCELERATION a,

$$a = \frac{F}{m} = \frac{6.4 \times 10^{-3} N}{0.5 \times 10^{-3} kg} = \underline{\underline{12.8 \ m/s^2}}$$

THE BAR WILL ACCELERATE, MOVING FASTER AND FASTER. AS IT DOES SO AN EMF WILL BE INDUCED, AND AS THE BAR MOVES FASTER THIS BACK EMF WILL INCREASE AS FLUX IS SWEPT OUT AT A FASTER AND FASTER RATE. FINALLY THE INDUCED EMF WILL BUILD UP UNTIL IT IS AS BIG AS THE BATTERY VOLTAGE. WHEN THIS HAPPENS THE INDUCED EMF WILL STOP ANY CURRENT FROM THE BATTERY FROM FLOWING. WHEN THE CURRENT DROPS TO ZERO THE FORCE ON THE BAR ALSO DROPS TO ZERO. THIS IN TURN MEANS ZERO ACCELERATION AND HENCE CONSTANT ("TERMINAL") VELOCITY. THIS WILL ALL OCCUR WHEN THE SPEED IS SUCH THAT

$$\mathcal{E}_{INDUCED} = v\ell B = \mathcal{E}_{BAT}$$

$$\text{OR} \quad v_T = \frac{\mathcal{E}_{BAT}}{\ell B} = \frac{12V}{(0.04m)(0.08T)} = \underline{\underline{3750 \ m/s}}$$

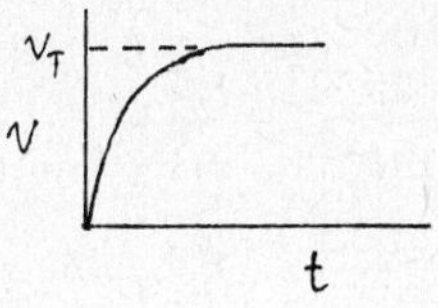

22.5 THE TOTAL FLUX IS $\Phi = N\Phi_1 = NBA$

$$= (200 \tfrac{1}{cm})(30cm)(0.08T)(0.02m)^2$$

$$\Phi = 0.192 \ T \cdot m^2$$

$$L = \frac{\Phi}{I} = \frac{0.192 \ T \cdot m^2}{2A} = \underline{0.096 \ H}$$

22.6 $U = \frac{1}{2} L I^2 = \frac{1}{2}(0.4H)(3A)^2 = \underline{\underline{1.8 \text{ JOULES}}}$

22.7 THE FLUX THROUGH THE COIL IS

$$\phi = NBA\cos\theta \qquad \theta = \omega t = \text{ANGLE BETWEEN B AND NORMAL TO COIL.}$$

$$= NBA\cos\omega t \qquad \omega = 2\pi f = \text{ANGULAR VELOCITY}$$

$$\mathcal{E} = -\frac{\Delta\phi}{\Delta t} \simeq -\frac{d}{dt}(NBA\cos\omega t) = \underline{\underline{\omega NBA\sin\omega t}}$$

HERE I USED THE DERIVATIVE OF $\cos\omega t$.

$$\frac{d}{dt}(\cos\omega t) = -\omega\sin\omega t$$

IF YOU ARE NOT FAMILIAR WITH DERIVATIVES YOU CAN FIGURE THIS OUT USING TRIGONOMETRY.

$$\Delta\phi = NBA[\cos(\theta+\Delta\theta) - \cos\theta]$$

$$= NBA[\cos\theta\cos\Delta\theta - \sin\theta\sin\Delta\theta - \cos\theta]$$

USING $\cos(a+b) = \cos a\cos b - \sin a\sin b$

IF $\Delta\theta \ll 1$, $\cos\Delta\theta \simeq 1$, $\sin\Delta\theta \simeq \Delta\theta$

THUS $\Delta\phi \simeq NBA(-\Delta\theta\sin\theta)$

$$\text{SO} \quad \mathcal{E} = -\frac{\Delta\phi}{\Delta t} = -\frac{NBA(-\Delta\theta\sin\theta)}{\Delta t} = \underline{\underline{\omega NBA\sin\omega t}} \qquad \text{USING } \theta = \omega t$$

$\omega = 2\pi f = (2\pi)\left(\frac{1800}{60s}\right) = 188.5$ RAD/s

$B = 0.5T$

$A = 160 \times 10^{-4}$ m

$N = 120$

$\mathcal{E} = (188.5/s)(120)(0.5T)(160 \times 10^{-4}) \sin \omega t$

$\mathcal{E} = 181 \sin \omega t$

Time Dependent Currents and Voltages; AC Circuits

23

• Summary of Important Ideas, Principles, and Equations

1. Consider what happens in the RC circuit here when the switch is thrown to position 1 to charge thecapacitor. Since the capacitor is initially uncharged there is at first a big rush of current on to the plates. As charge piles up on the capacitor and voltage builds up across it the rush of charge slowly dies down. The sum of the potential drop across the resistor plus the drop across the capacitor always adds up to the battery voltage ε, so as $V_G = Q/C$ builds up $V_r = IR$ decreases. Using Calculus one can find

$$V_c(t) = \mathcal{E}(1 - e^{-t/\tau})$$
$$V_R(t) = \mathcal{E}e^{-t/\tau}$$
$$i(t) = \frac{V_R(t)}{R} = \frac{\mathcal{E}}{R}e^{-t/\tau} \qquad (23.1)$$

Here τ = RC is the time constant (in seconds).

After a time τ the current has dropped to $1/e = 0.37$ of its initial value, and it has dropped to ½ its initial value in a time

$$\boxed{0.693\tau = t_{1/2}} \qquad (23.2)$$

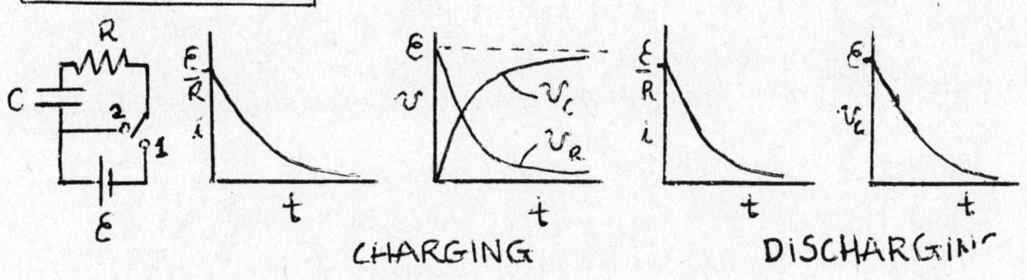

When the switch is then thrown to position 2 to discharge the capacitor the current and voltage decay exponentially.

$$V_R(t) = V_C(t) = \mathcal{E} e^{-t/\tau}$$

$$i(t) = \frac{\mathcal{E}}{R} e^{-t/\tau} \qquad (23.3)$$

2. Consider what happens when the switch in the RL circuit here is closed. The current starts to build up rapidly at first, and this causes a big back EMF in the inductor. This opposes the current at first, but as the current build up slowly, changing less rapidly, the induced back EMF gets weaker and weaker, until finally the current reaches a steady value and the induced EMF vanishes. The equations describing this are found to be

$$i(t) = \frac{\mathcal{E}}{R}(1 - e^{-t/\tau})$$

$$V_R(t) = \mathcal{E}(1 - e^{-t/\tau})$$

$$V_L(t) = \mathcal{E} e^{-t/\tau} \qquad (23.4)$$

The time constant is $\tau = \frac{L}{R}$ (in seconds)

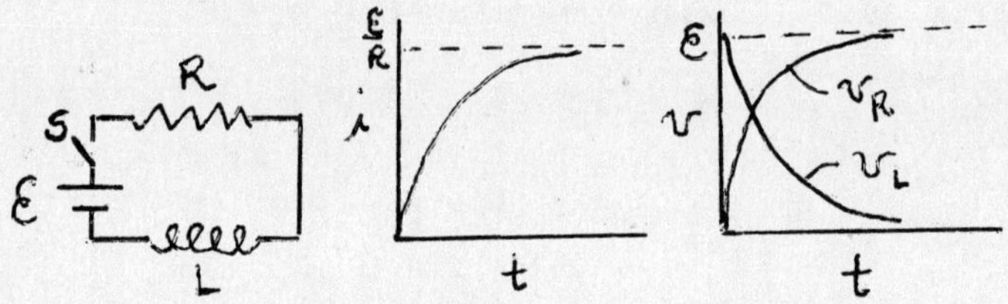

3. Alternating currents and voltages oscillate back and forth sinusoidally with frequency f, with

$$v(t) = v_o \sin(\omega t + \alpha)$$
$$i(t) = I_o \sin(\omega t + \beta) \qquad (23.5)$$

$\omega t + \alpha$ is called the <u>phase</u> of the voltage.

v varies from V_o to $-V_o$ and $\omega = 2\pi f$.

The "root mean square" (i.e., average) voltage or current is

$$V_{rms} = \frac{1}{\sqrt{2}} V_o = 0.707\, V_o \qquad I_{rms} = \frac{1}{\sqrt{2}} I_o = 0.707\, I_o \qquad (23.6)$$

The average power dissipated in a resistor carrying AC current across which the rms potential difference is V_{rms} is

$$P = I_{rms}^{\ 2} R = I_{rms} V_{rms} \qquad (23.7)$$

Thus we can treat resistive AC circuits using our DC equations, provided we use the average (rms) values for current and voltage. Power is dissipated only in resistors, not in capacitors or inductors.

4. In circuits containing capacitance the current across the capacitor is 90° in phase ahead of the voltage across the capacitor. The magnitudes of the current and voltage for the capacitor are related by

$$V_C = X_C I_C \qquad (23.8)$$

where $X_C = \frac{1}{\omega C}$ = capacitive reactance (ohms)

5. In circuits containing inductance the current lags the voltage in phase by 90°, and their magnitudes are related by

$$V_L = X_L I_L \qquad (23.9)$$

where $X_L = \omega L$ = inductive reactance (ohm.s)

6. In an RLC series circuit the current and voltage differ in phase by an amount which depends on X_C and X_L. The impedance Z (in ohms) of the circuit is defined as

$$Z = \frac{V}{I} \qquad (23.10)$$

Z depends on frequency, and $Z = \sqrt{R^2 + (X_L - X_L)^2}$.

Z is smallest (and the current largest for given V) when $X_L - X_C = 0$, or

$$\boxed{\omega L - \frac{1}{\omega C} = 0, \text{ or } \quad \omega = \frac{1}{\sqrt{LC}}} \qquad (23.11)$$

This frequency is called the resonant frequency and results in the largest current for given applied voltage.

7. A transformer consists of two coils wound so that all of the flux from one passes through the other. A transformer can be used to increase or decrease an AC voltage (it won't work with DC, since it works on the principle of induced voltages). If a voltage ε_1 is applied to a primary coil of N_1 turns, a voltage ε_2 will appear across the secondary coil of N_2 turns, where

$$\boxed{\frac{\varepsilon_1}{\varepsilon_2} = \frac{N_1}{N_2}} \qquad (23.12)$$

In an ideal transformer the power delivered to the primary is transmitted to the secondary without loss.

Thus $P = \varepsilon_1 I_1 = \varepsilon_2 I_2$, so

$$\boxed{\frac{I_1}{I_2} = \frac{N_2}{N_1}} \qquad (23.13)$$

8. Electrical injury is caused when current flows through your body, particularly through your heart (torso) or brain. AC and DC are both dangerous. In order for current to flow your body must form part of a closed electric circuit, so you must touch two electrical conductors or one "hot" wire and a grounded conductor (such as a water pipe or the earth). Water containing ions is a fairly good conductor, as is your blood (which contains lots of ions). A current of 1 mA at a frequency of 60 Hz will make you tingle. 10 mA will cause a hand to contract and be paralyzed (so that you can't let go). 100 mA for a second or so is enough to kill you. The current that flows in you depends on your resistance. Dry skin has fairly high resistance, perhaps 10^6 ohms, but wet skin does not, and with wet hands the resistance across your body may drop to 200-500 ohms. Thus 120 volts is enough to kill you in this instance. A high voltage in itself is not dangerous unless the source of voltage has enough charge (and low enough internal resistance) to drive a large current through you. Thus you can get charged up to 1000 volts by stoking a cat, but little charge has accumulated on your body.

• Qualitative Questions

23M.1 For each of the following indicate whether the device requires AC only, DC only, or either AC or DC.

A. Incandescent lamp.
B. Chrome plating of metal objects.
C. Electric stove.
D. Neon sign transformer.
E. Electric toaster.
F. Battery charger output.

23M.2 Often in working with electronic instrumentation it is necessary to use circuits with rapid response. In such applications it is important to take care that the instrumentation and circuitry does not decrease the speed of response of the apparatus. Which of the following is most likely to help achieve this aim?

A. Keep the capacitance in the circuits small.
B. Keep the resistance in the circuits large.
C. keep the voltages used large.
D. Keep the joule heating small.
E. Work in a room with low humidity.

23M.3 In this circuit the capacitor is initially uncharged. Which of the curves shown best describes the voltage between X and X' which would be observed after the switch is closed?

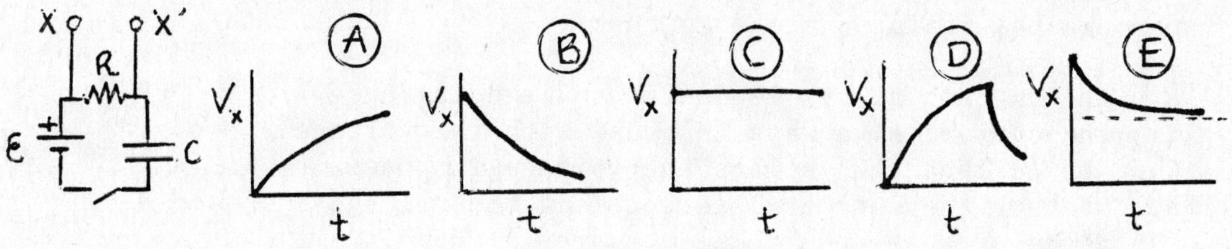

23M.4 The most critical factor determining injury due to electric shock is

A. the potential difference across your body.
B. the potential to which your body is raised.
C. the amount of electric charge on your body.
D. the induced EMF.
E. the current which flows in the body.
F. the polarity of the applied voltage.

23M.5 One can demonstrate the time required to charge or discharge a capacitor with the circuit shown here. If you actually show this to a class they will observe that the charging time (with switch in position 1) is longer than the discharging time (position 2). How would you best explain this?

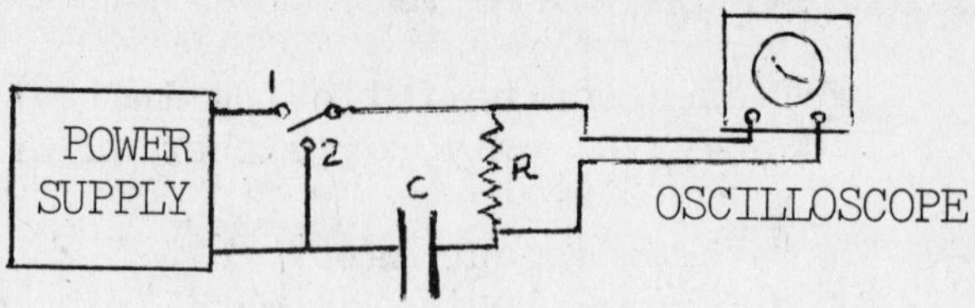

A. Circuit analysis shows that the time to charge a capacitor through a given resistor is less than the time needed for discharge through the same resistor.
B. The potential difference across the capacitor aided the discharging, whereas it opposed the charging.
C. The power supply had appreciable internal resistance.
D. In discharge the potential difference across R added to the potential difference across the capacitor, whereas this was not the case during the charging.
E. The switch was not left in position 1 long enough for the capacitor to become fully charged, hence one would not expect the time constants for charging and discharging to be equal.

23M.6 In an AC circuit the voltage across a resistor and the current through the resistor are such that

A. the current always leads the voltage in phase.
B. the current is always in phase with the voltage.
C. the current always lags the voltage in phase.
D. whether the current leads or lags the voltage in phase depends on what other elements are in the circuit.
E. the current is always 180° out of phase with the voltage, so one cannot say it is either leading or lagging, but it is not in phase.

23M.7 In an ideal series RLC circuit to which an AC voltage is applied,

A. power is dissipated equally in the capacitor, inductor and resistor.
B. all of the power is dissipated in the capacitor.
C. all of the power is dissipated in the inductor.
D. all of the power is dissipated in the resistor.
E. power is dissipated in all three elements, but in different proportions depending on the values of the parameters.

23M.8 A resistor and an inductance are connected in series to a battery of EMF E. When a switch is closed to complete the circuit

A. the current in the resistor will initially be greater than the current in the inductor.
B. the time required for the current to reach 50% of its final value will be dependent on E.
C. the current will rise rapidly at first, with the rate of increase decreasing with time.
D. the current will be large at first and will then decay exponentially.
E. the rate at which the current builds up will depend on the product RL.

23M.9 An important reason for using alternating current for domestic and industrial use is that

A. there is less energy wasted as heat with AC than with DC.
B. voltages can readily be changed with transormers for AC, but not for DC.
C. more energy can be obtained from AC.
D. the peak voltage is higher with AC.
E. motors won't work on DC.

23M.10 A transformer works on the principle that

A. an alternating current passed through a coil creates a changing magnetic field, and a changing magnetic field in a coil creates an EMF in that coil.
B. magnetic forces are always perpendicular to both current and magnetic field.
C. no current will flow between two points at the same potential.
D. a loop of current is magnetically equivalent to a small bar magnet.
E. a current in one coil of wire will exert a magnetic force on another current carrying coil.

23M.11 If the current through a circuit element is a maximum in an AC circuit when the voltage across that element is zero, then

A. something is wrong, because this cannot possibly happen.
B. the current and voltage are in phase.
C. the current and voltage are 90° out of phase.
D. the current and voltage are 180° out of phase.
E. it does not necessarily follow that the voltage will be a maximum when the current is zero.

• Multiple Choice: Answers and Comments

23M.1 — Electroplating and battery charging require DC. A transformer will work only with AC. A toaster, stove and lamp require only the heat generated and will work with either AC or DC.

23M.2 A — The time constant for charging or discharging a circuit containing capacitance depends on RC, so one wants low capacitance.

23M.3 B — Current rushes rapidly on to the plates at first and then decays to zero as the capacitor charges up.

23M.4 E

23M.5 C — If the power supply has internal resistance R_i, the capacitor is charged through a resistance $R + R_i$ with resulting time constant $(R + R_i)C$. Discharge occurs through only resistance R, so time constant is then RC.

23M.6 B — $V_R = IR$, and since R is just a constant multiplier, V_R and I are always in phase in every circuit.

23M.7 D — No power is dissipated in an ideal capacitor or inductor because the current and voltage are always 90° out of phase in them.

23M.8 C

23M.9 B

23M.10 A

23M.11 C — This will happen when $I = I_o \sin\omega t$ and $V = V_o \sin(\omega t + \frac{\pi}{2}) = V_o \cos\omega t$ (in which case I lags V by 90°) or when

$$I = I_o \sin\omega t \text{ and } V = V_o \sin(\omega t - \frac{\pi}{2}) = -V_o \cos\omega t$$

in which case I leads V by 90°.
To tell if I is lagging or leading V, we look at the phase. Thus if $I - I_o \sin\omega t$ and $V = V_o \sin(\omega t + \frac{1}{2}\pi)$, we see that at $t = 0$, $I = 0$ whereas $V = V_o$, a maximum. I will not reach its maximum until $t = \frac{\pi}{2\omega}$, a later time, so it is lagging V.

• Problems

23.1 A generator produces a voltage 170 sin 377t. What is the frequency, peak voltage and rms voltage?

23.2 A 2 μF capacitor is charged to 100 volts and then discharged through 4000 ohms. What is the time constant for discharge? How long will it take for the voltage to drop to 50 volts? To 25 volts? To 12.5 volts?

23.3 A large electromagnet has coils with inductance of 20 H and resistance of 5 ohms. If the magnet is connected to a 240 volt DC power supply, what will be the current in the coils after 4 seconds? After 8 seconds? After a very long time?

23.4 An inductance has a reactance of 200 ohms at a frequency of 60 Hz. What is the value of the inductance?

23.5 A 3 μF capacitor and a 200 ohm resistor are connected in series to a 120 volt rms generator with a frequency of 400 Hz. What power is dissipated in the circuit?

23.6 A series RLC circuit consists of a 5 ohm resistance, a 60 mH inductance, and a 2 μF capacitance. What is the maximum current which will flow in this circuit due to a 100 volts rms applied EMF? What should be the frequency of the applied EMF?

23.7 A transformer is used to step down the voltage in a transmission line from 13,200 volts to 220 volts. What is the ratio of turns needed between the primary and secondary windings? What current will flow in the secondary if the primary current is 30A?

• Problem Solutions

23.1 $V(t) = 170 \sin 377t = V_0 \sin \omega t$

THUS PEAK VOLTAGE $V_0 = \underline{170 \text{ VOLTS}}$ $V_{rms} = \frac{1}{\sqrt{2}} V_0 = \underline{120 \text{ VOLTS}}$

$\omega = 2\pi f = 377 \text{ RAD/S}$, $f = \frac{377}{2\pi} = \underline{60 \text{ Hz}}$

23.2 $\tau = RC = (4000\,\Omega)(2\times10^{-6}\,\text{F}) = 8\times10^{-3}\,\text{s} = \underline{8 \text{ ms}}$

$t_{1/2} = 0.693\tau = (0.693)(8 \text{ ms}) = 5.5 \text{ ms}$

THUS VOLTAGE WILL DROP FROM 100V TO 50V IN $\underline{5.5 \text{ ms}}$. IT WILL DROP FROM 50V TO 25V IN ANOTHER 5.5 ms OR $\underline{11 \text{ ms}}$ FROM $t = 0$. IT WILL AGAIN DROP A FACTOR OF 2, FROM 25V TO 12.5V, IN AN ADDITIONAL 5.5 ms. OR $\underline{16.5 \text{ ms}}$ FROM $t = 0$

23.3 FROM EQ. (24.4), $i(t) = \frac{\mathcal{E}}{R}(1 - e^{-t/\tau})$

$\tau = \frac{L}{R} = \frac{20\,H}{5\,\Omega} = 4\,s$ AT $t = 4\,s$, $i = 48(1 - e^{-4/4}) = \underline{\underline{30.3\,A}}$

AT $t = 8\,s$, $i = 48(1 - e^{-8/4}) = \underline{\underline{41.5\,A}}$

$\frac{\mathcal{E}}{R} = \frac{240\,V}{5\,\Omega} = 48\,A$ AT $t = \infty$, $i = 48(1 - e^{-\infty}) = \underline{\underline{48\,A}}$

23.4 $X_L = \omega L$, $\omega = 2\pi f$, $L = \frac{X_L}{2\pi f} = \frac{200\,\Omega}{(2\pi)(60)} = \underline{\underline{0.53\,H}}$

23.5 $Z = \sqrt{R^2 + (X_L - X_C)^2}$, $X_L = 0$, $X_C = \frac{1}{\omega C} = \frac{1}{(2\pi)(400)(3 \times 10^{-6}\,F)}$

$X_C = 133\,\Omega$

$Z = \sqrt{(200)^2 + (133)^2} = 240\,\Omega$

$V = IZ$, $P = I^2 R = \left(\frac{V}{Z}\right)^2 R = \left(\frac{120\,V}{240\,\Omega}\right)^2 (200\,\Omega) = \underline{\underline{50\,W}}$

23.6 $I = \frac{V}{Z}$, $Z = \sqrt{R^2 + (X_L - X_C)^2}$

I IS A MAXIMUM WHEN $X_L - X_C = 0$, SO $Z = R$

$X_L - X_C = \omega L - \frac{1}{\omega C}$, SO $\omega L = \frac{1}{\omega C}$, $\omega = \frac{1}{\sqrt{LC}}$

$\omega = \frac{1}{\sqrt{(60 \times 10^{-3}\,H)(2 \times 10^{-6}\,F)}} = 2887$ RAD/SEC

$f = \frac{\omega}{2\pi} = \underline{\underline{459\,Hz}}$

$I = \frac{V}{Z} = \frac{V}{R} = \frac{100\,V}{5\,\Omega} = \underline{\underline{20\,A}}$

23.7 $\frac{V_2}{V_1} = \frac{N_2}{N_1}$, $\frac{N_2}{N_1} = \frac{13200\,V}{220\,V} = \underline{\underline{60}}$ = TURNS RATIO

$\frac{I_2}{I_1} = \frac{N_1}{N_2}$, $I_2 = \frac{N_1}{N_2} I_1 = \left(\frac{1}{60}\right)(30\,A) = \underline{\underline{0.5\,A}}$

Electromagnetic Waves and the Nature of Light

24

• Summary of Important Ideas, Principles, and Equations

1. An accelerating charge will radiate electromagnetic radiation. These electromagnetic waves are oscillating electric and magnetic fields which can travel through vacuum or through many types of matter. They travel out from the source much as water waves or sound waves would. Far from the source the wave front is flat and the waves are called plane waves. The electric and magnetic fields are oriented at 90° to each other and lie in the plane of the wave front, which is perpendicular to the direction of propagation.

   The electric and magnetic fields oscillate back and forth at the frequency of the oscillating charges which generated them. The frequency and wavelength of the waves are related by

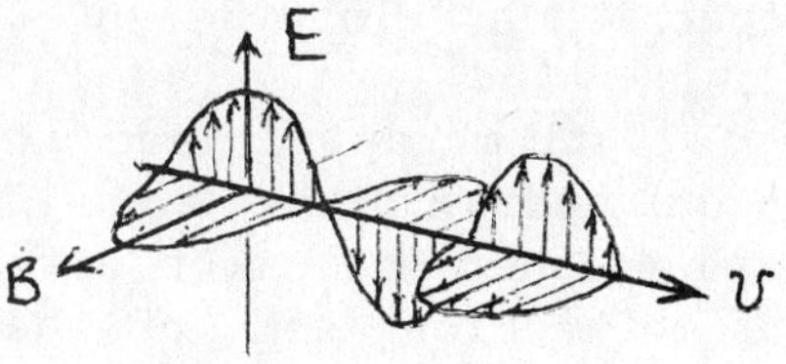

$$\boxed{v = f\lambda} \quad (24.1)$$

   where λ = wavelength (meters), f = frequency (Hertz) and v = velocity (m/s).

   In vacuum all frequencies travel at the same speed, $c = 3 \times 10^8$ m/s. c is an important constant of nature.

   In matter the waves travel more slowly, and different frequencies travel at slightly different speeds, with high frequencies travelling slightly slower than low frequencies in materials like glass or water.

   Electromagnetic waves are given different names, depending on their wavelength. Thus the longest wavelengths are called radio waves, and succeedingly shorter wavelengths are called microwaves, infrared rays, visible light, ultraviolet radiation. X-rays and finally gamma rays. Only a very small part of the spectrum, with wavelengths from 700 nm (red) to 400 nm (violet) is visible. 1 nm = 10^{-9} m.

2. The index of refraction of a material is defined as

$$\boxed{n = \frac{c}{v}} \qquad (24.2)$$

3. Consider a water wave travelling out from a pebble which has been dropped into the water. All the points on a given crest travelling out from the pebble form a wave front. A line drawn perpendicular to a wavefront is a ray. It points in the direction the wave is travelling (think of how a ray of sunshine looks piercing down through the clouds, or the ray of a flashlight beam). The spacing between two crests (or between two troughs or any two equivalent points) is a wavelength, λ. The number of crests passing a point in one second is the frequency, f. The wavelength stays constant as the wave moves along.

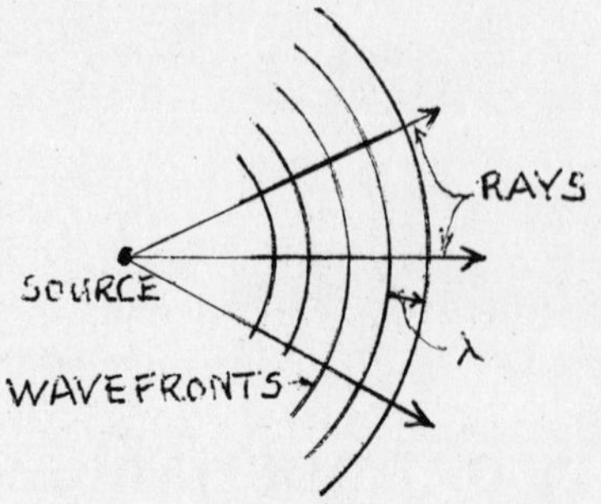

We can imagine the wave front as being a line of secondary sources which each radiate spherical waves, with the envelope of all of these little wavelets giving rise to the resultant wavefront. This idea is called Huygen's principle, and it is useful for understanding diffraction and interference.

4. When an electromagnetic wave travelling in one medium strikes the interface with another medium of different index of refraction, some of the wave may be transmitted and some may be reflected, with the relative amounts depending on the materials and the angle of incidence. The possible rays are sketched here, where θ_1 = angle of incidence, θ_r = angle of reflection, and θ_2 = angle of refraction. Note carefully how these angles are defined. The incident ray, the reflected ray and the transmitted (refracted) ray all lie in a single plane. The angles are measured with respect to a line drawn perpendicular to the surface (called the "normal" to the surface), and this line also lies in the plane of the three rays.

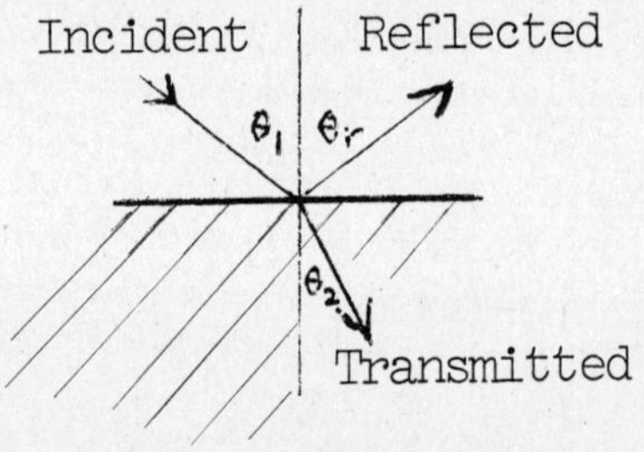

The law of reflection is:

Angle of incidence = Angle of reflection

$$\boxed{\theta_1 = \theta_r} \qquad (24.3)$$

5. Refraction is the change in direction of a ray when it passes from one medium into another of different index of refraction, i.e., where the speed of the wave is different. If the angle of incidence is θ_1 in a material with index n_1, the angle of refraction θ_2 in the second material of index n_2 will be given by Snell's law.

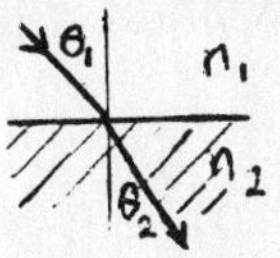

$$\boxed{n_1 \sin\theta_1 = n_2 \sin\theta_2} \qquad (24.4)$$

Refraction occurs for all kinds of waves, such as sound and water waves, as well as for electromagnetic waves.

It is easy to understand why refraction happens. Think of a band marching diagonally across a football field. Each row is two steps apart (one wavelength) and is a wavefront (like a wave crest). Everyone marches to the same beat of the drum everywhere on the field (i.e., the wave has the same frequency no matter where it is).

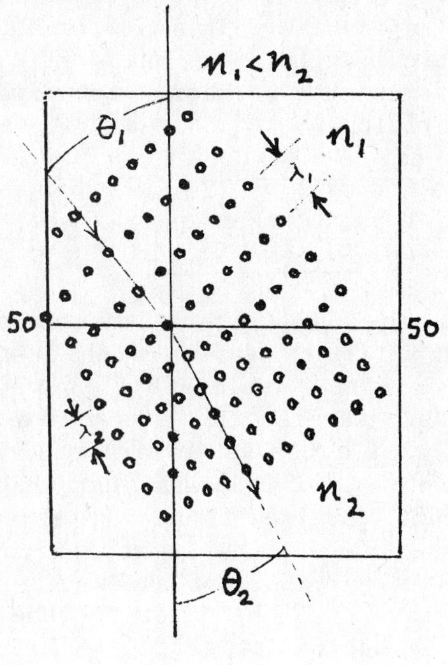

Suppose that the area north of the 50 yard line is air and the area south of the 50 is water (light goes faster in air than water by a factor of about 1.33. As soon as a person reaches the 50 yard line going south he is supposed to slow down. Since $v = f\lambda$, and f is always constant, the wavelength λ must get smaller when the wave slows down. In the drawing the persons on the right end of a row will hit the 50 first, and they slow down by shortening their stride as soon as they cross the line. This causes the entire row to swing to the right. Once the entire row is across the 50 everyone is once again moving with the same shortened stride, and the row then marches straight ahead, but its direction has been changed from what it was in medium 1 (air, north of the 50). Observe that in going from air into a more dense medium (where the wave slows down) the ray direction swings in toward the normal. In going in the opposite direction, just the opposite happens. Using geometry it is easy to derive Snell's law from this model.

6. When a wave is incident on the interface of a medium with a smaller index of refraction it may, for certain large angles of incidence, undergo total internal reflection. Suppose, for example, light in water is striking a water-air interface. In (a) some of the wave is transmitted and some is reflected.

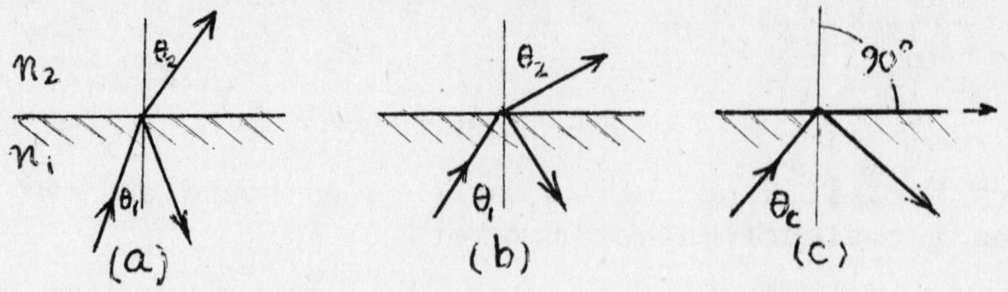

In (b) the angle of incidence is increased, and again some of the wave is reflected and some is transmitted, although the intensity of the transmitted wave is now relatively less than it was. In (c) the angle of incidence has increased to the point that any transmitted wave would just graze the surface of the interface. Its intensity has now dropped to zero, and all of the light is now in the reflected wave. This occurs at an angle of incidence called the critical angle, θ_c. Whenever $\theta_1 > \theta_c$ the light is totally internally reflected. When $\theta_1 = \theta_c$ we see that $\theta_2 = 90°$. Thus Snell's law yields $n_1 \sin\theta_c = n_2 \sin\theta_2 = n_2 \sin 90^\circ = n_2$ so

$$\boxed{\sin\theta_c = \frac{n_2}{n_1}} \qquad (24.5)$$

7. In many transparent materials the speed of light varies with frequency (i.e., with wavelength). Typically shorter wavelength rays move more slowly, i.e., have a larger index of refraction. This means that they will be bent more when passing through a prism. Thus white light, which consists of a mixture of all wavelengths (i.e., of all colors) will be dispersed into the colors of the rainbow when refracted by a prism. This is one way of getting a fairly monochromatic light beam. The variation of velocity with frequency (or wavelength) is called dispersion.

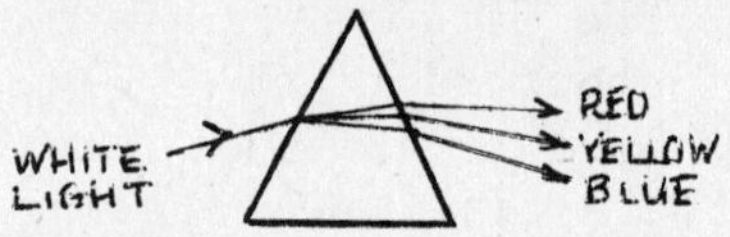

• Qualitative Questions

24M.1 When a plane electromagnetic wave travels through space with a velocity vector v,

A. v is not always perpendicular to the plane of E and B.
B. E is not always perpendicular to the plane of v and B.
C. B is not always perpendicular to the plane of E and v.
D. None of the three vectors E, B or v need be perpendicular.
E. v, E and B are all perpendicular.

24M.2 An electromagnetic wave

A. is a longitudinal wave.
B. is a transverse wave.
C. can't travel through a vacuum.
D. is different from a light wave.
E. caries charge.

24M.3 Ultraviolet rays are to microwaves as

A. a great white shark is to a minnow.
B. a tenor is to a baritone.
C. a whisper to a shout.
D. red is to green.
E. an orange is to an apple.

24M.4 When light passes from glass into air,

A. its frequency increases.
B. its velocity decreases.
C. its wavelength is unchanged.
D. some of the light will be reflected.
E. $\theta_1 \geq \theta_2$, where θ_1 = angle of incidence.

24M.5 Shown here are some possible paths followed by light going from air to glass or glass to air. You aren't told which way the light is going. Which is the most probable path followed?

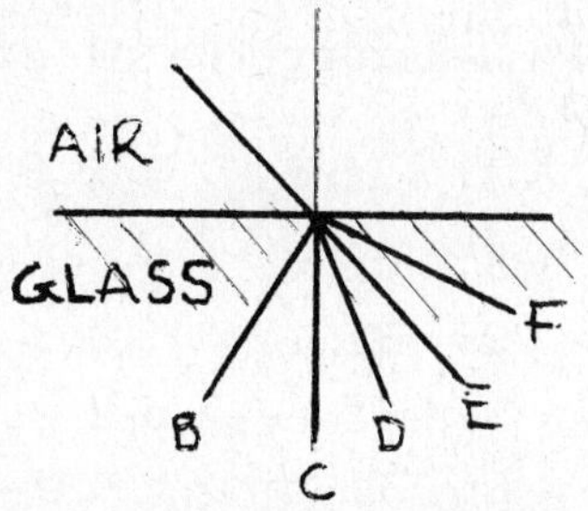

A. We can't tell which path was followed unless we know which way the light was going.

24M.6 In vacuum, gamma rays, X-rays and visible light all have the same

A. frequency.
B. wavelength.
C. velocity.
D. intensity.
E. energy.

24M.7 A glass prism separates white light into the colors of the spectrum because

A. of total internal reflection.
B. of interference between different frequencies within the prism.
C. electromagnetic energy is dissipated in the prism.
D. the white light, which had a single frequency, has its frequency shifted by varying amounts, depending on the thickness of glass traversed.
E. different frequencies of light travel at different velocities.

24M.8 Total internal reflection can occur when light in one medium is incident on another

A. which has a lower index of refraction.
B. which has a larger index of refraction.
C. which has the same index of refraction.
D. at an angle of incidence less than the critical angle.
E. in all cases, depending only on the angle of incidence.

24M.9 The refraction of light

A. occurs only at curved surfaces.
B. is a result in the change in frequency that occurs when light passes from one medium to another.
C. is a result of the change in speed that occurs when light passes from one medium to another.
D. is the result of many tiny reflections which can be imagined to occur as the light crosses the boundary between two media.
E. is described by the relation $\theta_i = \theta_r$.

24M.10 Consider two electromagnetic waves, X and Y.

A. If $f_X > f_Y$, then $\lambda_X > \lambda_Y$.
B. If $v_X > v_Y$, $\lambda_X > \lambda_Y$.
C. If $\lambda_X > \lambda_Y$, $f_Y > f_X$.
D. If $n_X > n_Y$, $v_X > v_Y$.
E. If $n_X > n_Y$, $f_X > f_Y$.

• Multiple Choice: Answers and Comments

24M.1 E

24M.2 B

24M.3	B	UV and microwaves are both the same kind of thing, but UV has a higher frequency (and shorter wavelength) than do microwaves. A tenor has a higher pitched voice than does a baritone. If A or D were reversed they would be acceptable answers by comparison with the wavelengths.
24M.4	D	
24M.5	D	It doesn't matter which way the light is going. It will travel closer to the normal line in glass.
24M.6	C	
24M.7	E	
24M.8	A	
24M.9	C	
24M.10	C	One can see this from the relation $f\lambda = v$.

• Problems

24.1 A radio station broadcasts at a frequency of 1.20 MHz. What is the wavelength of this radiation.

24.2 The wavelength of visible light ranges from about 400 nm (violet) to 700 nm (red). What is the range of frequencies associated with these frequencies?

24.3 How long does it take light to travel from the sun to the earth, a distance of 1.5×10^{11}m?

24.4 The index of acetone is 1.36. What is the speed of light in acetone?

24.5 A fish swims in a circular pond of diameter 2 meters. What is the deepest he can be and still see all of the outside world when he looks up?

24.6 An engineer wishes to design an optical light pipe using a straight cylindrical glass fiber. It is intended that once light enters the end of the fiber, it will not get out the sides, but will instead bounce back and forth until it reaches the other end. The end of the fiber is perpendicular to the length. What must the index of refraction of the fiber be?

• Problem Solutions

24.1 $f\lambda = c$, $\lambda = \frac{c}{f} = \frac{3\times10^8 \text{ m/s}}{1.2\times10^6 \text{ /s}} = \underline{\underline{250 \text{ m}}}$

24.2 $f_R = \frac{c}{\lambda_R} = \frac{3\times10^8 \text{ m/s}}{700\times10^{-9} \text{ m}} = \underline{\underline{4.3\times10^{14}}} \text{ Hz}$

$f_V = \frac{c}{\lambda_V} = \frac{3\times10^8 \text{ m/s}}{400\times10^{-9} \text{ m}} = \underline{\underline{7.5\times10^{14} \text{ Hz}}}$

24.3 $d = ct$, $t = \frac{d}{c} = \frac{1.5\times10^{11} \text{ m}}{3\times10^8 \text{ m/s}} = \underline{\underline{500 \text{ s}}} = \underline{\underline{8 \text{ MIN } 20 \text{ s}}}$

24.4 $n = \frac{c}{v}$, $v = \frac{c}{n} = \frac{3\times10^8 \text{ m/s}}{1.36} = \underline{\underline{2.21\times10^8 \text{ m/s}}}$

24.5 IF THE FISH IS TO SEE ALL OF THE OUTSIDE WORLD, HE MUST BE ABLE TO SEE RAYS COMING IN WITH AN ANGLE OF INCIDENCE θ_1 APPROACHING 90°. THIS MEANS THAT THE LIGHT REACHING HIM MAKES AN ANGLE θ_2 WITH THE NORMAL, WHERE

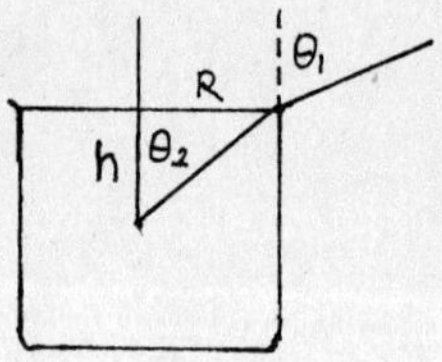

$$n_1 \sin\theta_1 = n_2 \sin\theta_2$$

$n_1 = 1$, $n_2 = 1.33$ (WATER), $\sin\theta_2 = \frac{(1)(\sin 90°)}{1.33}$, $\theta_2 = 49°$

$\tan\theta_2 = \frac{R}{h}$, so $h = \frac{R}{\tan\theta_2} = \frac{1 \text{m}}{\tan 49°} = \underline{\underline{0.88 \text{ m DEEP}}}$

24.6

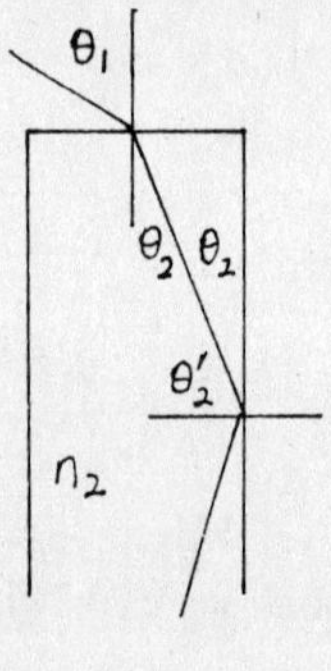

FOR TOTAL INTERNAL REFLECTION $\theta_2' \geq \theta_c$

WHERE $\sin\theta_c = \frac{1}{n_2}$ FROM 25.5

LIGHT IS MOST LIKELY TO ESCAPE THROUGH SIDES AS $\theta_1 \rightarrow 90°$,

$$n_1 \sin\theta_1 = n_2 \sin\theta_2$$

$$(1)(\sin 90°) = n_2 \sin\theta_2$$

so $\sin\theta_2 = \frac{1}{n_2}$

BUT $\sin\theta_2 = \cos\theta_2'$

so $\frac{1}{n_2} = \cos\theta_2'$ AND $\sin\theta_2' = \frac{1}{n_2}$ FOR TOTAL INTERNAL REFLECTION.

THUS $\cos\theta_2' = \sin\theta_2'$, so $\theta_2' = 45°$

AND $n_2 = \frac{1}{\sin 45°} = 1.414$

THUS IF $\underline{\underline{n_2 \geq 1.414}}$ NO LIGHT WILL ESCAPE.

Geometrical Optics; Optical Instruments

25

• Summary of Important Ideas, Principles, and Equations

1. A mirror or a lens can form an image of an object. The light rays one sees appear to come from this image. If they actually pass through the image point, the image is real. If they only appear to have passed through the image point, the image is virtual. If the image is right side up with respect to the object of which it is an image, it is upright. If it is upside down, it is inverted.

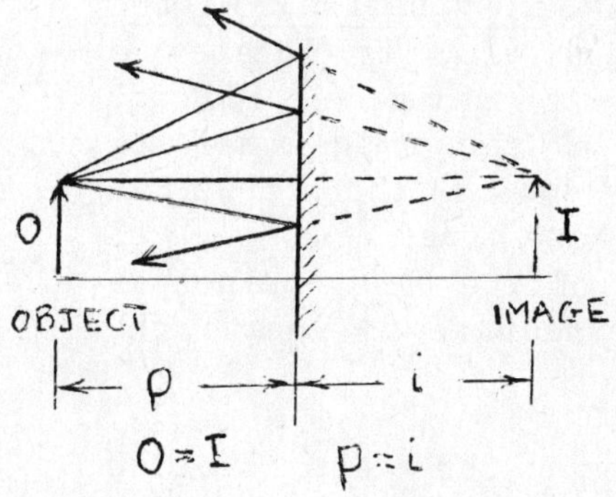

A plane mirror forms an image which is erect and virtual and the same size as the object. The object and image distances are equal.

2. A concave spherical mirror (it is "caved" in) causes all rays parallel to the axis (and not too far from it) to be reflected through the focal point of the mirror. The distance from the mirror to the focal point is the focal length, f. For a sperical mirror of radius R

$$f = \tfrac{1}{2}R \qquad (25.1)$$

One can graphically find the position of the image formed by a mirror by tracing out the paths of three principal rays. Consider some of the rays emanating from the tip of the object.

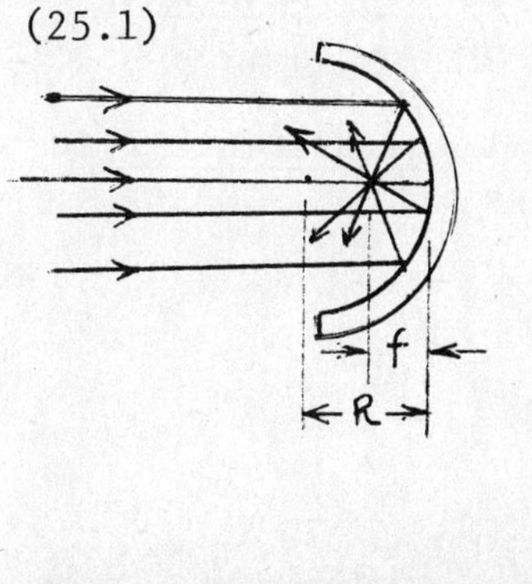

(#1) A ray along the line from the tip of the object to the center of curvature will be reflected straight back.

(#2) A ray parallel to the axis will be reflected back through the focal point.

(#3) A ray directed along the line from the tip of the object to the focal point will be reflected parallel to the axis.

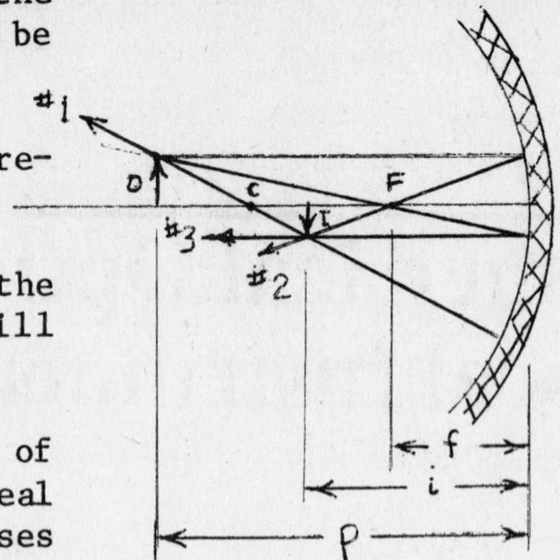

Where all of these rays intersect is the tip of the image. The image sketched here is a real image, because the light actually passes through it. One could place a piece of film there and expose it, whereas this could not be done with the image in a plane mirror.

A ray diagram is unchanged if the direction of every ray is reversed. Thus if an object at position A forms an image at position B, an object at position B will form an image at position A.

For a convex spherical mirror the focal point is located a distance f behind the mirror, and again the image may be found by using the principal rays. Note that the use of just two of the three principal rays is adequate to locate the image.

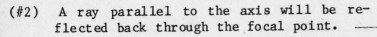

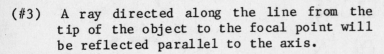

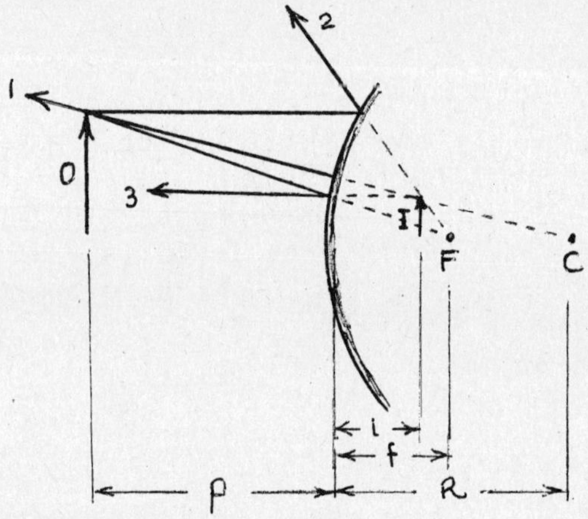

One can find the exact location and size of the image by using the mirror equation.

$$\frac{1}{p} + \frac{1}{i} = \frac{1}{f} \qquad (25.2)$$

To use this equation it is important that the proper signs for i and f be used.

The object distance p is always positive.

The focal length f is $+\frac{1}{2}R$ for a concave mirror, $-\frac{1}{2}R$ for a convex mirror.

The image distance i is positive for a real image and negative for a virtual image.

The ratio of the size of the image to the size of the object is the <u>magnification</u>, m.

$$m = \frac{I}{O} = -\frac{i}{p} \qquad (25.3)$$

If $m > 0$ the image is upright and if $m < 0$ the image is inverted.

To use the mirror equation, first sketch a ray diagram. Then you can see if the image is real or virtual, and this will make clear the sign of the image distance i.

3. When parallel light rays strike a <u>converging lens</u> they are brought to a focus on the opposite side of the lens. By reversibility, rays emanating from a focal point are refracted to travel parallel to the axis.

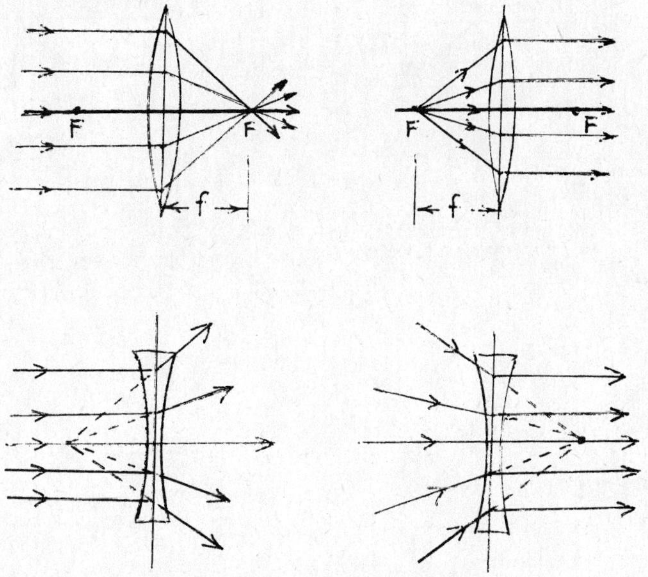

In a <u>diverging lens</u> rays parallel to the axis are refracted to diverge from the focal point, and rays directed in toward a focal point are bent to travel parallel to the axis, as illustrated here.

A lens which is thicker on the axis than on the edge is a converging (or "positive") lens. One which is thicker at the edge than on the axis is a diverging (or "negative") lens.

The three principal rays can again be used to locate the image. They are labeled 1, 2 and 3 in the drawing.

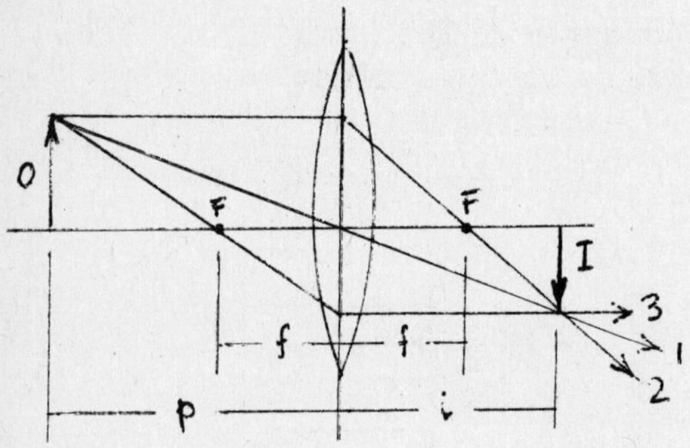

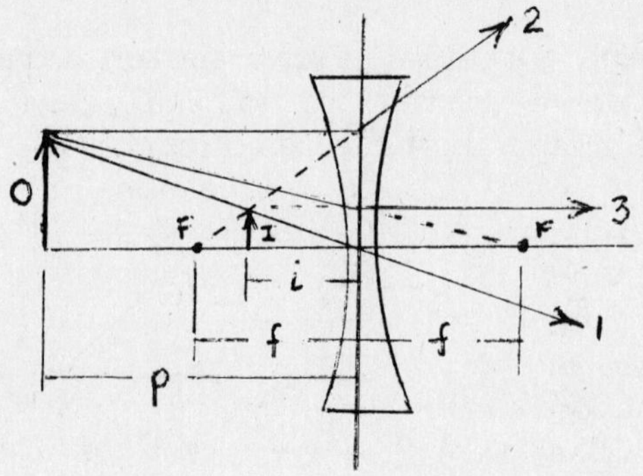

One can find the exact size and position of the image by using eq. (25.2), provided the following sign convention is followed:

Draw a sketch with light always travelling from left to right.

The object distance p is positive if the object is to the left of the lens. This is the case for every problem you will be facing.

The image distance i is positive if the image is on the right side of the lens and negative if the image is on the left side of the lens.

The focal length f is positive for converging lenses and negative for diverging lenses.

The magnification is given by eq. (25.3)

Always first draw a ray diagram and then use eq. (25.2) to obtain exact values.

A diverging lens always forms a virtual, upright, diminished image of a real object. A converging lens can form a real or virtual image, depending on object position.

4. Spherical lenses and mirrors have defects called aberrations. Spherical aberration causes blurring of images because rays parallel to but far from the axis are brought to a focus nearer to the mirror or lens than are rays near the axis. Different colors have different indices of refraction and hence different focal lengths. This results in chromatic aberration, a colored blurring of an image.

5. In the human eye light is refracted by the cornea and by the lens. The curvature, and hence the focal length, of the lens can be varied to form an image on the retina. Light is detected at the retina, from which electrical signals are sent to the brain. If the eyeball is too short or the cornea not sufficiently curved the image tends to be formed behind the retina, and the person is far sighted. This can be corrected with a converging glasses lens. If the eyeball is too long, or if the cornea is too sharply curved, the person is near sighted. This can be corrected with a diverging lens.

6. A camera has a lens of fixed focal length which can be moved to form a real, inverted image on a film.

7. A magnifying glass is a converging lens positioned with the object to viewed inside the focal point. A magnified, upright virtual image is formed at 25 cms, a comfortable distance for viewing.

The magnification is approximately D, where D is the power of the lens in diopters. Definition:

Power (in diopters) = $\frac{1}{f}$, f in meters.

8. A compound microscope consists of a short focal length objective lens and an eyepiece. The magnification is approximately $LD_oD_e/4$, where L is the separation of the objective and eyepiece, and D_o and D_e are the powers in diopters of the objective and eyepiece.

9. A telescope consists of an objective, which may be either a lens or a mirror, and an eyepiece lens. The objective gathers large amounts of light (to make a bright image) and forms a sharply resolved image. The eyepiece magnifies the image for examination.

10. A light pipe (or optical fiber) is a clear rod which transmits light from one place to another. It can bend light around corners, since the light is prevented from escaping because it experiences total internal reflection at the walls.

• Qualitative Questions

25M.1 Suppose a lighted candle is placed a short distance from a plane mirror, as shown here. Where will the image be located?

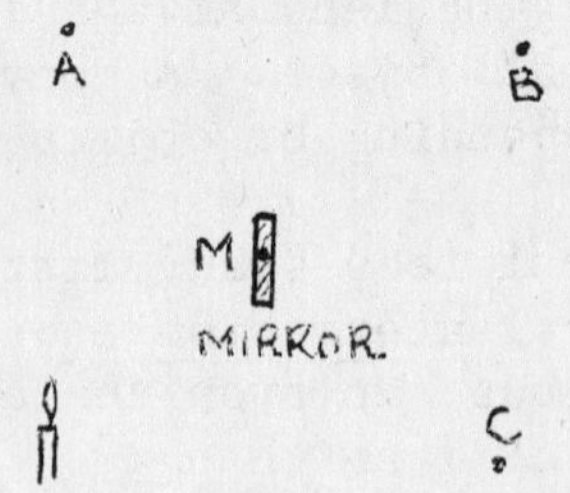

A. At A. B. At B.
C. At C. D. At M (at the mirror.
E. No image is formed since the object is not in front of the mirror.

25M.2 Sometimes when you look into a curved mirror you see a magnified image (a great big you) and sometimes you see a diminished image (a little you). If you look at the bottom (convex) side of a shiny spoon, what will you see? (Try it!)

A. You won't see an image of yourself, because no image will be formed.
B. You will see a little you, upside down.
C. You will see a little you, right side up.
D. You will see a little you, but whether you are right side up or upside down depends on how near you are to the spoon.
E. You will see either a little you or a great big you, depending on how near you are to the spoon.
F. You will see an image of yourself which is the same size as the real you.

25M.3 Which of the following is an accurate statement?

A. A mirror always forms a real image.
B. A mirror always forms a virtual image.
C. A mirror always forms an enlarged image.
D. A mirror always forms an image smaller than the object.
E. A mirror always forms an image the same size as the object.
F. None of the above is true.

25M.4 An object is placed a distance d in front of a plane mirror. The size of the image of the object will be

A. half as big as the size of the object.
B. dependent on the distance d.
C. dependent on where you are positioned when you look at the image.
D. twice the size of the object.
E. the same size as the object, independent of the distance d or the position of the observer.

25M.5 A bright lamp filament is placed one meter from a screen. Between the lamp and the screen is placed a converging lens of focal length 24 cms. As the lens position is varied with respect to the lamp,

A. no sharp image will be seen for any lens position.
B. a sharp image will be seen when the lens is halfway between the lamp and the screen.
C. a sharp image will be seen only when the lens is 40 cms from the lamp.
D. a sharp image will be seen only when the lens is 60 cms from the lamp.
E. a sharp image will be seen only when the lens is 24 cms from the lamp.
F. a sharp image will be seen when the lens is either 40 cms or 60 cms from the lamp, but not otherwise.

25M.6 A diverging lens ($f = -4$ cms) is positioned 2 cms to the left of a converging lens ($f = +6$ cms). A 1 mm diameter beam of parallel light rays is incident on the diverging lens from the left. After leaving the converging lens the outgoing rays

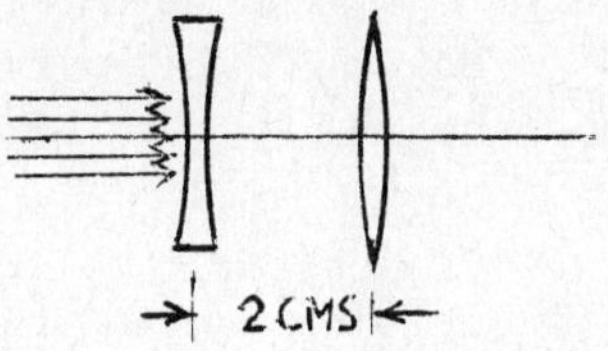

A. converge.
B. diverge.
C. form a parallel beam of diameter $D > 1$ mm.
D. form a parallel beam of diameter $D < 1$ mm.
E. will travel back toward the light source.

25M.7 Is it possible to see a virtual image?

A. No, since the rays which seem to emanate from a virtual image do not in fact emanate from the image.
B. No, since virtual images do not really exist.
C. Yes, the rays which appear to emanate from a virtual image can be focussed on the retina just like those from an illuminated object.
D. Yes, since almost everything we see is virtual because most things do not themselves give off light, but only reflect light coming from some other source.
E. Yes, but only indirectly in the sense that if the virtual image is formed on a sheet of photographic film, one could later look at the picture formed.

25M.8 Two thin convex-convex lens are placed in contact. If each has a focal length of 20 cms, how would you expect the combination to function?

A. About like a single lens of focal length 20 cms.
B. About like a single lens of focal length 40 cms.
C. About like a single lens of focal length slightly greater than 20 cms.
D. About like a single lens of focal length less than 20 cms.

25M.9 A fish's eye is well suited to seeing under water, but what can you say about his vision if he is taken out of water?

A. His acuity (sharpness of vision) would be greater out of water.
B. He would suffer from astigmatism out of water.
C. He would be very near-sighted out of water.
D. He would be very far-sighted out of water.
E. He vision would be limited to a cone of half-angle of about 49°.

25M.10 If the human eyeball is too short from front to back, this gives rise to a vision defect that can be corrected by using

A. convex-convex eyeglasses.
B. concave-concave eyeglasses.
C. cylindrical eyeglasses.
D. contact lenses, but no ordinary lenses.
E. shaded glasses (i.e., something that will cause the iris to dilate more).

25M.11 Which of the following is an accurate statement concerning the prisms used in the binocular shown here?

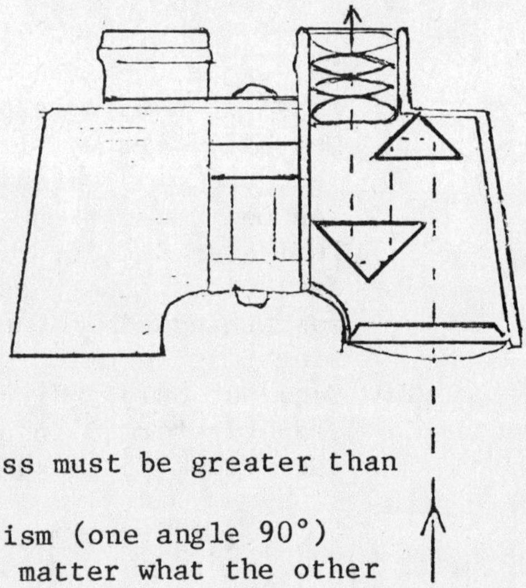

A. The prisms are used to refract light.
B. The prisms are used to form an image.
C. The index of refraction of the glass must be less than a certain maximum value.
D. The index of refraction of the glass must be greater than certain minimum value.
E. As long as the prism is a right prism (one angle 90°) oriented as shown here, it doesn't matter what the other two angles are.

25M.12 In which of the following ways is a camera different from the human eye?

A. The camera always forms an inverted image, the eye does not.
B. The camera always forms a real image, the eye does not.
C. The camera utilizes a fixed focal length lens, the eye does not.
D. For the camera the image magnification is greater than one, but for the eye the magnification is less than one.
E. A camera cannot focus on objects at infinity, but the eye can.

25M.13 In a single lens reflex camera the lens-film distance may be varied by sliding the lens forward or backwards with respect to the camera housing. If, with such a camera, a fuzzy picture is obtained, this means that

A. the lens was too far from the film.
B. the lens was too close to the film.
C. too much light was incident on the film.
D. too little light was incident on the film.
E. the object was too close to the camera.
F. the object was too far from the camera.
G. one cannot say which of the above reasons is valid.

25M.14 The image seen in a simple telescope is upside down. Suppose you cover the lower half of the objective lens on the front end while looking at an object. What will you see?

A. The top half of the image will be blacked out.
B. The lower half of the image will be blacked out.
C. The entire image will be blacked out, since the entire lens is needed to form an image.
D. The image will appear as it would if the objective were not blocked, but it will be dimmer.
E. There will be no noticeable difference in the appearance of the image with the objective partially blocked or not.

• Multiple Choice: Answers and Comments

25M.1 C If you don't believe this answer, try it and see for yourself.

25M.2 C Try it!

25M.3 F

25M.4 E As you back away from a mirror your image will look smaller and smaller. The same thing happens when a person moves away from you. She looks smaller when she is farther away because she subtends a smaller angle at your eye, but she of course is the same actual size no matter how far back she moves.

25M.5 F Suppose the lens is a distance p from the lamp. Then the screen is a distance 100 - p from the lens. The lens equation can be solved to find p.

$$\frac{1}{p} + \frac{1}{100-p} = \frac{1}{24}$$

$$24(100-p) + 24p = p(100 - p)$$

$$p^2 - 100p + 2400 = 0$$

$$(p - 40)(p - 60) = 0 \qquad p = 40 \text{ cms or } 60 \text{ cms.}$$

If $p = 60$ cms, $i = 100 - p = 40$ cms
If $p = 40$ cms, $i = 100 - p = 60$ cms

Thus the lens can be placed either 40 cms or 60 cms from the lamp. This illustrates the reversibility of a lens.

25M.6 C Note that the left hand focal points of both lenses coincide. Light striking the diverging lens thus diverges from this focal point. When it strikes the second lens it is thus bent back into a parallel beam, but it has been spread into a larger beam.

25M.7 C A is not a good answer, because while it is true that the rays do not actually emanate from a virtual image, this doesn't mean they can't be seen or focused. B is some kind of Zen Buddhist answer. D is a deperate wrong answer of a question maker who is running out of reasonable wrong answers. E is false. A virtual image can't expose film, since the light only appears to come from it.

25M.8 D The first lens will bend parallel rays in to a focus a distance 20 cms from the lens. 1 The second convex lens will bend them even more, so the combination lens will act like a single lens of shorter focal length. As the separation of the lenses approaches zero the effective focal length approaches $f = 10$ cms.

25M.9 C A fish's eyeball is sharply curved. This is necessary because the difference in index of refraction between his eye and the water is not very great, and it is this difference which determines the amount of refraction. When he is placed in air his eye bends the light too much, forming an image in front of his retina.

25M.10 A The glasses will bend the light in more and cause the image to be formed closer to the front of the eye.

25M.11 D The prisms act like mirrors to reflect the light back and thus shorten the needed length of the binocular. The reflection occurs because of total internal reflection, and this requires a minimum index of refraction in the glass.

25M.12 C

25M.13 G If the image is fuzzy this could be because it is formed in front of or behind the film. Thus we don't know if the lens is too close or too far from the film.

25M.14 D Any part of the lens will form the entire image.

• Problems

25.1 I recently had to replace the side view mirror on my truck. How tall should the mirror have been if I am to see all of a car 2 meters high when it is following 20 meters back (measured from the mirror)? The mirror is 0.5 meter from my eyes.

25.2 An object is placed 6 cms from a concave mirror of radius 4 cms. Graphically determine the image position and size. Is the image real or virtual?

25.3 An object is placed 3 cms from a convex mirror of radius 4 cms. Graphically determine the size and position of the image. If the image real or virtual?

25.4 A slide projector has a focal length of 200 mm. What size picture will it cast on a screen 12 meters distant when a slide 35 mm tall is used?

25.5 (a) How far from a 50 mm focal length lens, such as is used in many 35 mm cameras, must an object be positioned if it is to form a <u>real</u> image magnified in size by a factor of three?

(b) How far from the lens must the object be placed if it is to form a <u>virtual</u> image magnified in size by a factor of three (or is this even possible?)?

25.6 An object is placed at the origin. A converging lens of focal length 10 mm is placed at x = 40 mm on the x axis. A second converging lens of focal length 20 mm is placed at x = 90 mm. Graphically determine the size and location of the final image.

• Problem Solutions

25.1 LOOKING AT A MIRROR IS LIKE LOOKING THROUGH A LITTLE WINDOW. HERE THE IMAGE IS 2m TALL, SO THE MIRROR HEIGHT MUST BE X, WHERE

$$\frac{x}{2m} = \frac{0.5m}{20.5m}, \quad x = 0.049\,m = \underline{\underline{4.9 \text{ cms TALL}}}$$

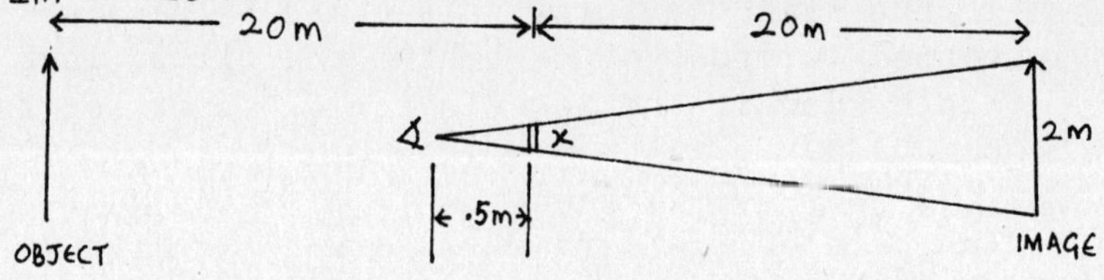

25.2

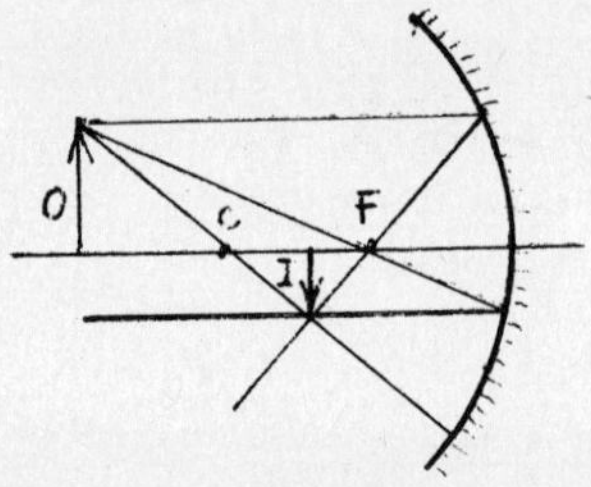

25.3

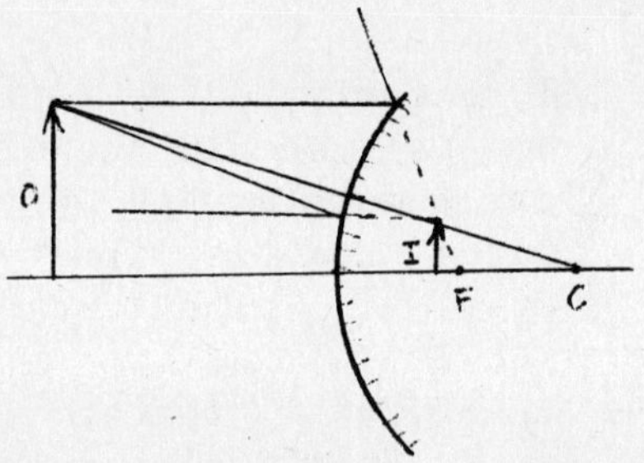

25.4 $\frac{I}{O} = |m| = \frac{i}{p}$ $\quad \frac{1}{p} + \frac{1}{i} = \frac{1}{f}$

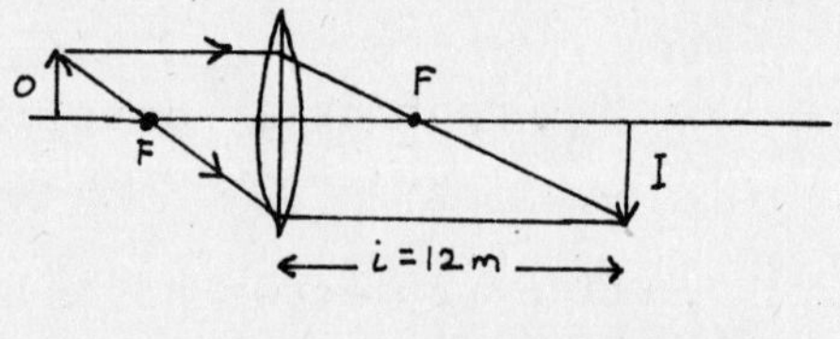

$$\frac{1}{p} + \frac{1}{1200} = \frac{1}{20}$$

$$p = \frac{1200}{59} \text{ cms}$$

$$|m| = \frac{i}{p} = \frac{1200}{1200/59} = 59$$

$$I = |m|\,O = (59)(35\text{ mm}) = 2065\text{ mm} = \underline{\underline{2.065\text{ m}}}$$

25.5 (a) $|M| = \frac{i}{p} = 3$, so $i = 3p$

$$\frac{1}{f} = \frac{1}{i} + \frac{1}{p}$$

$$\frac{1}{50} = \frac{1}{3p} + \frac{1}{p} = \frac{1}{3p} + \frac{3}{3p}$$

$$\frac{1}{50} = \frac{4}{3p} \;,\; p = \frac{200}{3}$$

$p = 66.7$ mm = OBJECT DISTANCE

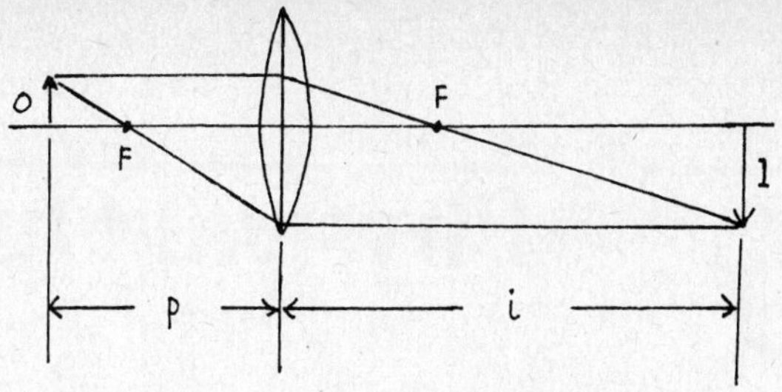

(b) OBJECT DISTANCE $p < f$ TO FORM A VIRTUAL IMAGE, AND $i < 0$

THUS $\frac{|i|}{p} = 3$

$$\frac{1}{f} = \frac{1}{p} - \frac{1}{|i|}$$

$$\frac{1}{50} = \frac{1}{p} - \frac{1}{3p}$$

$$\frac{1}{50} = \frac{3}{3p} - \frac{1}{3p} = \frac{2}{3p}$$

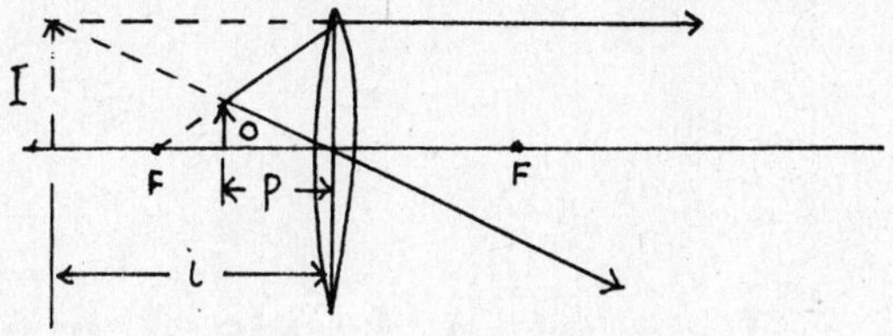

$3p = 100$, $p = 33.3$ mm

25.6 FIRST FIND THE IMAGE FORMED BY THE FIRST LENS. USE THIS AS THE OBJECT OF THE SECOND LENS.

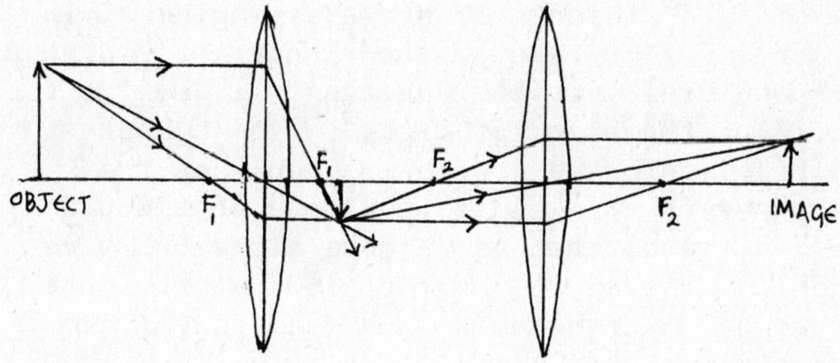

26 Physical Optics

• Summary of Important Ideas, Principles, and Equations

1. When light is propagating it exhibits wave properties. When it interacts with matter it behaves like a beam of particles. We say that light has a dual nature in this respect. Physical optics is concerned with some of the wave aspects of light.

   Many of the most interesting properties of light are associated with the oscillating electric field, as opposed to the magnetic field, of the light wave. To envision a light wave imagine a sine wave drawn in space. Its value represents the electric field at a point. The spacing between crests is the wavelength. Imagine the curve moved at a velocity v (the speed of light). The number of crests per second passing a given fixed point is f, the frequency of the light. If we call the phase of the wave zero at one of the places where the sine wave is also zero, then as we move along the wave the phase increases to 90° at a crest, then to 180° at the next zero of the wave, then to 270° where the wave has a trough, and finally to 360° at the next zero.

   Two waves are coherent if they maintain a fixed phase relationship as they travel along. Coherent waves may add together to reinforce each other or they may subtract to cancel each other. Incoherent waves add in some regions and cancel in others, and on the average have no effect on each other. Two waves which cancel each other are said to interfere destructively. Two waves which reinforce each other interfere constructively. Coherent waves must have the same frequency and wavelength.

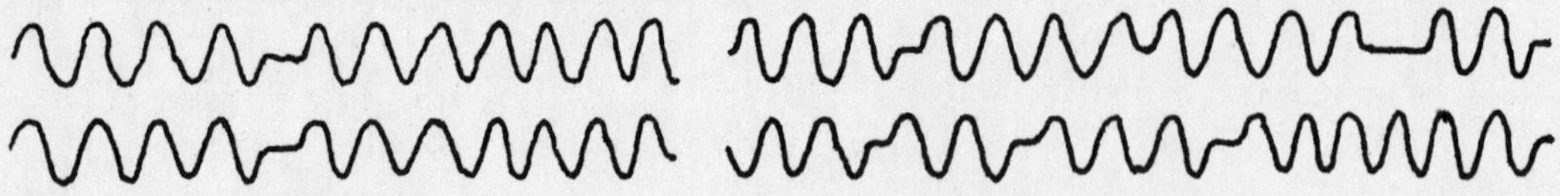

TWO COHERENT WAVES　　TWO INCOHERENT WAVES

Two coherent waves may be obtained by splitting a light beam, either by letting it impinge on two slits or by reflecting part of it out of the beam with a partially silvered mirror.

2. Two closely spaced slits of separation d illuminated by a monochromatic source produce an interference pattern on a screen a distance D>>d away. The interference maxima occur at

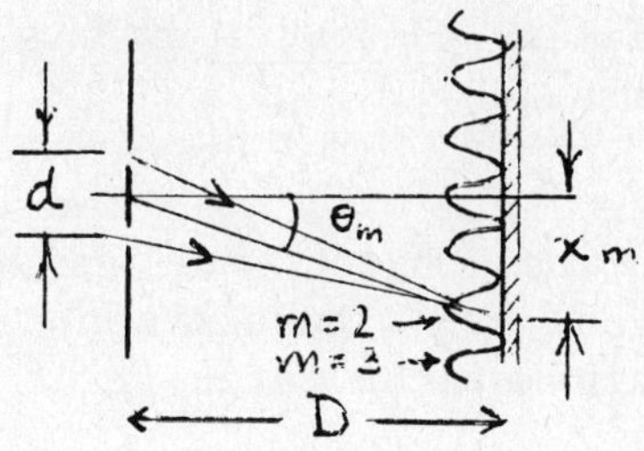

$$\sin\theta_m = m\frac{\lambda}{d} \qquad m = 0, 1, 2, 3 \ldots$$

or

$$\boxed{x_m \simeq \frac{m\lambda D}{d}} \qquad (26.1)$$

where x_m = distance to mth maximum from center of screen.
λ = light wavelength.

3. Light reflected from the two surfaces of a thin film can interfere. When light in one medium (e.g., air) is reflected from a medium with a larger index of refraction (e.g., water) it experiences a 180° phase shift. Light reflected from the front and back surfaces of an unsupported film (e.g., a soap film in air) will interfere constructively at normal incidence when

$$\boxed{t = \frac{(m + \frac{1}{2})\lambda}{2n}} \qquad (26.2)$$

where t = film thickness and n = film index of refraction.
λ = light wavelength in air.

A very thin film ($t \ll \lambda$) will appear black.

4. An interferometer is a device which utilizes the interference between two coherent beams of light produced by a half silvered mirror. It can be used to measure wavelengths, refractive indices, and very small distances.

5. Holography is a technique for producing a three dimensional image of an object. A hologram is a photograph of the interference pattern produced by a reference beam and the light scattered by the object. When the hologram is illuminated by a coherent source a three dimensional image is formed.

6. Diffraction is the same as interference, but this term is used when many beams (as opposed to only two) are interfering. A diffraction grating consists of a large number of uniformly spaced slits. It produces a pattern of many sharp intense lines when illuminated with monochromatic light. The maxima occur at angles θ_m, where

$$\boxed{\sin\theta_m = \frac{m\lambda}{d}} \quad m = 0, 1, 2, \ldots \qquad (26.3)$$

Here m is the order of the diffraction maximum and d is the slit separation.

White light will be dispersed into many colored lines by a grating, with the longest wavelengths (e.g., red) being deviated through the largest angle in a given order. Diffraction gratings are used to obtain monochromatic light beams from polychromatic ones.

7. Light passing near any object can be thought to interfere with itself, thereby producing a diffraction (i.e., interference) pattern. Thus objects do not cast sharp shadows. Light passing through a circular aperature and striking a screen will form a number of bright and dark rings, and not just a single sharp bright spot as one might expect for a beam of light particles. The angle at which the minima of the diffraction pattern due to a single slit occurs is θ_{min},

$$\boxed{\sin\theta_{min} = \frac{m\lambda}{d}} \qquad (26.4)$$

Here d = width of the single slit.

For a circular aperature of diameter d the first minimum occurs at

$$\boxed{\theta_m = \frac{1.22\lambda}{d}} \quad \theta \text{ in radians} \qquad (26.5)$$

In order for two point sources of light (like two stars) to be seen clearly when viewed through a circular aperature, the angle θ which they subtend at the aperature must be at least as big as θ_m given by eq. (27.5). Two sources which can be seen distinctly are said to be resolved.

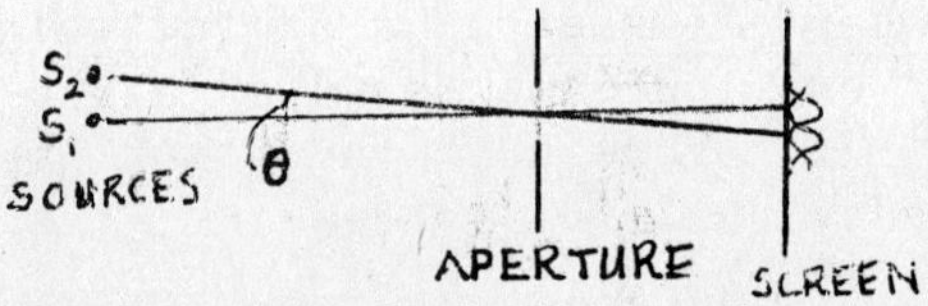

8. The plane formed by the velocity vector of a light wave and its electric field is called the plane of polarization. If a light wave always has its E vector in one plane it is said to be plane polarized. Most light sources (such as the sun or a light bulb) produce unpolarized light consisting of many short trains of light, each with a randomly oriented plane of polarization. Polarized light may be produced from unpolarized light by passing it through special materials (e.g., Polaroid) which transmit only one polarization. Light may also be polarized by reflection or by scattering

from small particles. Light reflected from the surface of a lake is partially polarized with the plane of polarization horizontal. Sunlight scattered by the atmosphere is also partially polarized.

9. Molecules with a spiral structure, including many important organic molecules, are optically active. They cause the plane of polarization to rotate as light passes through the material.

• Qualitative Questions

26M.1 If two light waves are coherent, which of the following is NOT necessary?

A. They must have the same frequency.
B. They must have the same wavelength.
C. They must have the same amplitude.
D. They must have the same velocity.
E. They must have a constant phase difference at every point in space.

26M.2 Why would it be impossible to obtain interference fringes in a double slit experiment if the separation of the slits is less than the wavelength of the light used?

A. The very narrow slits required would generate many different light wavelengths, thereby washing out the interference pattern.
B. The two slits would not emit coherent light.
C. The fringes would be too close together.
D. In no direction could a path difference as large as one wavelength be obtained, and this is needed if a bright fringe, in addition to the central fringe, is to be observed.

26M.3 What do we mean when we say that two light rays striking a screen are in phase with each other?

A. When the electric field due to one is a maximum, the electric field due to the other is also a maximum, and this relation is maintained as time passes.
B. They are travelling at the same speed.
C. They have the same wavelength.
D. They have the same frequency.
E. They alternately reinforce and cancel each other.

26M.4 White light is incident on the "black box" apparatus sketched here. The light striking the screen consists of a number of brightly colored bands. Within the black box there is probably

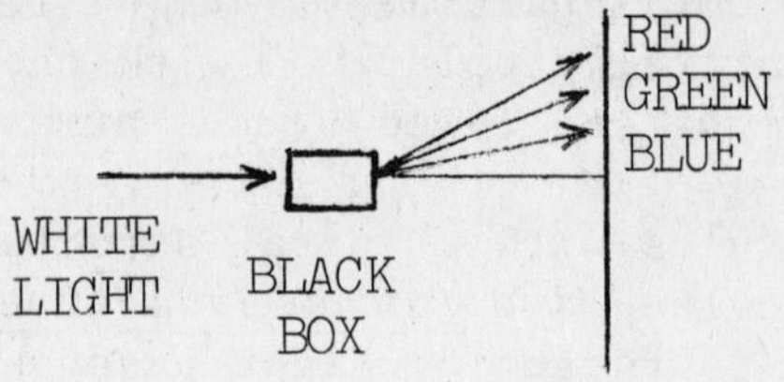

A. a diffraction grating.
B. an interference filter.
C. a polarizer.
D. a chromatic disperser.
E. a lens.
F. a prism.

25M.5 With what color light would you expect to be able to see the greatest detail when using a microscope?

A. Red, because of its long wavelength.
B. Red, because it is refracted less than other colors by glass.
C. Blue, because of its shorter wavelength.
D. Blue, because it is brighter.
E. The color makes no difference in the resolving power, since this is determined only by the diameter of the lenses.

26M.6 An important reason for using a very large diameter objective in an astronomical telescope is

A. to increase the magnification.
B. to increase the resolution.
C. to form a virtual image which is easier to look at.
D. to increase the width of the field of view.
E. to increase the depth of the field of view.

26M.7 A diffraction grating works on the principle that

A. waves polarized at right angles to each other will cancel each other.
B. the index of refraction varies with wavelength.
C. two waves initially in phase will cancel if one travels an extra odd number of half wavelengths before they are recombined.
D. the light passing through the center of a slit can either cancel or reinforce the light passing near the edges of the slit.
E. light passing through a grating forms essentially a shadow of the grating on a screen, with the number of bright lines observed approximately equal to the number of slits in the grating.

26M.8 Sunlight reflected from the surface of a lake

A. is unpolarized.
B. tends to be polarized with its electric field vector parallel to the surface of the lake.
C. tends to be polarized with its electric field vector perpendicular to the surface of the lake.
D. has undergone refraction by the surface of the lake.
E. None of the above is true.

26M.9 After a rain one sometimes sees brightly colored oil slicks on the street. These are due to

A. interference effects.
B. polarization effects.
C. diffraction effects.
D. selective absorption of different wavelengths by oil.
E. None of the above.

• Multiple Choice: Answers and Comments

26M.1 C In order for two waves to completely cancel when interfering destructively they must have the same amplitude, but this is not necessary for coherence.

26M.2 D

26M.3 A

26M.4 A A prism disperses blue more than red.

26M.5 C From eq. (26.5), a shorter wavelength results in less spreading of the image due to diffraction, so more closely spaced objects can be resolved with blue light.

26M.6 B

26M.7 C

26M.8 B

26M.9 A

• Problems

26.1 The threads in a nylon stocking can act as a diffraction grating. If light of wavelength 440 nm from a distant mercury arc is viewed through such a stocking, what is the angle between the central maximum and the first order fringe if there are 50 threads per cm?

26.2 Monochromatic light of 660 nm incident on a diffraction grating produces a series of fringes on a distant screen. Adjacent fringes are separated by 6 mm. When monochromatic light from another source is used the fringes are separated by 5 mm. What is the wavelength of the light from the second source?

26.3 Two flat glass plates 12 cms long are separated from each other at one end by a fiber whose diameter we wish to measure. Illumination from above with light of wavelength 600 nm produces interference bands separated by 0.8 mm. How thick is the fiber?

26.4 The world's largest refracting telescope is operated at the Yerkes Observatory in Wisconsin. It has an objective of diameter 102 cms. Suppose such an instrument could be mounted on a spy satellite at an elevation of 300 km above the surface of the earth. What is the minimum separation of two objects on the ground if their images are to be clearly resolved by this lens? Assume an average wavelength of 550 nm for white light.

• Problem Solutions

26.1 FROM EQN. (26.1) $\sin\theta_m = m\frac{\lambda}{d}$

$d = \frac{1}{50}$ cm $= 0.02$ cm $= 2\times10^{-4}$ m

$\sin\theta_m = (1)\frac{(440\times10^{-9}\text{m})}{(2\times10^{-4}\text{m})} = 2.2\times10^{-3}$, $\theta_m = 0.12°$

26.2 $\sin\theta_m = m\frac{\lambda}{d}$

$\sin\theta_m \simeq \frac{x_m}{D}$, so $x_m \simeq \frac{mD\lambda}{d}$

$\Delta x = x_{m+1} - x_m = \frac{(m+1)D\lambda}{d} - \frac{mD\lambda}{d} = \frac{D\lambda}{d}$

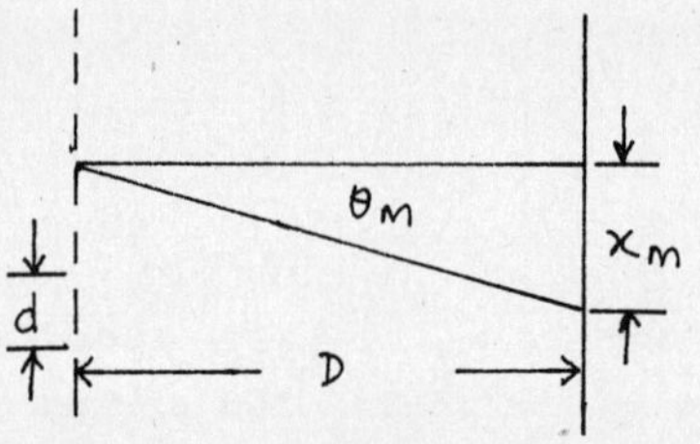

FOR A DIFFERENT WAVELENGTH λ',

$\Delta x' = \frac{D\lambda'}{d}$ THUS $\frac{D}{d} = \frac{\Delta x}{\lambda} = \frac{\Delta x'}{\lambda'}$,

$\lambda' = \left(\frac{\Delta x'}{\Delta x}\right)\lambda$

$= \left(\frac{5\text{mm}}{6\text{mm}}\right)(660\text{ nm})$

$\lambda' = 550$ nm

26.3 THE WIDTH OF THE WEDGE OF AIR MUST CHANGE BY $\frac{1}{2}\lambda = 300$ nm BETWEEN FRINGES (SO PATH LENGTH WILL CHANGE BY λ). THIS CHANGE OCCURS IN A LATERAL DISTANCE OF 0.8 mm.

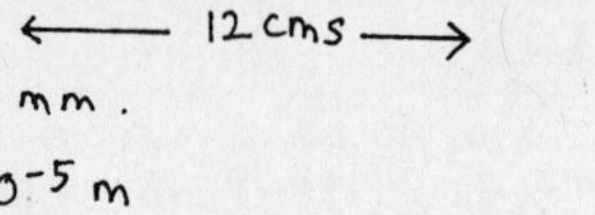

THUS $\tan\theta = \frac{x}{0.12} = \frac{300\times10^{-9}\text{m}}{0.8\times10^{-3}\text{m}}$, $x = 4.5\times10^{-5}$ m

26.4 FROM EQ. 26.5, $\theta_m = \frac{1.22\lambda}{d} = \frac{(1.22)(550 \times 10^{-9} m)}{1.02 m} = 6.6 \times 10^{-7}$ RAD

IF X IS THE SEPARATION OF THE TWO OBJECTS A DISTANCE D AWAY,

$$\theta \simeq \frac{x}{D}, \quad x \simeq \theta D = (6.6 \times 10^{-7})(3 \times 10^{5} m)$$

$$\underline{\underline{x = 0.20 m}}$$

CURRENT U.S. SPY SATELLITES AREN'T QUITE THIS GOOD, BUT THEY ARE WITHIN A FACTOR OF 10 OF THIS. THIS IS AN IMPORTANT FACTOR INFLUENCING ANY DECISION TO ENTER INTO DISARMAMENT TREATIES. ONE HAS TO BE ABLE TO CHECK ON WHAT THE OTHER PEOPLE ARE DOING. WE ARE ABLE TO DO THIS PRETTY EFFECTIVELY.

27 Relativity

• Summary of Important Ideas, Principles, and Equations

1. Before 1900 people believed that the universe was filled with a substance called the ether. They imagined electromagnetic waves to be some kind of oscillation taking place in this ether, and they envisioned light as travelling with a speed of 3×10^8 m/s with respect to this ether. No experimental evidence, however, has ever been obtained to substantiate this idea. In fact, experiments by Michelson and Morley with a sensitive interferometer gave a negative result for the existence of ether.

2. Einstein's Special Theory of Relativity replaced ideas based on the ether. It is based on the following two hypotheses:

   (a) The laws of nature have the same form in all inertial reference frames.

   (b) The speed of light in vacuum is always $c = 3 \times 10^8$ m/s, independent of any relative motion of the light source or the observer.

   The first idea is a reasonable one. Remember that inertial reference frames are all moving with constant velocity with respect to each other, so it may not seem unreasonable that a law valid in one (like $F = ma$) is also valid in another. Adding on a constant velocity will not change the acceleration, for example. The second hypothesis, however, is quite unreasonable from the viewpoint of common sense and everyday experience. Its only justification is that every experiment ever done to check it out has substantiated it. Its consequences are most noticeable for events involving velocities near c, and that is why most of us have never been aware of some of the strange effects which occur. There seems to be no "absolute" fixed reference frame in the universe.

3. One consequence of the special theory is that events which are simultaneous in one inertial reference frame are not simultaneous for an observer in another inertial reference frame (i.e. for a moving observer).

4. Moving clocks run slow. To see what this means, imagine a person in a moving rocket. She measures her pulse rate and finds that 100 beats occur in 90 seconds. If a person on earth could time the same 100 beats by some remote measuring device he might find an elapsed time of 91 seconds. He might then assert that the moving clock in the rocket was running slow. This is not so, and if the experiment were done in reverse the person in the rocket would deduce that the clock on earth was running slow. The rate at which time passes is simply not absolute, but depends on the reference frame in which it is measured. The best we can do is to keep track of the proper time, which is the time you measure on your own wristwatch (i.e. on a clock at rest with respect to the observer). In the preceding example the proper time is that measured by the woman in the rocket, i.e. 90 seconds. If we call this time τ, the time measured by an observer moving with speed u with respect to the woman is τ', where

$$\boxed{\tau' = \frac{\tau}{\sqrt{1 - \frac{u^2}{c^2}}} = \gamma\tau} \qquad (27.1)$$

where

$$\gamma = \frac{1}{\sqrt{1 - \frac{u^2}{c^2}}} > 1$$

This behavior of moving clocks (they run slow) is called time dilation.

5. Moving objects are shortened in their dimensions parallel to their direction of motion. This shortening is called the Lorentz-FitzGerald contraction. A rocket of length L moving with speed u will have a length L' as measured by a stationary observer.

$$\boxed{L' = L\sqrt{1 - \frac{u^2}{c^2}} = \frac{L}{\gamma}} \qquad (27.2)$$

Dimensions perpendicular to the motion are unchanged.

6. Suppose that a particle has velocity (v_x, v_y, v_z) as measured here on earth. A person in a rocket moving in the x direction with velocity u will observe the particle to have velocity (v_x', v_y', v_z'), where

$$v_x' = \frac{v_x - u}{1 - \frac{v_x u}{c^2}} \qquad v_x = \frac{v_x' + u}{1 + \frac{v_x' u}{c^2}}$$

$$v_y' = \frac{v_y}{\gamma\left(1 - \frac{v_x u}{c^2}\right)} \qquad v_y = \frac{v_y'}{\gamma\left(1 + \frac{v_x' u}{c^2}\right)}$$

$$v_z' = \frac{v_z}{\gamma\left(1 - \frac{v_x u}{c^2}\right)} \qquad v_z = \frac{v_z'}{\gamma\left(1 + \frac{v_x' u}{c^2}\right)}$$

The equations on the right are obtained from those on the left by replacing u by -u. Notice that although dimensions along y and z are not affected by motion along x, velocities are because of the time dilation.

7. Mass changes with speed. If an object has mass m_o when it is at rest, then when it is moving with velocity v we observe it to have mass m, where

$$m = \frac{m_o}{\sqrt{1 - \frac{v^2}{c^2}}} = \gamma m_o \qquad (27.3)$$

m_o is the rest mass of the object.

8. Mass and energy are equivalent.

$$E = mc^2 = \gamma m_o c^2 \qquad (27.4)$$

Mass, like light, sound and chemical energy, is just another form of energy. We can think of joules (energy) and kilograms (mass) as just two different units for the same thing, and c^2 is the conversion factor (like going from feet to meters).

$$E_o = m_o c^2 = \text{Rest energy} \qquad (27.5)$$

Kinetic energy $$KE = E - E_o = mc^2 - m_o c^2 = (\gamma - 1) m_o c^2 \qquad (27.6)$$

At low speeds we can approximate KE by KE $\simeq \frac{1}{2}m_o v^2$.

The rest energy of a proton is 931 MeV.

A little bit of mass can be converted into a lot of thermal energy, as demonstrated by the H-bomb.

9. Neither energy or matter can be transported at a velocity greater than 3 x 10^8 m/s, which is a kind of "cosmic speed limit" for our universe.

10. The general theory of relativity describes behavior in non-inertial reference frames. It rests on the principal of equivalence,

> An inertial frame S in a uniform gravitational field is equivalent to a reference frame that has constant acceleration (with respect to S) but no gravitational field.

The predictions of the general theory of relativity include black holes, the bending of light by a gravitational field, the gravitational red shift and the precession of the orbit of Mercury.

• Qualitative Questions

27M.1 According to the special theory of relativity,

A. space and mass are equivalent.
B. space and energy are equivalent.
C. time and energy are equivalent.
D. time and mass are equivalent.
E. energy and mass are equivalent.

27M.2 Suppose that you could travel at speeds approaching the velocity of light. What would you observe about yourself?

A. Your mass has increased.
B. Some of your dimensions have grown smaller.
C. Some of your dimensions have grown larger.
D. Your pulse rate has increased.
E. Your pulse rate has decreased.
F. Nothing has changed.

27M.3 A person standing on the ground observing a rocket ship moving past at a speed near c would find the ship to be

A. the same length as when it is at rest.
B. shorter than when it is at rest.
C. longer than when it is at rest.
D. shorter when it is approaching him and longer when it is going away.
E. longer when it is approaching him and shorter when it is going away.

27M.4 Two useful approximations for $x \ll 1$ are

$$\sqrt{1+x} \simeq 1 + \frac{x}{2}, \quad \sqrt{1-x} \simeq 1 - \frac{x}{2}$$

$$\frac{1}{1+x} \simeq 1 - x, \quad \frac{1}{1-x} \simeq 1 + x$$

Thus for $\frac{v}{c} \ll 1$, $\gamma = \frac{1}{\sqrt{1 - v^2/c^2}} \simeq 1 + \frac{1}{2}\left(\frac{v}{c}\right)^2$

In view of this, which of the following is largest?

A. $\frac{1}{2} m_o v^2$
B. $\frac{1}{2} m v^2$
C. Kinetic energy
D. It is not possible to tell without more information.

27M.5 From the equation $E = mc^2$ we conclude that

A. mass and energy when combined travel at twice the speed of light.
B. mass and energy when combined travel at the speed of light squared.
C. energy is actually mass travelling at the speed of light squared.
D. energy and speed are equivalent forms of the same thing.
E. energy and mass are equivalent forms of the same thing.

27M.6 Suppose that you measure the speed of a light beam from your flashlight and find that it travels 3×10^8 m/s. If you then shine it forward while travelling on a train moving with speed v,

A. you will find that it now travels at a speed slightly greater than 3×10^8 m/s.
B. you will now find that it travels at a speed slightly less than 3×10^8 m/s.
C. a person on the ground will observe that the light now travels at a speed slightly greater than 3×10^8 m/s.
D. a person on the ground will now observe that the light now travels at a speed slightly less than 3×10^8 m/s.
E. None of the above is true.

27M.7 We have seen that the theory of relativity predicts that all of the velocity components of a moving object will differ when measured in two reference frames which are moving with respect to each other, in contrast to the transformation of lengths, where only the length parallel to the direction of motion is altered. Which of the following best explains the origin of this result ?

A. Although the main length contraction occurs in the direction of motion, there is also a smaller, but not negligible contraction, perpendicular to the motion.
B. If something is squashed in one dimension, it must bulge out in another dimension. Hence if x is contracted it follows that y and z must be dilated.
C. It is through time dilation that velocity components can be affected in all directions.
D. This follows immediately from considerations of conservation of energy.
E. This is a consequence of the fact that according to Einstein everything is relative.

27M.8 If you were in a rocket ship speeding along in a straight line at a speed of 0.9c (about 170,000 miles per second) how might you detect this?

A. Your pulse rate would change.
B. You would feel more massive.
C. It would be very hard to breathe.
D. You would experience a great force crushing you.
E. There is no way to tell you are moving without looking out the window at some stellar object.

• Multiple Choice: Answers and Comments

27M.1 E

27M.2 F

27M.3 B

27M.4 B $m = \gamma m_0 > m_0$, SO A IS NOT CORRECT.

TO COMPARE $\frac{1}{2}mv^2$ AND $KE = mc^2 - m_0c^2$ CONSIDER THE CASE $\frac{v}{c} \ll 1$ AND EXPAND γ.

THUS $\frac{1}{2}mv^2 = \frac{1}{2}\gamma m_0 v^2 = \frac{1}{2}\left(1 + \frac{1}{2}\frac{v^2}{c^2}\right)m_0 v^2$

$$= \frac{1}{2}m_0 v^2 + \frac{1}{4}m_0\frac{v^4}{c^2}$$

COMPARE THIS WITH $KE = \gamma m_0 c^2 - m_0 c^2$

$$= \left(1 + \frac{1}{2}\frac{v^2}{c^2}\right)m_0 c^2 - m_0 c^2$$

$$KE = \frac{1}{2}m_0 v^2$$

THUS WE SEE $\frac{1}{2}mv^2 > KE$

27M.5 E

27M.6 E The speed of light is always found to be 3×10^8 m/s in vacuum for all observers, independent of the motion of the source. Strange but true!

27M.7 C

27M.8 E

• Problems

27.1 How fast would a rocket ship have to move so that it had contracted to half of its proper length (as observed by a stationary object?

27.2 By what factor has the mass of an electron increased when it is accelerated to a speed of 0.99 c?

27.3 A radar operator on earth sees two space ships moving straight at each other, each with speed 0.6c. With what speed does the pilot of one ship see the other ship approaching? Classically you might think 1.2 c, but this is not consistent with the theory of relativity (nor with what is actually observed to happen in nature).

27.4 A "tiny" atomic bomb which killed more than a hundred thousand people released energy equivalent to that released when 20,000 tons (20 kilotons) of TNT explosive is detonated. How much mass was converted to energy when this took place? 1 kT = 4.3×10^{12} Joules. Incidentally, modern H-bombs have energy yields in the 10-20,000 kT range (10-20 megatons).

• Problem Solutions

27.1 $L' = \sqrt{1 - \frac{u^2}{c^2}}\, L$, $\left(\frac{L'}{L}\right)^2 = 1 - \frac{u^2}{c^2} = \frac{1}{2^2}$, $u = \frac{\sqrt{3}}{2} c$

$u = 0.87c$

27.2 $m = \gamma m_0 = \frac{m_0}{\sqrt{1 - \frac{v^2}{c^2}}} = \frac{1}{\sqrt{1 - (0.99)^2}} m_0 = 7.1\, m_0$

27.3 LET SHIP #1 BE A MOVING REFERENCE FRAME WITH VELOCITY $u = +0.6c$ WITH RESPECT TO EARTH. VELOCITY OF #2 WITH RESPECT TO EARTH IS THEN $V_x = -0.6c$

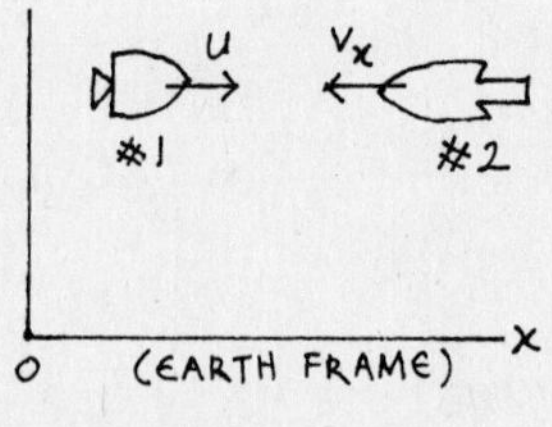

FROM EQN. 27.3

V' = VELOCITY OF #2 AS VIEWED FROM #1

$$= \frac{V_x - u}{1 - \frac{V_x u}{c^2}} = \frac{(-0.6c) - (0.6c)}{1 - \frac{(-0.6c)(0.6c)}{c^2}} = \frac{-1.2c}{1.36} = \underline{\underline{-0.88c}}$$

THUS #1 SEES #2 APPROACHING AT A SPEED LESS THAN c, AS REQUIRED BY RELATIVITY THEORY.

27.4 $E = mc^2$, $m = \frac{E}{c^2} = \frac{(20)(4.3 \times 10^{12} J)}{(3 \times 10^8 m/s)^2} = 9.6 \times 10^{-4}$ kg

$$\underline{\underline{m = 0.96 \text{ gm}}}$$

---

28 Origins of the Quantum Theory

• Summary of Important Ideas, Principles, and Equations

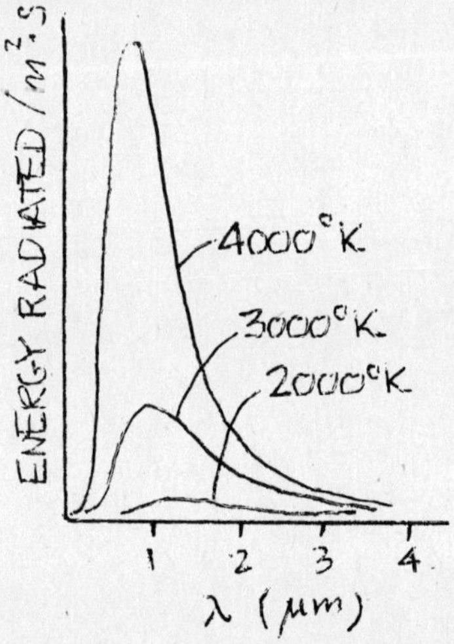

1. Blackbody radiation played an important role in the development of modern physics. An object which absorbs all of the radiation incident on it (and thus reflects none back) is called "black," because that is how it would look if you shined a flashlight beam on it. An object which is a good absorber is also a good radiator of radiation. The vibrating atoms in a hot material will radiate electromagnetic radiation with a wide range of frequencies. The energy radiated as a function of wavelength is sketched here for an ideal blackbody. notice two important features. First, the peak energy shifts to shorter wavelengths as the temperature is increased. If λ_{max} is the wavelength at which the radiated energy peaks, and T is the absolute temperature, then λ_{max} and T are related by Wien's law.

$$\lambda_{max}T = 2.90 \times 10^{-3} \text{ m}\cdot{}^{o}\text{K} \qquad (28.1)$$

A blackbody at 6000°K, about the temperature of the sun's surface, radiates most intensely near 480 nm in the visible region. By comparison, your body at 300°K radiates most in the infrared region of the spectrum, with a peak near a wavelength of 10 μm. The total amount of energy increases rapidly with temperature, and the power radiated from area A by a perfect blackbody is given by the

Stefan-Boltzmann law,

$$\boxed{P = \sigma A T^4} \qquad (28.2)$$

Here T = temperature, °K and $\sigma = 5.67 \times 10^{-8}\ W/m^2 \cdot {}^\circ K^4$

It was not possible to explain the blackbody radiation curves sketched above by using classical theory. In 1900 Planck proposed a description which fit the experimental results, but it required that the oscillating atoms in a hot object each emit radiation only in little chunks of energy (called quanta or photons). He envisioned a radiating solid to consist of millions of radiating oscillators with a distribution of frequencies. Einstein then deduced that each one radiated little particles of light with energy E,

$$\boxed{E = h\nu} \quad h = 6.63 \times 10^{-34}\ Js,\ \nu = \text{frequency, Hz} \qquad (28.3)$$

2. When light quanta strike certain metal surfaces they can knock loose electrons. This photoelectric effect can be explained using the photon picture of light, but not the wave model. Only one photon gives up its energy in knocking loose a particular electron, and if ϕ (called the work function) is the energy to break an electron loose, the KE of the electron will be related to the photon frequency ν by

$$\boxed{KE = h\nu - \phi} \qquad (28.4)$$

Thus if $h\nu < \phi$ no electrons whatever will be emitted, no matter how intense the incident light beam. Energies in such calculations are often measured in electron volts (1 e.V. = 1.6×10^{-19} Joules).

3. Einstein used the idea of quantized energy levels to explain the specific heat of solids at low temperatures. He treated a solid as a collection of oscillators, and he proposed that the energy of each oscillator was $nh\nu$, where n is an integer. The oscillator could change its amplitude of oscillation from one energy level to an adjacent one, say from 13 $h\nu$ to 12 $h\nu$. In so doing it would emit a quanta of energy $h\nu$. Notice that this idea of quantized energy levels is a very fishy sort of thing, quite different from what we are used to in everyday life. For example, we would assume that a mass attached to a spring could have any energy we wished. All we have to do is cause the amplitude of oscillation to have the appropriate value. For all intents and purposes this is true for macroscopic systems, since the differences in the quantized energy levels are so small. On an atomic level, however, the "graininess" of the energy levels shows up and is very important. Each oscillator has a characteristic frequency, ν. Physically a given oscillator is simply a mode of vibration in which all of the atoms vibrate in a particular pattern.

4. If an electron is accelerated through a potential difference V (in volts) it acquires $KE = \frac{1}{2}mv^2 = eV$. When it hits a solid it can give up this energy by emitting a quantum of electromagnetic energy, i.e., a photon. If the voltage V is several kilovolts the photon is an X-ray of frequency ν and wavelength λ.

$$h\nu_{max} = eV = \frac{hc}{\lambda_{min}} \quad \text{or} \quad \lambda_{min} = \frac{hc}{eV} \tag{28.5}$$

(using $\lambda\nu = c$)

5. A photon is a kind of particle in all respects, although a funny one. It has zero rest mass and it always travels at the speed c in vacuum. It has energy and momentum and can smash into things and knock them apart. If it hits an electron it loses energy to the electron and then becomes itself a photon of lower frequency and longer wavelength. In such Compton scattering the shift in wavelength is $\Delta\lambda$ for a scattering angle θ,

$$\Delta\lambda = \lambda' - \lambda = \left(\frac{h}{m_o c}\right)(1 - \cos\theta) \tag{28.6}$$

6. Just as light has both a wave nature and a particle nature, so matter particles have also a wave nature. The de Broglie wavelength of a particle of momentum $p = mv$ is

$$\lambda = \frac{h}{p} \tag{28.7}$$

This wavelength is extremely small for macroscopic objects, and so we don't notice the wavelike properties of baseballs. Not so for atomic particles.

7. The wave nature of particles means they are rather elusive will-o'-the-wisp kind of things. It is hard to pin them down, and in fact in a sense they do not have a simultaneous position and momentum. The best we can do is to say that if a particle is located in a small region Δ_x along the x-axis, it has momentum which is certain to within an amount Δp_x, where

$$\Delta x \Delta p_x \geq \hbar \tag{28.8}$$

$$\hbar = \frac{h}{2\pi} = 1.05 \times 10^{-34} \text{ J}\cdot\text{s}$$

This is the Heisenberg uncertainty principle.

In similar fashion we can never be certain of the energy of a particle. If we make a measurement of the energy of an electron, say, in a time Δt, then we can not have an accuracy greater than E, where

$$\boxed{\Delta E \Delta t \geq \hbar} \qquad (28.9)$$

HEISENBERG MAY HAVE BEEN HERE.

8. If these quantum mechanical ideas seem strange, what with waves which are particles and particles which are waves and energy levels which can only have certain values, don't feel bad. To this day no one truly understands what is going on, which is one reason the books sound kind of vague. Quantum mechanics is a complex set of rules which works beautifully in describing lots of interesting effects. But as for the underlying rationale, that is still to be discovered. As you edge closer and closer to the fresh edge of human knowledge things start to become ambiguous and fuzzy. There are still lots of unanswered questions. From studying physics in school you may get the mistaken idea that everything is all figured out. Not so. That is what makes it fun.

• Qualitative Questions

28M.1   If two adjacent levels of an oscillator have energies of 2.4 e.V. and 3.6 e.V., then when it makes a transition from one level to the other it can emit a quanta of energy

    A. 1.2 e.V.   C. 3.6 e.V.
    B. 2.4 e.V.   D. 6.0 e.V.

28M.2   The particle like nature of light is most clearly revealed by

    A. the interference pattern in a double slit experiment.
    B. the Doppler effect.
    C. the phenomenon of diffraction.
    D. the photoelectric effect.
    E. the polarization of reflected light.

28M.3   The radiation from a blackbody at a temperature T ($^\circ$K) is peaked at a wavelength of 900 nm. If the object is heated to 3T, the peak wavelength will be

    A. unchanged, since it is characteristic of the particular object.

    B. 11 nm.   D. 300 nm.
    C. 2700 nm.   E. 72.9 nm.

$\lambda_{max} T = 2.9 \times 10^{-3}$.

28M.4 Suppose that a light beam of wavelength 400 nm and power level P causes N photoelectrons per second to be emitted from a particular metal surface.

A. If the power level is doubled this will make no change in the number of electrons emitted per second.
B. If the power level is doubled and the wavelength is changed to 800 nm this will make no change in the rate at which electrons are emitted.
C. If the power level is unchanged and the wavelength is changed to 200 nm, the rate of electron emission will not change.
D. If the power level is doubled and the wavelength is changed to 200 nm, the rate of electron emission will not change.

28M.5 An important early application of the idea of quantized oscillator energy levels was Einstein's explanation of

A. diffraction by a narrow slit.
B. the phenomenon of beats between two light beams.
C. thermal expansion in solids.
D. the specific heat of solids at low temperatures.
E. the generation of radio waves by an antenna.

28M.6 When a photon is scattered from an electron, there will be an increase in its

A. energy. B. frequency. C. wavelength. D. momentum.

28M.7 When an electron is accelerated to a higher speed, there is a decrease in its

A. energy. B. frequency. C. wavelength. D. momentum.

28M.8 The reason the wavelike nature of a moving baseball is not noticed in everyday life is that

A. it doesn't have a wavelike nature.
B. its wavelength is too small.
C. its frequency is too small.
D. its energy is too small.
E. no one pays attention to such things except for the Mets, and they can't hit a curve ball anyway.

28M.9 In view of the wave nature of electrons, you might expect that instruments such as a microscope could be built with them. This has indeed been done with great success. One advantage you might expect to gain by using electrons instead of visible light is that

A. the shorter wavelengths possible will result in greater resolution.
B. the more energetic particles will be able to see deeper into samples.
C. electrons will not be scattered by the target, whereas light is.
D. chromatic aberration will not be a problem.
E. electrons are more readily obtained than are photons.

• Multiple Choice: Answers and Comments

28M.1 A The energy of the quanta emitted is equal to the difference in energy of the two levels.

28M.2 D

28M.3 D

28M.4 D A is false. Doubling the power level will double the electron rate.
B is false. If 800 nm is beyond cut-off (i.e., a wavelength too long) the rate will drop to zero. Otherwise the rate will increase when the power level is doubled.
C is false. If the wavelength is halved each photon has twice as much energy, and so for constant power the incident flux of photons is halved, resulting in a halving of the electron rate.
D. is true, since doubleing the power and halving the wavelength results in the same incident flux of photons. The added light energy will result in emitting the electrons with more KE now.

28M.5 D

28M.6 C The electron loses energy so wavelength increases.

28M.7 C Increased speed means increased momentum and hence decreased wavelength, $\lambda = h/p$.

28M.8 B

28M.9 A

$P = \sigma A T^4$

• Problems

28.1 A 560 nm laser beam with an intensity of 1000 W per m^2 is incident on an area of 1 mm^2. How many photons strike the surface each second?

28.2 What is the energy in Joules and in electron volts of a photon of wavelength 550 nm?

28.3 When light of wavelength 555 nm is shined on a metal surface it is found that electrons are emitted with KE of 0.25 e.V. (a) What is the work function of the metal? (b) What would the KE be if a wavelength of 400 nm were used? (c) What is the longest wavelength which will emit photoelectrons from this metal? Express energies in e.V.

28.4 In a hospital X-ray machine electrons are accelerated through 20,000 volts. What is the shortest wavelength X-ray produced?

28.5 Atoms in crystals are typically separated by distances of 0.1 nm. What KE must an electron have, in e.V., in order to have a wavelength of 0.1 nm?

28.6 Suppose that the speed of an electron travelling 2,000 m/s is known to an accuracy of 1 part in 10^5 (i.e., within 0.001%). What is the greatest possible accuracy within which we can determine the position of this electron?

• Problem Solutions

28.1 LET N = NUMBER OF PHOTONS STRIKING 1 mm^2 PER SECOND. EACH HAS ENERGY $E = h\nu = \frac{hc}{\lambda}$

$$\left(\frac{P}{A}\right) = \frac{Nhc}{A\lambda}, \quad N = A\left(\frac{P}{A}\right)\left(\frac{\lambda}{hc}\right) = (10^{-3}\,m)^2\left(1000\,\frac{W}{m^2}\right)\frac{560\times10^{-9}\,m}{(6.63\times10^{-34}\,J.s)(3\times10^8\,\frac{m}{s})}$$

$N = 2.8 \times 10^{15}$ PHOTONS/SEC.

28.2 $$E = h\nu = \frac{hc}{\lambda} = \frac{(6.63\times10^{-34}\,J.s)(3\times10^8\,m/s)}{(550\times10^{-9}\,m)} = 3.62\times10^{-19}\,J$$

$$E = (3.62\times10^{-19}\,J)\left(\frac{1}{1.6\times10^{-19}}\,\frac{eV}{J}\right) = 2.26\,eV$$

28.3 (a) $KE = h\nu_1 - \phi$

$$h\nu_1 = \frac{hc}{\lambda_1} = \frac{(6.63\times10^{-34}\,J.s)(3\times10^8\,m/s)}{(555\times10^{-9}\,m)} = 3.58\times10^{-19}\,J$$

$$h\nu_1 = \frac{3.58\times10^{-19}}{1.6\times10^{-19}}\,eV = 2.24\,eV$$

$$\phi = h\nu_1 - KE = 2.24\,eV - 0.25\,eV$$

$$\phi = 1.99\,eV$$

(b) FOR $\lambda_2 = 400$ nm, $h\nu_2 = \frac{hc}{\lambda_2} = \frac{(6.63\times10^{-34}\,J.s)(3\times10^8\,m/s)}{(400\times10^{-9}\,m)(1.6\times10^{-19}\,\frac{J}{eV})}$

$h\nu_2 = 3.11\ eV$

$KE = h\nu_2 - \phi = 3.11\ eV - 1.99\ eV = \underline{\underline{1.12\ eV}}$

(c) $h\nu_3 = \frac{hc}{\lambda_3} = KE + \phi$

FOR LONGEST WAVELENGTH $KE = 0$, SO $\frac{hc}{\lambda_3} = \phi$

$$\lambda_3 = \frac{hc}{\phi} = \frac{(6.63\times10^{-34} J.s)(3\times10^8 m/s)}{(1.99\ eV)(1.6\times10^{-19} \frac{J}{eV})} = 6.25\times10^{-7}\ m$$

OR $\underline{\underline{\lambda_{max} = 625\ nm}}$

28.4 $h\nu = \frac{hc}{\lambda} = eV$, $\lambda = \frac{hc}{eV} = \frac{(6.63\times10^{-34} J.s)(3\times10^8 m/s)}{(1.6\times10^{-19} C)(20{,}000\ \text{VOLTS})}$

$\underline{\underline{\lambda = 6.2\times10^{-11}\ m = 0.062\ nm}}$

28.5 $\lambda = \frac{h}{mv}$, $KE = \frac{1}{2}mv^2 = \frac{1}{2m}(mv)^2 = \frac{1}{2m}\left(\frac{h}{\lambda}\right)^2$

$$KE = \frac{1}{(2)(9.11\times10^{-31} kg)}\left(\frac{6.63\times10^{-34} J.s}{0.1\times10^{-9}\ m}\right)^2 = 2.41\times10^{-17}\ J$$

$KE = \frac{2.41\times10^{-17}}{1.6\times10^{-19}}\ eV = \underline{\underline{151\ eV}}$

28.6 $\Delta x\, \Delta p_x \simeq \hbar$

$\Delta p_x = m\Delta v_x$, $\Delta v_x = 10^{-5}\ v_x$

$$\Delta x \simeq \frac{\hbar}{m\Delta v_x} \simeq \frac{1.05\times10^{-34} J.s}{(9.11\times10^{-31} kg)(10^{-5})(2000\ m/s)}$$

$\Delta x \simeq 5.8\times10^{-3}\ m = \underline{\underline{5.8\ mm}}$

29 Atomic Structure and Atomic Spectra

• Summary of Important Ideas, Principles, and Equations

1. About 1911 Rutherford proposed the nuclear model of the atom. He envisioned a heavy positive nucleus, comprising most of the mass of the atom in a very small volume, about which circulated a swarm of negative electrons. Most of the atom is empty space. His deductions were based on experiments in which alpha particles (which are helium nuclei) are scattered from a gold foil. Today we still hold to this picture of an atom, but in Rutherford's time it presented serious problems. An electron moving in a curved orbit is experiencing centripetal acceleration, and according to well established electromagnetic theory an accelerated charge should radiate away energy. The electron should then spiral in to the nucleus, collapsing the atom in less than a microsecond. A continuous range of frequencies would be radiated in this process. Atoms do not, however, collapse like this, and they typically emit a series of sharp frequencies when they do radiate. It was the brilliant and radical ideas of Niels Bohr which gave a basis for the nuclear model and made a giant leap forward into the world of quantum mechanics.

2. Bohr based his atomic model on the following postulates:

   A. Electrons in an atom can exist only in special stationary orbits. They do not radiate while in one of these orbits.

   B. Newton's laws apply to the electrons and the nucleus.

   C. An electron can jump from one orbit to another, emitting or absorbing a photon of energy $h\nu = E$, where E is the difference in the electron's energy in the two orbits.

Bohr said that the stationary orbits of the electrons were to be determined by the condition

$L = n\hbar$, where L = angular momentum of electron and $\hbar = \frac{h}{2\pi}$

For circular orbits this condition is equivalent to requiring the orbit circumference to be an integral multiple of the de Broglie wavelength of the electron.

Thus only certain "quantized" orbit radii and velocities are allowed. Using $L = mrv$ and centripetal force $- mv^2/r = kZe^2/r^2$ we find for a hydrogen atom (an electron circulating around a single proton)

$$\boxed{r_n = \frac{n^2 \hbar^2}{ke^2 m} = n^2 a_o} \qquad (29.1)$$

$a_o = 5.29 \times 10^{-11}$ m = Bohr radius k = Coulomb's law constant

The resulting quantized anergy levels are obtained from $E = KE + PE$.

$$\boxed{E_n = -\frac{mk^2e^4}{2\hbar^2 n^2} = -\frac{E_o}{n^2}}, \quad E_o = 13.6 \text{ e.V.}, \quad n = 1, 2, 3, \ldots \qquad (29.2)$$

When an electron jumps to a lower energy orbit (nearer the nucleus) it emits a photon of freqeuncy ν. When it jumps to a higher energy level it absorbs a photon. The difference in energy ΔE between the two levels between which the transition is made is related to the photon frequency by

$$\boxed{\Delta E = h\nu} \qquad (29.3)$$

Using eq. (29.2) this can be written

$$\boxed{\frac{1}{\lambda} = \frac{\nu}{C} = R\left[\frac{1}{n_f^2} - \frac{1}{n_i^2}\right]} \qquad (29.4)$$

$$R = \frac{E_o}{hc} = 1.1 \times 10^7 \text{ m}^{-1}$$

n_f = final state, n_i = initial state

3. A more careful investigation of the possible orbits of an electron in a hydrogen atom shows that elliptical orbits are also possible. Thus in addition to n, the principal quantum number (which determines the energy) another quantum number ℓ is introduced to describe the orbital angular momentum, taking account of the fact that the orbits need not be circular. Thus the orbital angular momentum L is related to the quantum number ℓ by

$$\boxed{L = \sqrt{\ell(\ell + 1)}\,\hbar}, \quad \ell = \text{integer} \tag{29.5}$$

Classically ℓ would be a measure of how elliptical the orbit is, whereas n measures how near the orbit passes to the nucleus (the proton here).

An orbit can be tipped in different orientations, but it turns out that the possible orientations are also "quantized." The component of the angular momentum along an arbitrary z-direction can only have specific discrete values given by

$$\boxed{L_z = m\hbar}, \quad m = \text{integer} \tag{29.6}$$

Finally, an electron behaves in some respects as if it is spinning about its own axis with a spin angular momentum $S = \frac{1}{2}h$. The orientation of this spin angular momentum is also quantized, and

$$\boxed{S_z = m_s\hbar} \text{ where } m_s = \pm\tfrac{1}{2} \tag{29.7}$$

In summary, the "state" of an electron in a hydrogen atom is specified by four quantum numbers: n, ℓ, m, m_s.

Think of n as related to the size of the orbit and hence to the energy. ℓ is a measure of the orbital angular momentum. m is kind of a measure of how the orbit is oriented in space. m_s is a measure of how the spin of the electron is oriented with respect to the orbit.

It turns out that these numbers are limited to the following values:

n = integer = 1, 2, 3, ...

ℓ = integer but never greater than n, so $\ell = 0, 1, 2, \ldots, n-1$

m = integer ranging from $-\ell$ to $+\ell$, $m = -\ell, -\ell + 1, \ldots, 0, 1, \ldots \ell$

$m_s = +\frac{1}{2}$ or $-\frac{1}{2}$

The picture I have described here is not quite accurate, because electrons are not in fact like little BB's speeding around in orbits like the earth around the sun. The electrons are more like a kind of pulsating cloud or wave, which in some ways acts like a particle. Still, the Bohr model gives one something which can be visualized, and it fits fairly well with what seems to happen in nature.

4. It is a fact of nature that certain kinds of particles (such as electrons) obey the Pauli exclusion principle, which states that two electrons cannot occupy the same state in a given atom. This principle helps us to understand how the periodic table of the elements is built up.

All atoms are electrically neutral. The element with atomic number Z has a nucleus containing Z protons (with total charge +Ze) around which swarm Z electrons (with total charge -Ze). One can imagine building up the atoms by starting with one proton and one electron (hydrogen). The electron will occupy the state of lowest energy, i.e., one with n = 1, quite near the nucleus. The next element, helium, has two electrons and two protons. The second electron will occupy the lowest energy state available to it consistent with the Pauli exclusion principle (which happens to be n = 1 also). The process continues, and as each state or collection of states (called a shell) fills up, succeeding electrons will go into higher and higher energy states. The number of electrons in the outermost orbits are mainly responsible for the chemical properties of an element, and elements with the same number of electrons outside a "closed shell" (e.g., hydrogen, lithium, sodium, potassium, ...) constitute a chemical family.

5. Bombardment of an atom with an energetic electron may dislodge an electron from a low lying energy state in a heavy atom. If an atomic electron from a higher energy state (n = 3, 4, ...) then drops into the low-lying (n = 1, 2) state, energy is conserved by the emission of a photon. Typically the photon energy is in the kilovolt range and the wavelength is in the X-ray region. These are called characteristic X-ray lines for a particular atom.

6. In a collection of atoms or molecules most electrons are in the lowest energy states available to them. However, at any finite temperature there will be a few atoms with electrons in higher energy states. There is a distribution of atoms with electrons in various states, with the number in a given state decreasing as the energy of the state gets larger and larger. This is the normal thermal equilibrium distribution. One can invert this distribution (i.e., have more atoms with electrons in high energy states) by "pumping" energy into the system with light irradiation, through collisions or by electrical discharges. An inverted population

distribution is unstable, but if it is moderately long lived it may be possible to stimulate electrons with a light wave and induce them to fall back to lower energy levels, emitting photons in the process. Interestingly, such stimulated emission of radiation is all coherent (the electrons all fall back down in synchronization, so to speak). This stimulated emission can in turn stimulate more coherent emission, and one has a kind of avalanche effect. A device designed to utilize this light amplification by stimulated emission of radiation (hence the acronym LASER) can produce a light beam which has a very pure frequency and is very coherent. This allows one to focus it to a very small point and obtain huge light intensities.

• Qualitative Questions

29M.1 One of Niels Bohr's brilliant insights, which led to our modern understanding of atoms, was that

A. the charge on the electron must be quantized.
B. the angular momentum of an electron in an atom had to be an integral multiple of Planck's constant divided by 2π.
C. the energy of a photon was Planck's constant times the frequency.
D. the energy of an electron in an atom could vary continuously from zero to some maximum cut-off value.
E. the energy of an electron in a hydrogen atom was an integral multiple of the rest mass of the electron.

29M.2 A key observation which led to Bohr's model of the atom was the fact that

A. the spectra of all atoms are essentially identical.
B. the spectra of isolated atoms consist of discrete frequencies, as opposed to a continuous distribution.
C. alpha particles were found to be transmitted through thin gold foils with no scattering whatever.
D. almost all of the mass of an atom is localized in the nucleus.
E. electrons in an atom are moving at velocities near that of light, and consequently relativistic effects are important.

29M.3 Which of the following was an important factor which led Bohr to postulate his model of the atom?

A. A given atom seems able to emit light of any frequency.
B. Classical theory would predict that electrons in atoms would radiate their energy and the atom would collapse.
C. Newton's laws were found to be inapplicable to atomic particles.
D. The neutron was found to be unstable.
E. Light exhibits wave properties when interacting with an atom.

29M.4 The light absorption spectra of pure water and of water containing bacteriochlorophyll (BChl) are shown here. From this data we would infer that the light sensitive process in BChl involves an electronic transition from the ground state to an excited level whose energy above the ground state is about

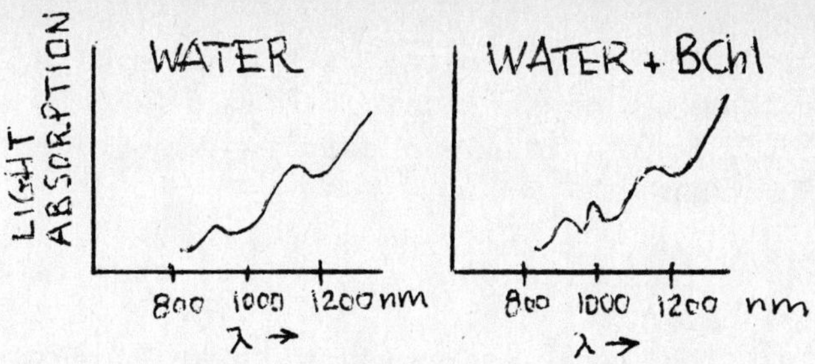

A. 0.8 e.V. C. 1.2 e.V. E. 2.0 e.V.
B. 1.0 e.V. D. 1.6 e.V. F. 2.3 e.V.

29M.5 An atom emits a photon when one of its electrons

A. collides with another one of its electrons.
B. undergoes a transition to a state of lower energy.
C. undergoes a transition to a state of higher energy.
D. loses its charge.
E. None of the above is true.

29M.6 A hydrogen atom is in its ground state when its orbital electron

A. is within the nucleus.
B. has escaped from the atom.
C. is in its lowest energy level.
D. is stationary.
E. is ionized.

29M.7 If an atom had only three distinct energy levels between which electrons could make transitions, how many spectral lines of different wavelengths could the atom emit?

A. 1 B. 2 C. 3 D. 4 E. 5 F. 6

29M.8 Before an atom in its normal state can emit radiation it must

A. be ionized.
B. be irradiated with X-rays.
C. increase its atomic number.
D. have energy supplied to it.
E. None of the above is true.

29M.9 Atoms are

A. positively charged.
B. negatively charged.
C. electrically neutral.
D. either positively charged, negatively charged or neutral, depending on the particular atom.

29M.10 Although de Broglie had not yet made his discovery of the wavelength associated with electrons at the time Bohr advanced his model of the atom, one can see how such an idea fits in with Bohr's ideas concerning electron orbits. The basic idea is that

A. since electrons are waves, they can be anywhere in the atom.
B. the wave associated with an electron must interfere constructively with itself, which suggests an orbit circumference should be a multiple of the electron wavelength.
C. the diameter of an orbit will be equal to the electron wavelength.
D. since electrons are waves, their frequency of revolution will be given by v/λ, where v is the electron velocity in an orbit.
E. electrons would be expected to emit light of well-defined sharp wavelengths, since they themselves have definite wavelengths.

29M.11 In understanding how the periodic table is built up, we recognize that

A. in the neutral atom all electrons have the same energy.
B. in an atom no two electrons have the same energy.
C. lowest energy states are occupied first.
D. all electrons in a given atom have the same values for their quantum numbers.
E. all of the energy levels in an atom are characterized by positive energy.

29M.12 The chemical properties of an atom are primarily determined by

A. the total number of electrons in the atom.
B. the number of electrons in closed shells.
C. the number of electrons outside closed shells.
D. the number of nucleons in the nucleus.
E. the size of the atom.

29M.13 Quantum numbers are

A. used to indicate the number of photons in a beam.
B. used to label the various states in an atom.
C. used in place of the continuous spatial variable "x" when describing atomic phenomena.
D. the names given quantities such as Planck's constant, the de Broglie wavelength and the Heisenberg uncertainty parameter.
E. described by none of the above.

29M.14 The most significant feature which the atoms neon and xenon have in common is that

A. they have the same number of 1s electrons.
B. they are both gases.
C. they both have a strong electron affinity.
D. they are both very strongly electropositive.
E. they both have all of their electrons in closed shells.
F. they both have a single electron missing from their outer shell.

29M.15 An important principle which guides us in building up the periodic table is our belief that

A. electrons can never have a positive net energy.
B. electrons in an atom can have any energy, provided only that the total energy is negative.
C. in an atom no two electrons with parallel spins can occupy the same orbital.
D. no two atoms can have electrons in the same type of orbitals.
E. the uncertainty principle prohibits us from assigning an energy to an electron in an atom.
F. all electrons in a given atom occupy states of the same energy.

• Multiple Choice: Answers and Comments

29M.1 B

29M.2 B

29M.3 B A is definitely not true, as is C. D is true but irrelevant here. The neutron had not yet been discovered when Bohr did his work. Light exhibits particle nature, not wave nature, when interacting.

29M.4 C We see that a peak has appeared about 1000 nm, so a transition has been made with energy E,

$$\Delta E = h\nu = \frac{hc}{\lambda} = \frac{(6.63 \times 10^{-34} \text{J}\cdot\text{s})(3 \times 10^{8} \text{m/s})}{(1000 \times 10^{-9} \text{m})}$$

$$= 1.99 \times 10^{-19} \text{J}$$

$$\Delta E = \frac{1.99 \times 10^{-19}}{1.6 \times 10^{-19}} \text{ e.V.} = 1.2 \text{ e.V.}$$

29M.5 B

29M.6 C

29M.7 C

29M.8 D To emit a photon an electron must fall down into an empty state, but if the atom is in the ground state the electron is already in the lowest available state.

29M.9 C 29M.10 B 29M.11 C 29M.12 C

29M.13 B 29M.14 E 29M.15 C

• Problems

29.1 What is the wavelength of the light emitted when a hydrogen atom makes a transition from a state with n = 3 to one with n = 1?

29.2 About how much enegy, in e.V. is needed to remove an innermost electron from lead (Pb)?

29.3 What element has as its ground state configuration $1s^2 2s^2 2p^2 2s^2$?

29.4 Write the ground state electron configuration of calcium.

• Problem Solutions

29.1 FROM EQN. (30.4), $\frac{1}{\lambda} = R\left(\frac{1}{n_f^2} - \frac{1}{n_i^2}\right) = (1.1 \times 10^7\ m^{-1})\left(\frac{1}{1} - \frac{1}{9}\right)$

$\lambda = 102\ nm$

29.2 THERE ARE TWO 1s ELECTRONS IN THE INNERMOST SHELL. IF WE ASSUME THAT THE CHARGE ON ONE ELECTRON SCREENS OFF A CORRESPONDING CHARGE ON THE NUCLEUS, THE SECOND ELECTRON SEES AN EFFECTIVE NUCLEAR CHARGE OF $(Z-1)e$. FROM EQ. (30.2) WE SEE THAT THE ENERGY OF THE LOWEST STATE IN HYDROGEN (WITH Z = 1) IS −13.6 eV. IF THE NUCLEAR CHARGE IN HYDROGEN HAD BEEN Ze THE IONISATION ENERGY OF HYDROGEN WOULD HAVE BEEN BIGGER BY A FACTOR OF Z^2 (THE Z COMES ALONG WITH THE FACTOR K IF YOU GO THROUGH THE DERIVATION). FOR LEAD, Z = 82, SO Z − 1 = 81 AND THE ENERGY TO REMOVE AN INNER ELECTRON WOULD BE $(81)^2(13.6\ eV) = 89.2\ keV$.

29.3 WE SEE THAT THIS ELEMENT HAS 2 + 2 + 6 + 2 = 12 ELECTRONS, SO Z = 12. FROM THE PERIODIC TABLE WE SEE THAT THIS ELEMENT IS Mg, MAGNESIUM.

29.4 $1s^2 2s^2 2p^6 3s^2 3p^6 4s^2$ IS CALCIUM, ELEMENT 20.

Aggregates of Atoms; Molecules and Solids

30

• Summary of Important Ideas, Principles, and Equations

1. Atoms usually stick together to form aggregates called molecules. Molecules can be classified according to the kind of bond which holds them together. In all cases the energy of the system is lowered when a bond is formed.

   When extra electrons are added to an atom, or when some electrons are removed, the atom then carries a net electric charge and is called an ion. In an ionic molecule (such as sodium chloride, NaCl) electrons leave one constituent, leaving it positively charged (Na^+) and attach themselves to another (Cl^-). The resultant ions are attracted together by the Coulomb force. A typical bond energy is 5 e.V. per molecule.

   In a covalent molecule some outer electrons (valence electrons) are shared by two atoms, thereby giving both atoms effectively filled electron shells, as in Cl_2. A typical bond energy is a few e.V., slightly less than for the ionic bond.

   Hydrogen bonded molecules result when a proton (a positively charged hydrogen atom) attracts two negatively charged ions and bonds them. Such weak bonds are often in organic molecules.

   Van der Waals molecules are very weakly bonded molecules in which random fluctuations in charge on one molecule induces an electric dipole on another in a mutual way, and the resulting two dipoles are weakly attracted. Because this bond is very weak, normal thermal agitation can break it. It is important for the low temperature condensation of some molecules such as H_2, N_2 and He.

2. Molecules have a more complex spectra than do atoms because there are more possible ways they can vibrate. First imagine the energy levels of an atom associated with different electron energy levels.

These might have separations of the order of one electron volt. If the atoms are part of a molecule these electron energies are split by amounts of the order of 0.1 - 0.01 e.V. due to vibrations of the atoms against each other. Thus each electron level is split into many finely spaced sublevels. Furthermore the molecule can rotate and the rotational modes further split the levels by amounts of the order of 10^{-4} to 10^{-5} e.V. This great multitude of energy levels gives rise to very many possible transition, with frequencies ranging from microwaves through visible light. Analysis of such spectra has been vital to understanding and identifying molecules.

3. The bonds described above can cause atoms to condense into solids. One additional bond, the metallic bond, appears only in condensed matter and is associated with the ability of electrons to move freely in conductors. The metallic bond is weaker than the ionic or covalent bonds.

The electron energy levels in solids broaden into energy bands (as opposed to the sharp discrete levels in atoms and molecules). A filled energy level in an atom corresponds to a filled energy band in a solid. A material in which all of the electrons are in filled bands is an insulator if there are no empty bands nearby in energy. Such a material will not conduct an electric current because in order for an electric field to accelerate an electron there must be available a nearby empty energy level.

If the energy bands are all filled, but an empty band (called the conduction band) is not too far removed in energy (perhaps 1.5 e.V. or less), some electrons will have enough thermal energy to jump from the filled band (the valence band) up into the empty band, where they can then conduct a current. Such materials are called intrinsic semiconductors. As they are heated more and more electrons are able to get into the conduction band, and so the conductivity increases as the temperature is increased. This is because the increase in the number of conduction electrons outweighs the added scattering of the electrons due to vibrations of the lattice (which raises resistivity and hence decreases conductivity).

It is possible to dope a semiconductor with appropriate impurities which contribute added electrons to the conduction band. Such materials are extrinsic n-type (for negative charge carriers) semiconductors.

When an electron jumps up to the conduction band from the completely filled valence band it leaves behind a single empty electron state near the top of the valence band. This empty state can then be filled by other electrons, and the resultant behavior is very much as if a positive electron (called a "hole") were present. The holes can also conduct an electric current, and in intrinsic semiconductors there are equal numbers of holes (near the top of the valence band) and electrons in the bottom of the conduction band.

A semiconductor can be doped with an impurity which can capture electrons from out of the top of the valence band, leaving behind holes which can conduct a current. Such doped semiconductors are called p-type (for positive carriers).

In a metal one of the bands is only partially filled, and the electrons can thus accelerate and gain energy without problem. Such materials are metals. For example, in sodium the outermost orbit, the 3s level, contains only one electron. Correspondingly, in sodium metal the highest energy band is only half full, and metallic behavior results.

In summary, if all energy bands are full and the nearest empty band (the conduction band) is 3 e.V. or more above the highest filled band (the valence band) the material is an insulator. If this energy gap is less than about 3 e.V. the material is a semiconductor. The material may also be a doped semiconductor if addition of impurities results in added energy levels near the edge of one of the bands. Finally, if a band is only partially full the material will be a metal.

• Qualitative Questions

30M.1 If two electrons are removed from a manganese atom, the resulting atom will be

A. cobalt.
B. vanadium.
C. Mn^{2-}
D. Mn^{2+}
E. $2Mn^{+}$

30M.2 A proton is the same as

A. an alpha particle.
B. a hydrogen atom.
C. H^{-}
D. H^{+}
E. an ionized electron.

30M.3 If the atoms in a solid were all alike, which of the following would not be a possible bond holding them together?

A. The metallic bond.
B. The ionic bond.
C. The covalent bond.
D. The Van der Waals bond.
E. The indentured bond.

30M.4 Potassium has a single 4s electron in its outermost shell in the atomic state. Thus you would expect solid potassium to be

A. an insulator.
B. an intrinsic semiconductor.
C. an extrinsic semiconductor.
D. a metal.
E. a semimetal.

30M.5 A large bond strength in a solid would be associated with

A. high electrical conductivity.
B. low density.
C. high melting point.
D. instability.
E. a high degree of ionization of the atoms.

30M.6 The n=2 shell can hold 8 electrons ($2s^2 2p^6$). Carbon has only 4 electrons in this shell. In diamond a strong tetrahedral (i.e. four cornered) bond is formed. We can thus deduce that this is a

A. ionic bond.
B. covalent bond.
C. metallic bond.
D. hydrogen bond.
E. Van der Waals bond.

30M.7 Suppose that E, V and R are of the order of magnitude of the energy level separations associated with electron energies, vibrational energies and rotational energies in a molecule. Then typically

A. $E > V > R$
B. $V > R > E$
C. $R > E > V$
D. $E > R > V$
E. $V > E > R$
F. $R > V > E$

30M.8 If the temperature of a sample of an intrinsic semiconductor (such as pure silicon) is increased

A. the number of holes increases, but the number of conduction electrons does not.
B. the number of conduction electrons increases, but the number of holes does not.
C. the forbidden energy gap also increases.
D. the electrical conductivity increases.
E. the electrical properties are unchanged.

• Multiple Choice: Answers and Comments

30M.1 D

30M.2 D

30M.3 B For an ionic bond one needs two kinds of ions, one positive and one negative, and if all of the ions are alike, this will not happen.

30M.4 D The 4s band in the metal would be only half full, so solid potassium is a metal.

30M.5 C A large bond strength means it will be necessary to get the material very hot (i.e. so that it is vibrating vigorously) in order for them to pull apart and become liquid.

30M.6 B An electron from each of four neighbors joins a given atom to effectively fill the n = 2 shell. By sharing electrons every atom manages to effectively "fill" its own n = 2 shell. Such a bond is called covalent.

30M.7 A

30M.8 D Increasing the temperature increased the number of charge carriers (both electrons and holes).

31 Nuclear Physics and Elementary Particles

• Summary of Important Ideas, Principles, and Equations

1. An atom consists of a very heavy nucleus with charge +Ze surrounded by a cloud of Z electrons with charge -Ze. The atom is electrically neutral and it is mostly empty space. Virtually all of the mass of the atom is concentrated in the nucleus.

   The nucleus consists of Z protons and N neutrons. Z is the atomic number and N is the neutron number. A neutron and a proton are about the same size, about 1840 times as massive as an electron. The neutron is electrically neutral and the proton has a charge +e. The neutron and the proton are two forms of a single particle called a nucleon.

   The number of nucleons in the nucleus is called A, the mass number. A = N + Z. A nucleus of element X is designated by the symbol

$$\boxed{{}^{A}_{Z}X} \qquad (31.1)$$

   For example, ${}^{235}_{92}U$ indicates the nucleus of uranium, element number 92.

   In this nucleus there are 235 neutrons and protons. Since the atomic number Z = 92 here, there are 92 protons in the nucleus and thus N = A - Z = 235 - 92 = 143 neutrons. The symbol U and the number 92 are redundant, since from the periodic table we see that the element 92 is uranium, but this is still the way it is done for convenience.

   A given element always has a definite atomic number Z (i.e. a definite number of protons). Isotopes are nuclei of a given element, i.e. given number of protons Z, but with varying numbers of neutrons. Thus ${}^{235}_{92}U$ and ${}^{238}_{92}U$ are two isotopes of uranium. They are sometimes written U-235 and U-238.

A parenthesis around a symbol means that the nucleus is unstable.

We can calculate the energy absorbed or given off in the reaction by comparing the masses of the reactants and the products.

8. Unstable nuclei will undergo radioactive decay in which they change into a different form.

In beta decay an electron is emitted, e.g. ${}^{214}_{82}Pb \rightarrow {}^{214}_{83}Bi + \beta^-$.

In alpha decay an alpha particle is emitted, ${}^{226}_{88}Ra \rightarrow {}^{222}_{86}Rn + {}^{4}_{2}He$.

A gamma emitter is a nucleus which is in an excited state. The nucleons are vibrating vigorously and can fall into a lower energy state with emission of a gamma ray (a photon) with no change in the atomic number Z or the mass number A.

Nuclei with $Z > 83$ (bismuth) are all naturally radioactive.

A nucleus can be made artificially radioactive by bombarding it with particles. This is the way radioactive isotopes (radioisotopes) are made with a proton bombardment in a cyclotron.

9. Radioactive decay varies exponentially with time. If N is the number of nuclei present at some initial time $t = 0$ (which may be chosen whenever we want) and $N(t)$ is the number left after time t (i.e. which have not yet decayed), then

$$\boxed{N(t) = N_o e^{-\lambda t}} \qquad (31.5)$$

λ is the decay constant.

$\tau = \frac{1}{\lambda}$ is the decay time.

The half-life $t_{\frac{1}{2}}$ is the time required for the number of nuclei which have not undergone radioactive decay to be reduced by a factor of ½.

$$\boxed{t_{\frac{1}{2}} = \frac{0.693}{\lambda} = 0.693\,\tau} \qquad (31.6)$$

The activity of a sample is the number of radioactive decays per second.

$$\boxed{\text{Activity} = -\frac{\Delta N}{\Delta t} = \lambda N} \qquad (31.7)$$

Activity is measured in Curies (Ci) or in microcuries ($1\mu Ci = 10^{-6} Ci$).

10. Fission is the disintegration of a heavy nucleus, such as the nucleus of uranium or plutonium, into two fragments of roughly equal masses. Certain unstable isotopes can be induced to undergo fission by bombardment with neutrons of appropriate energy. In the fission of U-235 about 200 MeV of energy plus about two neutrons are released in each event. This makes possible a chain reaction, for the released neutrons can cause other nuclei to disintegrate. This is the process utilized in an atomic bomb or a power nuclear reactor. The neutrons released are going too fast to be effective in causing fission, so they are slowed down ("moderated") by collision with an appropriate substance such as deuterated water or carbon. The rate of reaction is controlled by the use of a "control rod" (e.g. cadmium) which absorbs neutrons.

    In a power reactor the energy released is used to make steam which drives an electric generator.

    A breeder reactor utilizes fission of the plentiful isotope U-238 which is turned into plutonium. The plutonium is then used as fuel for a conventional fission reactor. The device is called a "breeder" because it turns plentiful U-238 into another useful fuel, P-239, which will undergo a fission chain reaction such as U-235 does. Unfortunately, pure plutonium is produced. This is a horrible material which is extremely toxic, is an alpha emitter with the long half life of 24,000 years, and can be used to make atomic bombs (ordinary reactor uranium isn't pure enough).

11. Fusion is the process in which protons and neutrons combine to form an alpha particle, releasing large amounts of energy (about 27 MeV per alpha). This is the process by which energy is released in the sun or in a hydrogen bomb.

12. Many kinds of nuclear detectors are used. They include photographic emulsions, Geiger counters, scintillation counters, semiconducting p-n junctions, cloud and bubble chambers, and spark chambers.

13. There are numerous other elementary particles in addition to electrons, neutrinos, neutrons, and protons; among them are mesons and hyperons. Mesons have rest masses between that of an electron and that of a proton; hyperons have rest masses greater than that of a neutron. Elementary particles are classified according to the dominant interactions in which they can participate. Leptons are particles that interact via weak force; hadrons are particles that interact via the strong nuclear force. According to current theory, all hadrons are composed of a combination of particles known as quarks, which carry charges of ±1/3 and ±2/3 of an electronic charge.

• Qualitative Questions

31M.1 The nucleus of an atom

A. contains most of the charge of the atom.
B. occupies most of the volume of the atom.
C. contains most of the mass of the atom.
D. is electrically neutral.
E. consists of an aggregation of neutrons.

31M.2 Two nuclei which contain the same number of protons are

A. not necessarily nuclei of a single element.
B. called isomers.
C. called isotopes.
D. called isonuclei.
E. isotopes with the same mass number.

31M.3 A nucleon is

A. any stable nucleus.
B. a general term describing any subatomic particle.
C. either a proton or a neutron.
D. an ionized neutron.
E. a composite particle consisting of a neutron plus a proton.

31M.4 The helium nucleus is a very stable nuclear unit. It occurs frequently in nuclear reactions and is called

A. a gamma ray.
B. an alpha particle.
C. a beta particle.
D. deuterium.
E. tritium.

31M.5 For a given separation of two nucleons the nuclear force is

A. strongest between two protons.
B. strongest between two neutrons.
C. strongest between a neutron and a proton.
D. equally strong between two protons, two neutrons or between a proton and a neutron.

31M.6 Which of the following is <u>NOT</u> a characteristic of the nuclear force.

A. For intranuclear distances it is much stronger than either the gravity force or the electrostatic force.
B. Its magnitude is independent of electric charge.
C. It is short range and drops to zero rapidly beyond a critical distance.
D. It is attractive between some particles and repulsive for others.
E. It is the force which holds the nucleus together.

31M.7 The "saturation" of the nuclear force refers to the fact that

A. a nucleon interacts significantly only with other nucleons more or less far removed from it.
B. a nucleon interacts significantly only with other nucleons more or less touching it.
C. once a nucleon interacts with one other nucleon it cannot interact with additional nucleons.
D. the nuclear force can bind two protons and two neutrons to form a stable unit (an alpha particle), but it cannot bind additional particles, and these must be held by the electrostatic force.
E. the nuclear force increases with decreasing particle separation up to a maximum value, and then remains constant for smaller separations.

31M.8 Which of the following is <u>NOT</u> one of the rules of radioactive decay?

A. Electric charge must be conserved.
B. The nucleon number A must be conserved.
C. The atomic number Z must be conserved.
D. Energy must be conserved.
E. Linear momentum must be conserved.
F. Angular momentum must be conserved.

31M.9 Which of the following is most nearly the same as a gamma ray?

A. Visible light.
B. Beta rays.
C. Alpha rays.
D. Helium nucleus.
E. Protons.
F. Neutrons.
G. Ions.

31M.10 For each of the following indicate if the reaction is possible.

A. $^{239}_{93}Np \rightarrow ^{239}_{94}Pu + \beta^-$

B. $^{7}_{3}Li + ^{1}_{1}H \rightarrow 2\ ^{4}_{2}He$

C. $^{2}_{1}H + ^{2}_{1}H \rightarrow ^{3}_{1}H + ^{1}_{0}n$

D. $^{12}_{6}C + ^{1}_{1}H \rightarrow ^{12}_{6}C + \beta^+$

E. $^{107}_{48}Cd + ^{0}_{-1}e \rightarrow ^{106}_{47}Ag$

31M.11 What is the missing particle in the reaction

$$^{235}_{92}U + ^{1}_{0}n \rightarrow ^{140}_{54}Xe + ^{94}_{38}Sr + ^{1}_{0}n + ?$$

A. A proton.
B. A neutron.
C. An electron.
D. A positron.
E. An alpha particle.

31M.12 An electron is called

A. an alpha particle.
B. a nucleon.
C. a gamma ray.
D. a beta ray.
E. a sting ray.

31M.13 When a nucleus undergoes radioactive decay

A. its atomic number cannot change.
B. its atomic number cannot increase.
C. its mass cannot change.
D. its mass cannot increase.
E. it usually splits into two fragments of about the same size.

31M.14 If the half-life of a certain radioactive isotope is $T_{\frac{1}{2}}$, this means that

A. after a time $T_{\frac{1}{2}}$, 37% of the original number of radioactive nuclei in a particular sample will still be present.
B. after a time $T_{\frac{1}{2}}$, 63% of the original number of radioactive nuclei in a particular sample will still remain.
C. after a time $\frac{1}{2}T_{\frac{1}{2}}$, all of the radioactive nuclei in a given sample will have decayed.
D. after a time $T_{\frac{1}{2}}$, the number of radioactive nuclei will have decayed to half of their original number.
E. after a time $T_{\frac{1}{2}}$, the level of radioactivity (in counts per minute) will have decayed to half its original value.
F. More than one of the above are true.
G. None of the above is true.

31M.15 A major radioactive constitutent in the fallout from nuclear weapons is $^{90}_{38}Sr$. This is particularly hazardous because

A. it has a very short half-life and hence a high level of radioactivity.
B. it decays into uranium, which can in turn undergo a chain reaction.
C. it enters the food chain, and behaves chemically, much like calcium.
D. when absorbed in the body it causes other atoms to become radioactive.
E. it emits neutrons which can trigger chain reactions.
F. it is very toxic chemically.
G. it can be collected and used by terrorists in building bombs.

31M.16 What is the source of the energy which the sun radiates to us?

A. Chemical reactions.
B. Nuclear fission reactions.
C. Nuclear fusion reactions.
D. Magnetic explosions.
E. Cosmic rays.

31M.17 A "breeder reactor"

A. essentially creates fuel out of its own energy without using any other input material.
B. creates fuel by changing $^{238}_{92}U$ into plutonium.
C. is considered more dangerous than other reactors because it generates unusually high radiation levels.
D. is a device for creating many little reactors from one large one.
E. uses plutonium as a starting material.

31M.18 In an accident in a nuclear power plant an area was contaminated with radioactive material. Health physicists monitored the radioactivity over three weeks and obtained the data given here. On the basis of this data what would you expect the radioactivity (in counts per minute) to be 30 days after the accident (at t = 30)?

A. 62 cpm
B. 320 cpm
C. 405 cpm
D. 1200 cpm
E. 1296 cpm
F. 1316 cpm
G. 1396 cpm

Time (days)	Activity (cpm)
0	5265
2	4800
6	3990
17	2400
21	1995

31M.19 The rates of radioactive decay of the two isotopes sludgium and gunkium are described by the equations

$$\frac{\Delta N_S}{\Delta t} = -100\ N_S \qquad \text{(Sludgium)}$$

$$\frac{\Delta N_G}{\Delta t} = -200\ N_G \qquad \text{(Gunkium)}$$

From this information

A. we cannot deduce the half-lives of the two isotopes.
B. we can deduce the the half-life of sludgium is twice that of gunkium.
C. we can deduce that the half-life of gunkium is twice that of sludgium.
D. we can deduce the half-lives of the two isotopes, but the half-lives are not in the ratio of 1:2 or 2:1.
E. we can make none of the above deductions.

• Multiple Choice: Answers and Comments

31M.1 C

31M.2 C

31M.3 C

31M.4 B

31M.5 D

31M.6 D The nuclear force is always attractive (like gravity in this respect).

31M.7 B

31M.8 C Look at some of the examples of radioactive decay and you will see that elements change into other elements.

31M.9 A Visible light and gamma rays are both photons, but the gamma photon has much more energy (higher frequency).

31M.10 A is O.K. because a β^- is ${}^{0}e$, so 93 = 94 - 1 and 239 = 239 + 0.

B is O.K.
C is wrong because we have charge +2 on left and only +1 on right.
D is wrong because we have 12 + 1 = 13 nucleons on left and only 12 on the right.
E is wrong because we have 107 nucleons on the left and only 106 on the right.

31M.11 B From Z values we see that the missing particle has 92 + 0 = 54 + 38 + 0 + Z, so Z = 0.
From A values we see 235 + 1 = 140 + 94 + 1 + A, so A = 1.
Thus the particle is a neutron (Z = 0, A = 1).

31M.12 D

31M.13 D

31M.14 F Both D and E are true.

31M.15 C From the periodic table we see that strontium and calcium are in the same column, so they behave very similarly chemically.

31M.16 C

31M.17 B

31M.18 F From the data notice that the activity dropped by a factor of two between t = 2 days and t = 17 days, so the half-life is 15 days. Thus in 30 days (two half-lives) the activity will drop by (½)(½) = ¼, or to a level of (¼)(5265)=1316 cpm.

31M.19 B

• Problems

31.1 How much energy is required to split a ${}^{12}_{6}C$ nucleus into three alpha particles? The atomic mass of carbon-12 is 12.0000u and the atomic mass of helium-4 is 4.0026u.

31.2 ${}^{232}_{90}Th$ decays in a series of steps in which six alpha particles and four beta particles are emitted. What is the resulting isotope?

31.3 The atomic mass of $^{7}_{3}Li$ is 7.01601 (including the electrons). What is the biding energy per nucleon?

31.4 The muon is an unstable lepton (in the same family as the electron) whose mass is 207 electron masses. Express this mass in MeV.

31.5 How much energy is released in the reaction $^{1}_{1}H + ^{2}_{1}H \rightarrow ^{3}_{2}He$? The atomic masses of the three isotopes are 1.00783u, 2.01410u, and 3.01603u.

31.6 The radioactivity of carbon from trees felled in Wisconsin glaciation is about 12.5% as intense as the activity from trees cut today. How long ago did this last ice age occur, according to carbon dating?

31.7 In an exciting discovery some astronauts recover some material from the asteroid belt. When analyzed the rock fragments show rubidium-80 and strontium-80 present in the ratio of 3:1 (more rubidium is present). It is known that rubidium-80 decays radioactively to the stable isotope strontium-80 with a half-life of 4.7×10^{10} years. Further, geologists believe no strontium would have been present in this type of rock unless it was the result of radioactive decay of rubidium-80. If this is the case, how old would you estimate the rocks to be?

31.8 In a nuclear reactor using U-235 as fuel each fission yields 200 MeV. How much mass is lost each day in a reactor operated at 1 GW (10^9 W)? How many fissions occur per second?

• Problem Solutions

31.1 $^{12}_{6}C \longrightarrow 3\,^{4}_{2}He$

NOTE THAT ATOMIC MASSES INCLUDE ELECTRON MASSES, BUT SINCE THESE ARE INCLUDED ON BOTH SIDES OF THE EQUATION THEIR CONTRIBUTION TO THE MASS CANCELS OUT.

FOR $^{12}_{6}C$, $M = 12.0000\ u$

FOR $3\,^{4}_{2}He$, $m = (3)(4.0026u) = 12.0078\ u$

MASS DEFECT $\Delta m = 0.0078\ u$

$E = \Delta mc^2$. 1u IS 931.5 MeV. SO $E = (0.0078u)(931.5\ MeV/u)$

$E = 7.27\ MeV$

31.2 $^{232}_{90}Th \longrightarrow 6\,^{4}_{2}He + 4\,^{0}_{-1}e + ^{208}_{82}Pb$

THE MASS NUMBER MUST BE $232 - (6)(4) = 208$

THE ATOMIC NUMBER Z MUST BE $90 = (6)(2) - (4)(1) + Z$, OR $Z = 82$

FROM THE PERIODIC TABLE I FIND ELEMENT 82 IS LEAD, Pb.

31.3 THE MASS OF THE $^{7}_{3}Li$ NUCLEUS IS THE ATOMIC MASS LESS THE MASS OF THE THREE ELECTRONS. OR $7.01601 - (3)(0.00055) = 7.01436\ u$

THE NUCLEUS CONTAINS 3 PROTONS AND 4 NEUTRONS. THEIR INDIVIDUAL MASSES YIELD

$$(3)(1.0072) + (4)(1.0087) = 7.0564 \text{ u}$$

THE BINDING ENERGY IS THUS

$$BE = (7.0564 \text{ u} - 7.01436 \text{ u})(931.5 \text{ MeV/u}) = 39.2 \text{ MeV}$$

THERE ARE 7 NUCLEONS IN ${}^{7}_{3}Li$, SO THE BINDING ENERGY PER NUCLEON IS $\frac{39.2}{7}$ = 5.6 MeV/NUCLEON

31.4 $m = 207\, m_e = (207)(0.511 \text{ MeV}) = 106 \text{ MeV}$

THE ENERGY EQUIVALENCE OF THE ELECTRON MASS IS OBTAINED FROM TABULATED VALUES.

31.5 FOR ${}^{1}_{1}H$, ATOMIC MASS = 1.00783 u

FOR ${}^{2}_{1}H$, ATOMIC MASS = 2.01410 u

TOTAL REACTANT MASS = 3.02193 u

SUBTRACT ${}^{3}_{2}He$ MASS − 3.01603 u

.00590 u

THUS ENERGY RELEASED IS $E = (0.00590 \text{ u})(931.5 \text{ MeV/u})$

$$E = 5.50 \text{ MeV}$$

31.6 IN ONE HALF-LIFE THE RADIOACTIVITY OF THE CARBON-14 DROPS TO 50% OF WHAT IT WAS ORIGINALLY. IN ANOTHER HALF-LIFE IT DROPS TO 25%. AND IN A THIRD TO 12.5%. THUS THE TREE DIED ABOUT (3)(5730 YRS) = 17,190 YEARS AGO. HERE I ASSUME THAT THE FRACTION OF CARBON IN THE FORM OF C-14 IN A LIVING TREE WAS THE SAME 17,000 YEARS AGO AS NOW. ONCE THE TREE DIES IT TAKES IN NO MORE CARBON.

31.7 WE DEDUCE THAT 25% OF THE ORIGINAL RUBIDIUM HAS DECAYED (LEAVING 3 PARTS RUBIDIUM FOR 1 PART STRONTIUM). IF N_0 WAS THE NUMBER OF RUBIDIUM NUCLEI ORIGINALLY PRESENT, THE NUMBER PRESENT AFTER TIME t IS

$$N = 0.75\, N_0 = N_0 e^{-\lambda t}$$

OR $0.75 = e^{-\lambda t}$, $\ln(0.75) = -\lambda t$

$-\lambda t = -0.288$ (ON CALCULATOR PUNCH 0.75, THEN "ln")

WE ARE GIVEN $t_{1/2} = 0.693/\lambda = 4.7 \times 10^{10}$ YRS.

THUS $t = 0.288/\lambda = (0.288)\left(\frac{4.7 \times 10^{10}}{0.693} \text{ YRS}\right) = 1.95 \times 10^{10}$ YRS.

31.8 ENERGY USED DAILY IS $E = Pt = (10^9 \text{ W})(24 \text{ HRS})(3600 \text{ SEC/HR})$

$$mc^2 = E = 8.64 \times 10^{13} \text{ J}$$

$$m = \frac{E}{c^2} = \frac{8.64 \times 10^{13} \text{ J}}{(3 \times 10^8 \text{ m/s})^2} = 9.[illegible] \times 10^{-4} \text{ kg} = 0.9[illegible] \text{ gm}$$

EACH FISSION RELEASES 200 MeV = $(200 \text{ MeV})(1.6 \times 10^{-13} \text{ J/MeV}) = 3.2 \times 10^{-11}$ J

ENERGY USED IN ONE SECOND IS $E_1 = Pt = (10^9 \text{ W})(1 \text{ s}) = 10^9$ J

NUMBER OF FISSIONS PER SECOND IS THUS $N = \frac{10^9 \text{ J}}{3.2 \times 10^{-11} \text{ J/FISSION}} = 3.1 \times 10^{19}$ PER SEC

INDEX